Smart Nanomaterials Technology

Series Editors

Azamal Husen, Department of Biotechnology, Smt. S. S. Patel Nootan Science & Commerce College, Sankalchand Patel University, Gujarat, Ethiopia

Mohammad Jawaid, Chemical and Petroleum Engineering, United Arab Emirates University, Al Ain, United Arab Emirates

Nanotechnology is a rapidly growing scientific field and has attracted a great interest over the last few years because of its abundant applications in different fields like biology, physics and chemistry. This science deals with the production of minute particles called nanomaterials having dimensions between 1 and 100 nm which may serve as building blocks for various physical and biological systems. On the other hand, there is the class of smart materials where the material that can stimuli by external factors and results a new kind of functional properties. The combination of these two classes forms a new class of smart nanomaterials, which produces unique functional material properties and a great opportunity to larger span of application. Smart nanomaterials have been employed by researchers to use it effectively in agricultural production, soil improvement, disease management, energy and environment, medical science, pharmaceuticals, engineering, food, animal husbandry and forestry sectors.

This book series in Smart Nanomaterials Technology aims to comprehensively cover topics in the fabrication, synthesis and application of these materials for applications in the following fields:

- Energy Systems—Renewable energy, energy storage (supercapacitors and electrochemical cells), hydrogen storage, photocatalytic water splitting for hydrogen production
- Biomedical—controlled release of drugs, treatment of various diseases, biosensors,
- Agricultural—agricultural production, soil improvement, disease management, animal feed, egg, milk and meat production/processing,
- Forestry—wood preservation, protection, disease management
- Environment—wastewater treatment, separation of hazardous contaminants from wastewater, indoor air filters.

Madhusudhan Alle · Ruchir Priyadarshi ·
Shiv Dutt Purohit
Editors

Functional Nanomaterials and Nanocomposites for Biodegradable Food Packaging

Springer

Editors
Madhusudhan Alle
Department of Chemistry
The University of Memphis
Memphis, TN, USA

Shiv Dutt Purohit
Department of Biomedical Engineering
and Biotechnology
College of Medicine and Health Sciences
Khalifa University of Science
and Technology
Abu Dhabi, United Arab Emirates

School of Chemical Engineering
Yeungnam University
Gyeongsangbuk-do, Korea (Republic of)

Ruchir Priyadarshi
BioNanocomposite Research Center
Department of Food and Nutrition
Kyung Hee University
Seoul, Korea (Republic of)

ISSN 3004-8273 ISSN 3004-8281 (electronic)
Smart Nanomaterials Technology
ISBN 978-981-97-9086-9 ISBN 978-981-97-9084-5 (eBook)
https://doi.org/10.1007/978-981-97-9084-5

This Springer imprint is published by the registered company Springer Nature Singapore Pte Ltd.
The registered company address is: 152 Beach Road, #21-01/04 Gateway East, Singapore 189721, Singapore

Preface

The discovery of plastics in the 1950s introduced the world to a revolutionary material that was tough, strong, lightweight, easily processable, and moldable. This material quickly became indispensable across various sectors, replacing traditional materials like metals, paper, and glass. However, the initial enthusiasm soon gave way to concerns as the non-biodegradable nature of plastics led to significant environmental and health hazards. The food industry, a major consumer of plastics, has contributed substantially to the generation of non-biodegradable waste, with food packaging being a primary culprit. This situation has created a dual challenge: managing the vast amounts of plastic waste and addressing the issue of food wastage due to spoilage during transportation and storage.

In recent years, biopolymer nanocomposites have emerged as a promising solution to these challenges. By integrating various nanomaterials with biopolymers, these advanced composites offer potential benefits for food packaging, including enhanced barrier properties, antimicrobial activity, and improved mechanical strength, all while being biodegradable. This book aims to provide a comprehensive overview of the latest developments in the field of biopolymer nanocomposite materials and their implications for the food packaging industry.

Functional Nanomaterials and Nanocomposites for Biodegradable Food Packaging is organized into 13 chapters, each addressing a critical aspect of this emerging technology:

Chapters "Introduction to Biodegradable Food Packaging" and "Plastics and Microplastics—Environmental and Health Hazards" of the book provide a background of problems associated with non-biodegradable plastic-based packaging, and the emergence of biopolymers as a possible solution to this problem. Chapter "Introduction to Biodegradable Food Packaging" provides an overview of the need for and benefits of biodegradable food packaging solutions. Chapter "Plastics and Microplastics—Environmental and Health Hazards" examines the detrimental impacts of traditional plastics, and the resulting microplastics, on human health and the environment over the long run. The insights into the extraction and characteristics of several biopolymer types, their sources, and their use in packaging have been provided in Chapter "Biopolymers—Isolation and Properties from a Food Packaging Perspective".

Following the background information, the succeeding chapters discuss the role of important nanomaterials, biopolymers, and their composite structures, in functional food packaging research. Chapter "Metal/Metal Oxide Nanomaterials in Biodegradable Food Packaging" explores the role of inorganic metal or metal oxide-based nanomaterials as functional fillers in food packaging polymer materials and their importance in enhancing packaging properties. Whereas, Chapter "Carbon-Based Nanomaterials in Food Packaging and Preservation" analyzes the applications and benefits of carbon-based nanomaterials as the class of sustainable nanostructures, that make the future of food packaging.

Chapter "Cellulose-Based Nanomaterials in Food Packaging" discusses the use of cellulose, the most abundant plant-based renewable biopolymer, for sustainable packaging. The chapter also covers the role and application of cellulose-based nanomaterials in this research area. While Chapter "Chitosan-Based Films for Food Packaging Applications" reviews the properties and applications of chitosan, which is the second most abundant biopolymer on earth after cellulose, and the most abundant animal-based polysaccharide, in food packaging.

Chapters "Nanoemulsions in Food Packaging"–"Quantum Dots in Food Packaging" shed light on other advanced (natural, nature-derived, or composite) nanomaterials which have been widely used in functional food packaging. Chapter "Nanoemulsions in Food Packaging" provides investigations into the use of nanoemulsions for improved packaging performance. Also, insight into the benefits and applications of nanoclays in packaging is provided in Chapter "Nanoclays in Food Packaging". Further, Chapter "Quantum Dots in Food Packaging" explores the potential uses of sustainable quantum dots, which are the latest nanomaterials being used in functional food packaging, in enhancing packaging functionalities.

An overview of three important dimensions of advanced functional packaging technology, i.e., active, intelligent, and smart packaging, and the role of functional bionanocomposites, have been provided in Chapter "Functional Food Packaging—Active, Intelligent, and Smart Bio-nanocomposites". Safety and migration issues are paramount when food contact materials are considered. Chapter "Migration and Safety Issues in Food Packaging Bionanocomposites" examines the safety concerns related to the migration of nanomaterials from packaging to food. Finally, the discussion on the practical challenges of integrating these novel materials into production processes is provided in Chapter "Converting Challenges for the Novel Materials in Food Packaging Production".

This book is intended for researchers, professionals, and students in the fields of materials science, food technology, and environmental science. By covering the intersection of nanotechnology and biodegradable materials, we aim to highlight the innovative approaches being developed to tackle the pressing issues of plastic waste and food spoilage.

We hope that the insights and knowledge shared in this book will inspire further research and development, ultimately leading to more sustainable and effective food packaging solutions.

Memphis, USA
Seoul, Korea (Republic of)
Abu Dhabi, United Arab Emirates/
Gyeongsangbuk-do, Korea (Republic of)

Madhusudhan Alle
Ruchir Priyadarshi
Shiv Dutt Purohit

Contents

About the Editors

Dr. Madhusudhan Alle is currently working as a Postdoc-fellow at The University of Memphis, TN, USA. Earlier he worked as a Research Professor at Department of Biomedical Science & Institute of Bioscience and Biotechnology, Kangwon National University, South Korea and also worked as an Associate Professor at Department of Chemistry, Gondar University, Gondar, Ethiopia (2014–2018). During this period, he served as a Coordinator of MSc (Chemistry) Programs and as the Head of the Chemistry Department. Four students submitted their master's thesis under his supervision. He received his doctorate from Department of Chemistry, Osmania University, India (2013), where he worked on the research projects sponsored by University Grants Commission, India. He has more than 18 years of experience in teaching and research in both academia and industry. He has published more than 70 scientific papers in SCI/SCIE-grade journals (>2700 citations; h-index = 29) as a first and corresponding author and published several books and book chapters (Springer and Elsevier). He is also listed as a potential reviewer for many reputed international journals. He is also a member of Ethiopian chemical society and Indian chemical society. He has presented his work in several national and international conferences in India, Ethiopia, South Korea, and U.S.A. He has conducted two research projects sponsored by University Grants Commission (UGC), New Delhi, India. His current research interests include liquid biopsy, cancer biology, polymeric nano-drug delivery systems, biomaterials for sustainability and catalysis. Extraction and chemical modification of nanocellulose from lignocellulosic biomass, green synthesis of metal and metal oxide nanoparticles on nanocellulose support and their novel applications in biomedicine, biosensors, and catalysis.

Dr. Ruchir Priyadarshi is currently working as a Brain Pool Fellow (2019–present) at the Department of Food and Nutrition, Kyung Hee University, Seoul, Republic of Korea. He obtained his Ph.D. (2018) from the Indian Institute of Technology Roorkee, India, in Biodegradable Food Packaging. He served as Head of Biopolymer Innovations at Yash Papers Ltd. India (2018–2019), where he worked on the development of bagasse-based molded tableware products. His research interests include active food packaging, functional materials, sustainable materials, biodegradable polymers, and bionanocomposites. He has published more than 70 research and review articles published in high-impact international journals. He has also published various book chapters for reputed international publishers. He has more than 5000 citations on his name with an h-index of 34. Dr. Priyadarshi is also a reputed reviewer for several peer-reviewed international journals in the Food Technology and Materials Science research areas. He has been awarded as a Highly Cited Researcher (HCR) in Agricultural Sciences for the year 2024 by Clarivate^{TM}, based on the Web of Science citation indices.

Dr. Shiv Dutt Purohit is currently working as a postdoctoral fellow at Khalifa University, Abu Dhabi. Prior to this, he worked as an International Research Professor (Assistant professor) at the School of Chemical Engineering, Yeungnam University, Gyeongsan, South Korea. He obtained his Ph.D. from the Department of Polymer and Process Engineering, Indian Institute of Technology Roorkee, in 2021. He is a materials scientist, and his research interests lie in Biomedical and related areas such as Tissue Engineering and Wound Healing. Besides, he is also intrigued to work in other areas, such as Food Packaging materials. He has published more than 35 research articles and book chapters and has more than 1000 citations on his name with an h-index of 18. Dr. Purohit is also a reputed reviewer for several peer-reviewed international journals and working as an Academic Editor for Wiley-VCH journals (IET Nanobiotechnology and Scientifica). He is also involved in the editing of a few special issues in SCIE Journals. In addition, he has edited a couple of books for Springer Nature.

Introduction to Biodegradable Food Packaging

Ana Caroline Ferreira Carvalho, Caroline Meinert, Fernanda Raquel Wust Schmitz, Sarah Finardi, Tuany Gabriela Hoffmann, Ruchir Priyadarshi, and Carolina Krebs de Souza

Abstract Food packaging plays a crucial role in maintaining and improving the quality of foods/beverages by protecting them against chemical, physical, biological, mechanical, and environmental damage. Plastic (petroleum-based) packaging is appreciated for its characteristics such as softness, strength, lightness, and transparency, besides being economical and effectively minimizing food waste. However, due to their non-biodegradation, single-use nature, and improper disposal, they create serious environmental problems, including excessive waste generation. Research has been directed toward the development of sustainable and biodegradable alternatives to conventional plastics, such as bioplastics, derived from renewable sources (agricultural products, food industry waste, and lignocellulosic biomass), which can be applied in coatings, films, composites, nanocomposites, etc. The type of starting material classifies polymers as either extracted from biomass, produced from microorganisms, or synthesized from monomers. Renewable, environmentally friendly polysaccharides and their polymer composites (cellulose, nanocellulose, hemicellulose, starch, chitosan, and pectin) have been extensively used for biodegradable food packaging. Food packaging has evolved from a simple primary container to a more specialized and complex one with other desirable properties (biodegradable, edible, intelligent, active, and smart). The growing consumer awareness of food safety and quality has driven innovations in food packaging.

Keywords Biopolymers · Biodegradable packaging · Food safety innovations · Smart food packaging · Sustainable materials

A. C. F. Carvalho · C. Meinert · F. R. W. Schmitz · S. Finardi · T. G. Hoffmann · C. K. de Souza (✉)
Department of Chemical Engineering, University of Blumenau, Blumenau 89030-000, Brazil
e-mail: carolinakrebs@furb.br

R. Priyadarshi
Department of Food and Nutrition, BioNanocomposite Research Center, Kyung Hee University, Seoul 02447, South Korea

M. Alle et al. (eds.), *Functional Nanomaterials and Nanocomposites for Biodegradable Food Packaging*, Smart Nanomaterials Technology,
https://doi.org/10.1007/978-981-97-9084-5_1

1 Introduction to Food Packaging

Considering the continuous increase in the world population, which results in an inescapable rise in the demand for food and, consequently, in the production of packaging—a more sustainable view of the context is urgent and unavoidable. Growing concern about food safety and changing consumer attitudes toward healthy eating drives the food industry to keep up with trends and meet customer demands, including food packaging as one of the many factors to consider.

Around 1.3 billion tons of food are wasted annually worldwide. Post-harvest losses and those that occur during the production chain significantly affect the generation of this waste, especially in low-income countries, for which it represents about 40% of their total production [1]. Indeed, from an industrial point of view, the development and application of packaging are some of the foremost parameters to be considered due to large-scale production, which must be efficient in ensuring mechanical properties, barriers, and resistance to extend the shelf life of products until consumption [2].

Food packaging performs different functions from filling to consumption, including marketing, communication of necessary information on preparation, storage, ingredients, allergens, and nutritional value [3, 4]. Furthermore, it is essential that proper packaging promotes the protection, enhancement, and maintenance of food/beverage quality during transportation, distribution, and storage [5, 6]. The purpose of packaging is also directly related to food waste reduction strategies [7], as it protects food and beverages from potential chemical, physical, biological, mechanical, and environmental damage [8, 9].

There are a variety of materials that are used in the storage, transportation, and preservation of food (glass, polymers, metals, wood, paper, cardboard, or other combinations of materials), which constitute packaging that can be classified as primary, secondary, and tertiary (Fig. 1). While primary packaging materials are those that come into direct contact with food, secondary packaging are those that contain the primary packaging and usually labeling information, and tertiary packaging (wood pellets, cardboard boxes, or plastic wrap) are those that help transport multiple units of products contained in the previously mentioned packages [10].

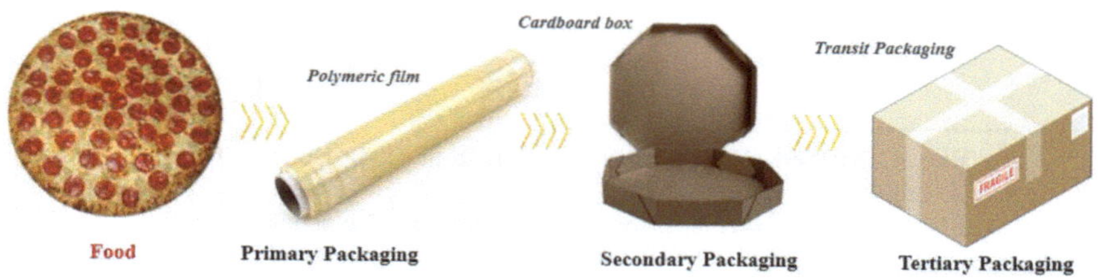

Fig. 1 Model packaging (primary, secondary, and tertiary) of a food product stored and distributed frozen

For a long time, polymers have supplied the most common packaging materials because they present several desired features like softness, transparency, and lightness [11]. Roughly 95–99% of plastic materials utilized for various purposes are produced by petrochemical industries using non-renewable sources (synthetic plastics).

Traditionally, plastic materials have been used in the food industry due to their wide availability, appreciable mechanical performance, low cost, and proven efficient protection [3]. However, problems such as non-biodegradability, recycling, and leaching of harmful chemicals create severe environmental and human health concerns [12, 13]. Although synthetic plastic-based packaging does not undergo physical, biological, and chemical degradation and therefore contributes to the increasing amount of waste generated, it has been an economical tool to minimize food waste [14]. Often, these non-biodegradable plastics are only used once, and when disposed of improperly and irresponsibly, they dramatically impact the environmental system (water, air, soil, and living beings) [15].

Researchers, packaging producers, and food industries have been developing and applying alternative materials (abundant, low-cost, renewable, sustainable, and biodegradable) to conventional petroleum-based materials, among the significant sources of solid waste [12]. To address this scenario, using biodegradable and sustainable compounds in food packaging has become a topic of great interest for many researchers [16–21].

Food packages, with ideal characteristics, should be easy to process, inexpensive, transparent, mechanically resistant, have good barrier properties to prevent water transfer with the external environment, protect the food from moisture and oxygen, be made from renewable sources, recyclable after use, and biodegradable [2].

Biodegradable polymers or biopolymers have emerged as an interesting and efficient alternative for many industrial applications, contributing to reducing the problems caused by non-biodegradable plastic [22].

Food packaging has evolved in the last decades from a simple container to a more specialized, complex, and differentiated one with other desirable properties added and guaranteed. Biodegradable packaging can be used in modified atmosphere packaging, active packaging systems (antimicrobial and/or antioxidant), intelligent [23, 24], and edible packaging [19, 25] to increase the shelf life of different high-quality and value-added food products.

According to the type of raw material, origin, and specific production route, biopolymers are categorized as polymers extracted from biomass/isolated directly from biomass or natural material (polysaccharides and proteins); polymers produced by classical chemical synthesis from bio monomers (Polylactic acid, Polycaprolactone, Polybutylene succinate, Polylactide aliphatic copolymer, Polyglycolide, Polybutylene adipate-co-terephthalate, Polyvinyl-alcohol, and Polypropylene carbonate), and polymers produced from microorganisms (Polyhydroxylalkanoates) [26, 27]. The quality of biopolymers depends on their mechanical, thermal, physical, and barrier properties [7].

The main challenge in introducing bio-based packaging is to combine its durability with the shelf life of the food. Intrinsic and extrinsic food factors, such as temperature, relative humidity, microbiological and enzymatic deterioration, and ultraviolet exposure, working individually or in interaction, are the primary means of deterioration and degradation of food quality and safety [28, 29]. These factors that deteriorate food are the same factors that impact the degradation of plastic material of biological origin. Thus, special care must be ensured while developing and generating bio-based plastic materials [3].

Bio-sourced plastics in the food industry are mainly used for packaging vegetables, fruits, chocolate varieties, and beverages in the form of cutlery, bags, clear and rigid containers, films, coatings, bottles, trays, plates, pouches, composites, nanocomposites, etc. [19, 28].

Increasing consumer awareness of food safety and quality has driven research and innovation in the food packaging industry. The biopolymer-based food packaging market is expected to grow with changes in consumer preference for recyclable and "eco-friendly" materials.

2 Biodegradable Packaging and Sustainability

Biodegradable materials for food packaging applications have been investigated for decades due to their sustainability-related properties such as renewability, abundant available raw materials, low cost, biodegradability, non-toxicity, biocompatibility, and functionality [21]. Biodegradable packaging or biopolymers, sustainable and with superior properties, have been developed using organic waste [30] from agricultural by-products or renewable resources such as forest lignocellulose material (sugarcane bagasse, beet pulp, and molasses), leather gelatin [31], agar, alginate, carrageenan from seaweed [16], tea residue [32], chitosan from crustaceans, and fish scale [33].

Biodegradable plastics are not necessarily based on renewable raw materials, as they can be produced from petrochemicals or plant-based, and the biodegradation is related to the chemical structure of the compound, not its source. Materials such as cellulose, starch, lignin, gelatin, and keratin are natural carbon-containing polymers that are biodegradable or compostable [34, 35]. Some materials used to form films are cellulose, chitosan, alginate, polyvinyl alcohol, methylcellulose, sodium alginate, sodium caseinate, and carrageenan [36].

Biodegradability is a term that defines the ability of a material to decompose into natural chemical elements after interaction with biological components, such as enzymes, in a relatively short time [37–39]. Biodegradable materials are interesting alternatives for replacing derived materials or materials that use oil to be produced, as they favor protecting the environment by reducing carbon dioxide emissions [39].

The term sustainability has several definitions in the literature, with the first dating back to the 1990s. According to Elkington [40], sustainability designates the ability to resist or adapt over time to changes, whose change to obtain improvements corresponds to the population's needs. For Elkington [40], the definition would be the balance between environmental, economic, and social categories, and companies should employ sustainability to generate stable business [41]. However, a more

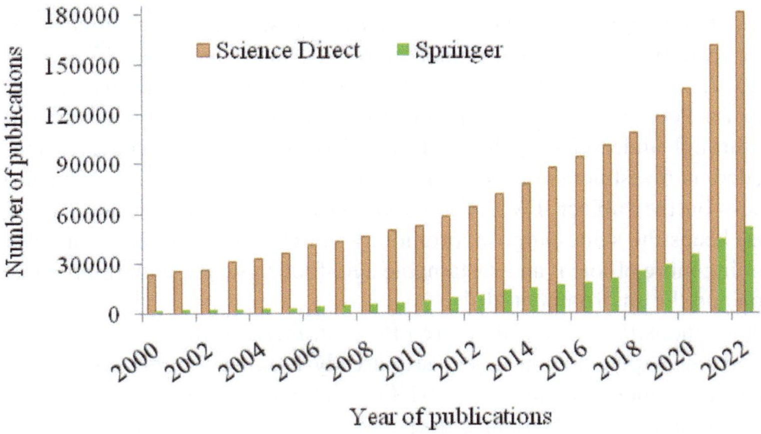

Fig. 2 Publication with term sustainability in the past 22 years in Science Direct platform and Springer

recent definition states that sustainability is a process to achieve desired sustainable development [42].

Many researchers mistakenly use the term sustainable development to express the same meaning as sustainability. However, the former term refers to the common interest of society based on its principles of industrial ecology, reuse, environmental management, and zero waste, among others [41, 43, 44]. Thus, studies on these terms have been increasing, both to try to explain better the definition of the terms sustainability and sustainable development and to develop more efficient ways to apply these proposals in everyday life. Figure 2 shows the continuous growth in the number of publications and scientific papers related to sustainability.

In the food sector, there is a constant search for the application of sustainable actions and materials in its products and processes, mainly due to climate change, scarcity of fresh drinking water, air, land, and sea pollution, and food-related diseases. In this way, the food industry has sought to develop and apply sustainable and efficient systems [45, 46], including in the packaging sector, since about 75% of food packaging is produced from non-biodegradable plastics conventional, obtained from petroleum derivatives, which cause severe damage to the environment and human health [47].

According to Jambeck et al. [48], and based on current trends, by 2050, there is expected to be more plastic waste than fish in the oceans, and by 2025, considering only PET packaging, consumption growth of 27 million tons is estimated. Several commissions worldwide started programs to contain these environmental impacts in this context. In 2015, the European Commission (EC) estimated that by 2025, 55% of plastic packaging should be reused or recirculated and that these should be recyclable by 2030 [49]. This is because, in some European countries (Malta, Luxembourg, Denmark, and Hungary), less than a third of this material is recycled [50]. The same was performed by the government of China, which in 2009 passed the circular economy law to encourage increased research and development to curb resource scarcity, decrease environmental impact, and reduce costs [51].

In 2022, the EC presented a proposal for "Partnership for Sustainable Food Systems for People, Planet and Climate" [52], aiming to achieve a sustainable food system through understanding the food chain. In this way, global studies and research are needed as investments in innovation in the sector, in addition to education and awareness on the subject. FAO (Food and Agriculture Organization of the United Nations) keeps the same purpose, promoting in the 2030 Agenda incentives aimed at obtaining more efficient and sustainable agri-food systems, with improvements in nutrition and the environment [46].

In other industry sectors, there are efforts to increase the use of washable and biodegradable wipes, as consumers are already looking for this type of product. European and American consumers (52%) are already purchasing this product, and the expectation for the Brazilian, Chinese, and Indian populations is an increase in this product by 29% [53].

One of the changes already implemented by some companies is reusable packaging, although this is still a significant challenge for most companies. These packages are focused on restocking and, consequently, acquiring fewer packages of the same product for some time. This simple strategy can help to reduce energy consumption, carbon dioxide, and greenhouse gas emissions [54].

Considering "green chemistry," studies present packages developed from biomass, such as proteins, natural polysaccharides, and petroleum-based synthetics [47, 55, 56], aiming to apply sustainable and biodegradable alternatives. In this vision, Europe coordinates a project with diverse organizations, including local, regional, national, and global organizations (public and private), to create a sustainable food system [57].

Biodegradable plastic has drawn attention because it can be transformed into substances that do not harm the environment (inorganic compounds, water, and biomass) through microbiological action [13]. Among the essential factors to be considered in the degradation of biodegradable plastics are the availability of oxygen, temperature, microorganisms, and humidity [58, 59], which, under specific conditions, allow these materials to degrade quickly. In contrast, conventional plastics, the most used worldwide (about 370 million tons), are accumulated in the environment, as only 9% of this material is recycled and 12% is incinerated [60–62].

About 40% of that comes from packaging, of which approximately 175 million tons are sent to landfills [50, 63], and more than 8 million tons are found/discarded per year alone in the ocean, causing a tremendous environmental impact [62]. According to [50], another critical factor is the recycling of plastic material, which consumers often need to dispose of correctly, as they need to learn or need clarification with the classification of waste due to the different instructions found. In this way, the production of biodegradable materials as substitutes for conventional plastics has presented a considerable increase, as shown in Fig. 3.

Shaikh et al. [13] reports that a 17% increase in the biodegradable packaging market is expected by 2025, from 2.11 million tons to about 2.43 million tons. This

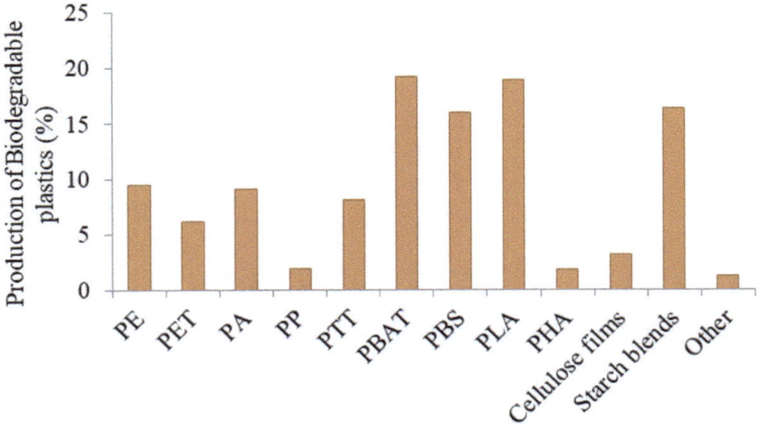

Fig. 3 Global production of biodegradable plastics in the year 2021. Polyethylene (PE), polyethylene terephthalate (PET), polyamide (PA), polypropylene (PP), polytrimethylene terephthalate (PTT), polyhydroxyalkanoate (PHA), polybutyleneadipate-co-terephthalate (PBAT), polybutylene succinate (PBS), and polylactic acid (PLA). Adapted from [64]

increase is due to the environmental concern of governments, which have carried out several campaigns to reduce the use and waste of plastic.

Recent research shows that biodegradable packaging has been gaining ground in the food sector, not only to package foods but also to help maintain food product quality through coating films. This is because conventional packaging could be more technically and economically viable for mechanical and chemical recycling [65]. Thus, there is a growing effort by world organizations and by the industry to promote greater sustainability in the consumption chain, as well as to explore the possibilities of producing different biodegradable packaging, from the simplest, applied to pack only one product, to the composite ones with other elements to help extend the shelf life of food products.

3 Biodegradable Materials for Food Packaging

Three types of raw materials can produce biopolymers: extracted from natural sources, in general plants (starch, cellulose, among others), microorganisms or genetically modified bacteria (xanthan, polyhydroxy alkanoates, among others), and synthetic polymers made from renewable polymers (polyethylene lactic acetate acid monomers) [49], as shown in Fig. 4. Although all biopolymers are biodegradable materials, not all are considered biopolymers. Polycaprolactone (PCL), polyglycolide (PGA), and polybutylene succinate adipate (PBSA) are biodegradable materials that are not classified as biopolymers as they are produced from non-renewable resources, i.e., they are chemicals of fossil origin [66].

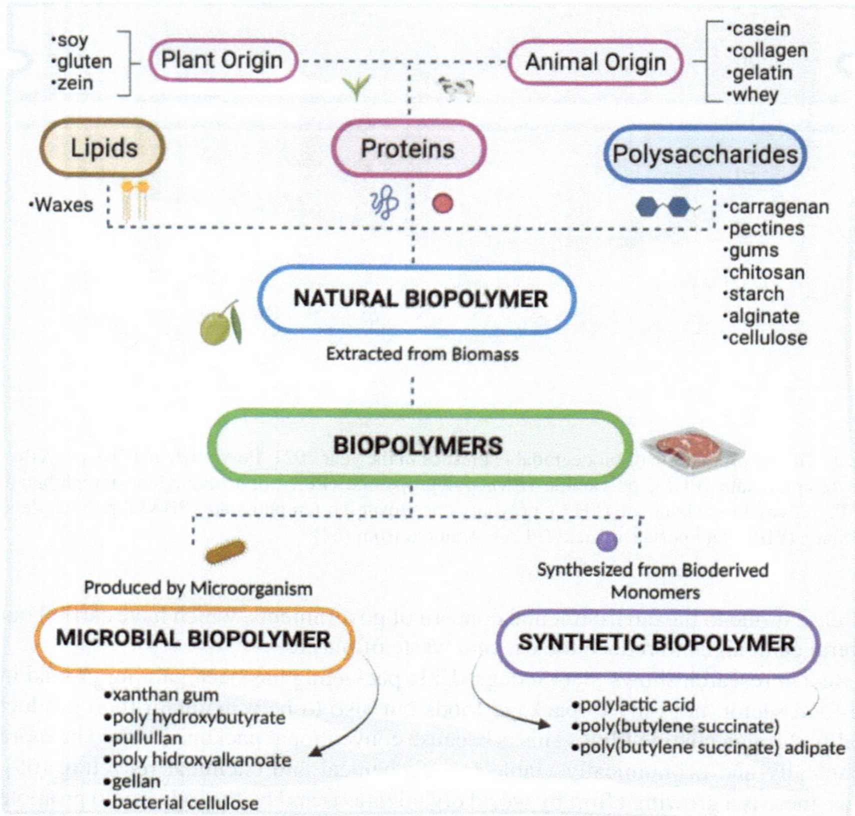

Fig. 4 Feedstock for biopolymers. Inspired from: [13, 27]

Therefore, biofilms should be biodegradable and able to extend food shelf life, made of biodegradable materials to reduce the use of plastic packaging [67]. Starch can be combined with PLA to reduce the cost of PLA films, improve biodegradation, and contribute to the barrier properties of PLA films [68].

These films can also be produced based on polysaccharides, proteins, and lipids, added to a solvent for complete dissolution and a plasticizer (glycerol, monoglycerides, glucose, or polyethylene glycol), as well as optionally include compounds with functional activities that can be antioxidants and antimicrobials [69].

3.1 Polysaccharides

Among the natural biopolymers, polysaccharides (starch, carrageenan, chitin, chitosan, cellulose, and their derivatives) are non-toxic materials with the ability to generate colorless films with an excellent barrier to carbon dioxide and oxygen [69]. In addition, their great accessibility and easy processing make natural polymers attractive for use in the plastics production industry [7].

Starch is a renewable biopolymer, widely used among natural biopolymers, composed of amylose and amylopectin. It is an agricultural raw material usually used to manufacture low-cost biofilms; it is easy to handle and fully biodegradable because it is metabolized by a wide range of organisms [70], and it is easy to process and shows thermoplastic behavior. However, this biopolymer is composed of two polysaccharides (linear amylose and highly branched amylopectin), whose amounts are relative and dependent on the plant source, which affects the properties of this material and its gelatinization behavior [70].

Starch is the final product of the photosynthetic process and constitutes the main carbon reserve of plants [71]. Typically found in potatoes, sweet potato, ginger, and cassava [72], deposited in the form of semi-crystalline granules (Fig. 5), composed of linear and branched macromolecules [71]. Starch-containing films can provide an adequate barrier to oxygen, depending on the relative humidity conditions of the environment [69], which contributes to researchers focusing on starch and cellulose as viable alternatives to conventional plastics [7].

Cellulose is a natural biopolymer composed of a single monomer, glucose under its β-D-glucopyranose, with a regular structure and matrix of hydroxyl groups, tending to form hydrogen-bonded microfibers and crystalline fibers, most found in the form of paper and cardboard [66]. This compound is extracted from wood at around 40–50% by weight, the primary source for obtaining this material [7]. Commercially, cellulose has several derivatives, such as methylcellulose (MC), carboxymethyl cellulose (CMC), ethyl cellulose (EC), hydroxyl propyl cellulose (HPC), hydroxyethyl cellulose (HEC), and cellulose acetate (CA) [73]. Biopolymers produced from pure cellulose, with low mechanical and water barrier properties, poor processing, brittleness, and highly crystalline structures [7], can be improved by combining them with other products, such as lipids, which provide better water vapor barrier properties [74].

Chitin is an acetylated polysaccharide composed of N-acetyl-D-glucosamine, obtained by chemical extraction from shrimps and crabs or by enzymatic hydrolysis or fermentation [66]. Through deacetylation in the presence of the alkali of chitin, chitosan (CH) is obtained, a compound with excellent properties for biofilm formation [74]. This (1–4)-linked 2-amino-deoxy-β-D-glucan polysaccharide is classified as one of the most abundant and non-toxic polysaccharides, with numerous antimicrobial properties. It is commercially inexpensive and easy to dissolve in solutions with a pH below 6.3 [75].

The high degree of biocompatibility makes chitosan a vector for minerals, vitamins, polyphenols, and functional substances, and although it has low water

Fig. 5 Microscopic image of starch in different plant sources [72]

solubility, it is interesting for packaging applications [69]. It enables the formation of strong, hard-to-break, flexible films with similar mechanical properties to conventional films but with moisture sensitivity [73].

3.2 *Proteins*

Different proteins are in evidence in the development of biopolymers, such as corn zein, wheat gluten, soy protein, whey protein, casein, collagen/gelatin, pea protein, rice bran protein, cottonseed protein, peanut protein, and keratin [76]. Films developed from proteins have superior characteristics to polysaccharides, as they have good gas barrier properties and low water vapor permeability due to the extensive intermolecular forces of the protein chains. However, it is still necessary to incorporate plasticizers in their composition to improve flexibility [69]. Casein-based films are an excellent example of application for their high nutritional quality and sensory properties [76].

3.3 Lipids

The most common lipids used in biopolymers are beeswax, candelilla wax, carnauba wax, triglycerides, acetylated monoglycerides, fatty acids, fatty alcohols, resins, fats, and sucrose fatty acid esters. These compounds are commonly used for their hydrophobicity, which, by reducing moisture transfer, controls the dehydration of fresh fruits and vegetables [66]. Lipids favor improved barrier properties when combined with proteins and polysaccharides; isolated ones provide a thicker film than other raw materials [69]. Lipid and protein compounds can exhibit active properties, which offer antimicrobial and antioxidative protection in packaging while ensuring food quality [75].

3.4 Plasticizers

Due to the fragile structure of some materials and poor mechanical strength and moisture sensitivity, it is necessary to incorporate plasticizers into the formulation (glycerol, sorbitol, and polyethylene glycol), along with some polymers' mechanical and thermal processing [7, 69].

These plasticizing agents are additives used to improve the physicochemical characteristics of polymers, responsible for increasing flexibility and distensibility. They are usually liquid, not very volatile, and composed of small molecules. When added to these polymers, the plasticizer modifies the three-dimensional organization, decreases the attractive intermolecular forces, and increases the free volume and the chain mobility [77].

Polyols such as sorbitol, glycerol, and polyethylene glycol are commonly employed as plasticizers in polysaccharide-based films. These plasticizers aim to prevent cracking and enhance the flexibility of the films throughout their processing, handling, and storage stages [78, 79]. The choice of a specific plasticizer is guided by its distinct attributes, encompassing compatibility, which indicates the degree of attraction between the polymer and the plasticizer; permanence, influenced by factors like volatility and susceptibility to extraction; and efficiency, which relies on the plasticizer's ability to solvate or dissolve within the material.

Therefore, in the presence of plasticizers, there is an increase in flexibility by changes in molecular organization and functional properties of films, decreasing the rigidity of the structure. However, adding polyols in the formulation should be done cautiously since excessive dosages result in sticky surfaces and complex applications [77].

3.5 Microbial Polymers

Genetically modified microorganisms or bacteria are capable of producing biopolymers known as polyhydroxyalkanoates (PHA), mainly polyhydroxy butyrate (PHB), hydroxybutyrate (Hb), and hydroxyvalerate (HV) copolymers [76]. PHAs are alkanoic acid biopolyesters, which can be fabricated into films similar to conventional polymers, produced by extrusion and casting, and are used for oily, frozen, and organic foods [73]. A wide range of microorganisms can create this linear aliphatic polyester through the polymerization process of lactic acid, obtained by microbial fermentation of sugars extracted from agricultural products [7].

PHB is considered the most prominent representative of the PHA family, which, besides being completely biodegradable, has physical properties like polypropylene [80]. Obtained by fermentation of microorganisms such as *Cupriavidus necator*, *Bacillus megateriumal*, and *Alcaligenes eutrophus*, it may be combined with PLA to develop films with good physical, mechanical, and thermal properties [7].

Produced by bacteria like *Bacillus subtilis*, the polyglutamic acid (PGA) is a water-soluble and biodegradable biopolymer. It has been explored for packaging applications due to its ability to form films with good gas barrier properties, essential for maintaining product freshness and extending shelf life [34].

Produced by the yeast-like fungus *Aureobasidium pullulans*, pullulan is a polysaccharide with film-forming properties. It is used in packaging due to its biodegradability, excellent oxygen barrier properties, and potential for use in edible coatings [30]. In addition, synthesized by diverse bacterial strains, dextran is a polysaccharide that has garnered attention for its potential in crafting biodegradable packaging films and coatings. Its inherent ability to retain moisture makes it apt for precise humidity regulation [81, 82].

Emerging from cyanobacteria, renowned as blue-green algae, the biopolymer cyanophycin beckons with its ecological origins. Further enhancing its appeal, its adeptness in forming films with excellent mechanical properties fuels excitement in packaging technology. This combination of biodegradability and mechanical properties can revolutionize packaging paradigms, offering a sustainable route to safeguard products while minimizing environmental impact [83].

Derived from the bacterium *Xanthomonas campestris*, xanthan gum plays a pivotal role beyond its bacterial origins. Its film-forming prowess positions it as a promising candidate for eco-friendly biodegradable packaging solutions, further expanding its impact [84].

Exopolysaccharides are complex compounds released by diverse microorganisms, carrying profound importance. These polymers shape shielding frameworks around cells, assisting in creating biofilms and endurance within challenging surroundings. Due to their distinct rheological traits, exopolysaccharides are stabilizers, thickeners, and emulsifiers within industry and scientific domains. Furthermore, their promise in healthcare and ecological restoration highlights their versatile significance, illustrating the dynamic interconnection between microbial mechanisms and broader global utilities [85].

Elastin-like Polypeptides (ELPs) exemplify a remarkable achievement in genetic manipulation, drawing inspiration from the elastin protein inherent to humans. Engineered as synthetic biopolymers, they are nurtured within microbial systems, capitalizing on the efficiency of these platforms for manufacturing. Notably, ELPs exhibit a unique sensitivity to temperature shifts, a quality that unlocks novel possibilities such as coatings capable of self-repair and packaging materials attuned to diverse environmental circumstances. These strides harmonize biology and materials science, presenting a vista of inventive and eco-conscious resolutions [86].

Chitosan, originating from chitin sourced in crustacean shells, can also be synthesized by fungi. Beyond its diverse origins, this biopolymer boasts notable antimicrobial attributes, extending its utility. By blending its antimicrobial potency with its film-forming capability, chitosan offers a potent avenue for enhancing the shelf life of consumables while aligning with eco-friendly packaging trends [87].

These biopolymers highlight the diverse range of microbial-origin materials being developed to address the environmental challenges associated with conventional plastics and packaging materials.

3.6 Synthetic Polymers

Among the synthetic polymers, there's polylactic acid (PLA), a substance derived from corn. This corn-derived PLA is produced by fermenting or other environmentally sustainable agricultural materials to extract starch [88]. PLA stands out for its ecological profile, which includes the inputs used to produce corn seed, fertilizer, lime, electricity, fuels (natural gas, diesel, propane, and gasoline), as well as the use of atmospheric carbon dioxide through photosynthesis and the irrigation water applied to the field [89]. PLA is an advantageous biopolymer due to its biocompatibility, renewability, minimal strength consumption, and mechanical and thermal processability, and it can be produced by extrusion, film casting, and fiber spinning [27].

Similar to materials made from PLA, polyvinyl alcohol (PVA) is a polymer obtained commercially from polyvinyl acetate by alkaline hydrolysis and possesses the ability to mold artificial water-soluble films, with chemical resistance, top biodegradability, low permeability to oxygen and carbon dioxide, and easy to manufacture [90].

4 Functional Aspects in Biodegradable Packaging

To maintain the quality and safety of food, it is necessary to use packaging with appropriate materials and technologies since during food storage, microorganisms (pathogenic and deteriorating), chemical reactions, enzymatic reactions, physical impacts, and environmental parameters can cause undesirable changes [23, 91].

In addition to the traditional functions of packaging, correlated to conservation, transportation, and information about the food, in the past decades, there has been a significant use of concepts related to biodegradable, sustainable, renewable, edible, active, and intelligent properties [21, 92].

4.1 Active and Intelligent Packaging

Active packaging is characterized by a technology where the packaging material interacts directly with the packaged food aiming to maintain or extend the shelf life, and ensure food quality and safety. In addition, there are also packages with properties provided by bioactive compounds that confer antimicrobial, antioxidant, CO_2 and O_2 emitting, and ethylene control characteristics [23].

Active packaging can be classified into absorber systems, which aim to remove undesirable compounds from the free space of the package (oxygen, ethylene, carbon dioxide, water, etc.) or around the food, and that can somehow alter or compromise the quality of the food. Another classification that can be assigned to active packaging is the emitting systems, which, unlike absorbers, incorporate substances into the packaging material (carbon dioxide, ethanol, antioxidant compounds, antimicrobial substances, preservatives, etc.) and are gradually released into the food [93].

Ethylene is a compound produced and released from plant metabolism, resulting from a biochemical process. It is responsible for plant tissue's growth, ripening, aging, and death. When a climacteric vegetable is packaged, and as the concentration of ethylene increases within the package, the vegetable's respiration rate also increases and accelerates the senescence process. However, when the ethylene concentration inside the package is reduced using an absorber (Fig. 6a), the senescence process becomes slower, and therefore, the vegetable has an extended shelf life [94].

Water represents a significant mass fraction in fruits and vegetables, and its dynamics are the driving force for the transfer from the inside of the food to the

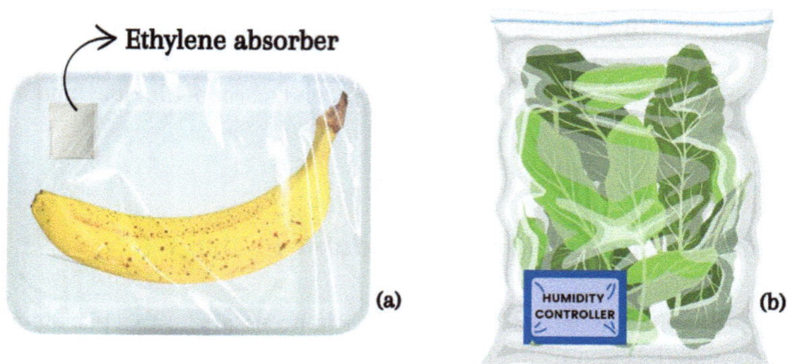

Fig. 6 Packaging with ethylene absorber (**a**) and humidity controller (**b**)

outside environment. A negative factor of mass transfer in preserving fruits and vegetables is the dehydration of these foods, which changes the nutrient content, flavor, and texture. In addition, the loss of mass when it reaches 3–10% causes the product to begin to shrivel and shrink irreversibly, making the product's appearance less attractive to the consumer, and commercial acceptance is lost [19, 95].

The frequent condensation inside fruit and vegetable packages makes it difficult to visualize the product, which is fundamental in determining the purchase. Active packaging with moisture absorbers (Fig. 6b) minimizes this problem and reduces quality changes in these foods. Techniques to reduce the humidity inside the packaging are employed to obtain a drier environment, ideal for vegetable preservation [96, 97].

Several compounds with antimicrobial potential are used as alternatives in the formulation of materials applied in food packaging or inside it. The antimicrobial agent incorporated into the active packaging can be in a solid or gaseous state [28]. Active packaging may contain antimicrobial active compounds impregnated into films, which, during storage, inactivate or prevent microbial proliferation, increasing the shelf life of the food [98].

Besides microbiological spoilage, spoilage by lipid oxidation can also affect the sensory and nutritional characteristics of foods, especially those with a high content of unsaturated lipids, which are the most susceptible to oxidation. Lipid oxidation reactions cause changes in the flavor of the food, bringing out the rancidity, which significantly affects the acceptability of the product [99]. There are two types of active packaging with antioxidant action: packaging that releases the antioxidant compounds into the food and packaging that removes undesirable compounds, such as oxygen, which causes oxidation [28, 99]. Fresh acerola, a fruit rich in a natural antioxidant that quickly degrades (vitamin C) when coated with a whey-based film, showed reduced microbiological contamination, improved coloration, and more significant conservation of the vitamin C content [19].

Foods that are susceptible to mechanical damage during harvest, transport, and storage processes, such as strawberries, need special care to avoid alterations in the quality of their properties. Study shows that strawberries coated with material containing propolis showed more excellent stability in vitamin C content [28].

Intelligent packaging does not directly contribute to increasing the shelf life of food since its main objective is to monitor changes that occur in the food during storage. The purpose of the intelligent packaging application is to ensure the provision of information about the quality of the food, contributing to its safety during transportation and storage [23]. Intelligent packaging works as sensors and indicators about the conservation and freshness of food, presenting the product's conditions in real time [100].

Some fruits do not show color change and other signs of ripeness that are easily noticeable when ripe make identifying the degree of ripeness challenging. However, they produce volatile compounds during ripening that can be identified from ripeness indicators, which warn the consumer through color change [101]. Thus, by indicating the color, the consumer can choose his preference according to the degree of ripeness [102].

As high-protein foods like fish or meat lose their freshness over time during storage, enzymes and bacteria degrade the food. This degradation leads to the accumulation of volatile compounds inside the packaging, including ammonia and di/trimethylamine (organic amines). These compounds can be easily detected by pH and colorimetric indicators, which change color when exposed to volatile compounds, showing consumers the quality of these foods [103].

Intelligent packaging, which has time and temperature indicators, presents the actual situation of what is happening inside the packaging regarding storage conditions. These devices are easy to use, as they can be integrated as part of the packaging and allow the consumer to check the quality of the food through a color response corresponding to the quality of the food at a given temperature. In addition, this system assists in the transport conditions of the food, ensuring safety during the shelf life until consumption [104].

4.2 Smart Packaging Systems

The quality of most foods is affected by sudden changes in storage temperature, and at every step of the supply chain, this parameter is a critical point for food distributors and manufacturers. The use of sensors even assists in checking the time and temperature history of refrigerated foods. In addition to monitoring freshness, microbiological growth, and chemical changes in food, they are vital devices for maintaining consumer safety [105].

Smart packaging monitors changes in the product or its environment through intelligent packaging combined with compounds that act on the changes. Smart packaging uses biosensors or chemical sensors that monitor food for quality and safety from producers to consumers [106].

Food monitoring can be done, for example, through the detection and analysis of freshness, oxygen, carbon dioxide, pH levels, time or temperature, leaks, and pathogens. Smart packaging allows the tracking of the product throughout its life cycle and the analysis of the conditions outside and inside the packaging, communicating with the manufacturer and consumer about the state of the product at any time [107].

Incorporating functional substances in packaging formulation can add value to the product and increase the shelf life of food due to the added active properties. These substances, besides helping in the preservation, can facilitate the consumer to identify the freshness state of the food through intelligent properties (packaging smart). Incorporating functional substances in packaging can add value to the package and increase the shelf life of food through packaging with active properties. Furthermore, besides assisting in food preservation, it can facilitate the consumer to identify the freshness state of the food through packages with intelligent properties.

Anthocyanin, a bioactive compound found in flowers, vegetables, and fruits, has different types of coloration (orange, red, purple, and blue). This compound has antioxidant and antimicrobial properties, in addition to the ability to change color at different pH. The color change results from modifying the chemical structure of the phenolic substances present in the anthocyanin due to the pH, thus making its application in pH-sensitive packages to evaluate the freshness and quality of the food [23].

5 Advances in Research and Applications

Food packaging has a long and varied history. It has been used to preserve and store food for long periods, making it more appealing to consumers [27]. Early forms of food packaging with almost no refinement included wrapping food in leaves, storing it in hollowed-out logs, and using clay containers [108]. In the 1800s, with the Industrial Revolution, there was an evolution to storage in metal cans with Nicholas Appert's preservation method, in addition to the glass jars and bottles that were common at that time [108, 109].

In the early 1900s, after the Second World War, the development of various plastic polymers used for food applications was recorded. Today, food packaging is developed from multiple materials (plastic, metal, and glass), most notably bio-based packaging [109, 110].

Applying bio-based materials in the sustainable packaging industry shows strong growth due to recent consumer trends toward greener (environmentally friendly) packaging and lower waste generation [111–113]. As of 2020, only 1% of plastic produced in Europe annually was derived from bio-based materials and was biodegradable. However, an expected increase of 9.45 million tons of bio-based materials for packaging is expected by 2023 [113].

Throughout the years, there have been advances in research into food packaging, as presented in Table 1. In the early-1900s post-war, petroleum-based industries focused on plastic products for the food industry, mainly for package purposes [108]. With globalization and advances in import/export, the necessity of different packaging materials started to grow [114]. In the late 1990s and early 2000s, research was improved towards greener, environmentally friendly, biodegradable, and compostable materials to minimize pollution impacts and benefit food preservation with active and intelligent packaging [110, 115].

As the research advanced throughout the years, it is possible to observe an increase in specific trends in using bio-based materials and expected results in food packaging applications [133]. Table 2 presents a few relevant studies in developing bio-based, intelligent, and active packaging, with biomaterials, properties obtained, and the food product in which the packaging was applied.

Studies have evolved since the early 2000s, considerably increasing in the past decade. From an industrial point of view, there is still much to be considered, despite the innovation, for intelligent and innovative bio-based packaging to become a viable product [145]. An interdisciplinary approach is vital to overcome uncertainties and global challenges. However, it is in our near future to address such impasses and overcome difficulties related to eco-friendly practices in the food industry [133, 146].

Table 1 Timeline of studies on bio-based food packaging

Study	Overview	Year	References
Primary, secondary, and tertiary packaging for shipment	Indicates the necessity of research advances in packaging for overseas export that are resistant, compact, and environmentally friendly	1943	[114]
Application of bio-based materials in food packaging	Challenges involved in the reduction of food packaging materials and substitution for biologically based packaging	1999	[116]
Development of bioactive and bio-based food packaging	Study of bioactive cellulose-based packaging material enhanced with bacteriocin and its effects on foods	2000	[117]
Industrial production of bio-based materials	Review of industrial achievements in the development of bioplastics from renewable sources and their applications	2001	[118]
Alternative bio packaging for foods	Experimental study on PHA/PHB as food packaging to replace petroleum-based plastics	2005	[119]
Improvement of starch-based films	Analysis of natural biopolymers on physical and chemical properties of biofilms	2006	[120]
Innovations in food packaging	Overview of packaging in food and beverage industries with some future trends and innovations	2008	[121]
Packaging materials from renewable sources	Review on the chemistry of hydrogels in the development of biopolymers for food packaging materials	2009	[122]
Environmental impacts of bio-based packaging	Comparison between conventional and bio-based materials used in food packaging, and impacts assessment in ecosphere	2010	[123]
Application of nanocomposites in food packaging	Review on the enhancement of food packaging performance made from bio-based materials and its properties	2013	[124]
History and challenges of bio-based packaging materials	Comparative study on the substitution and evolution of conventional plastics for bio-based plastics	2014	[125]
Development and testing of hydrogel films	Advances in the creation and improvement of active hydrogel films with nanoparticles	2015	[126]
Research on developing packaging with new biopolymers	Experimentation with different biopolymers from polysaccharides of chitosan and quinoa protein extracts	2016	[127]
Recent trends in food packaging techniques	Review research on future trends and innovative packaging technologies	2018	[128]
Study on properties of different bio-based films	Investigation of properties of bio-compostable films enhanced with different nanomaterials	2019	[129]
Biodegradable packaging from bio-based materials and active packaging	Review on alternatives for petrochemical-based materials and advances in bio-based active food packaging	2020	[111]

(continued)

Table 1 (continued)

Study	Overview	Year	References
Biopolymers and biodegradability in food packaging	A review on food packaging, plastics substitutes, biodegradable, environmentally friendly, and biopolymers packaging	2021	[130]
Advances in bio-based smart active packaging	A review of the main raw materials used for bio-based smart active packaging and its application in food preservation	2022	[131]
Biopolymeric barrier materials for food packaging	A review of advances in the development, production, and application of biopolymers in packaging such as films and paper coating	2022	[132]

6 Final Considerations

The concept of sustainability has countless definitions, but the advances in research on this topic have been of great interest. In their processes, companies incorporate sustainability concepts concerned with environmental issues [45].

The use of biodegradable food packaging addresses the environmental issue of the accumulation of conventional plastic waste (based on non-renewable resources) and, consequently, the reduction of global warming and toxic gas emissions. In this regard, biopolymers present a potential sustainable application in food packaging and can also assign functional compounds in their structures [16].

The food industry seeks to develop and apply sustainable and efficient systems, especially in the packaging sector, in which most of the materials used are based on non-renewable raw materials that cause damage to the environment and human health, going against the concept of sustainability.

In this sense, it is essential that government agencies and companies work together to implement and efficiently implement programs that contain environmental impacts [48], generating a sustainable food system for the planet [52].

Several studies emphasize the development and improvement of packaging obtained from biomass [47, 55, 56], which aim to apply biopolymers as sustainable alternatives to non-biodegradable packaging.

Innovative technologies in food packaging are extrinsic ways to ensure food safety, quality, freshness, and maintenance. Recent research on multifunctional systems, such as active and smart packaging, has been prominent because of its benefits for food preservation and quality [18, 21, 23].

These studies consolidate and reinforce the promising trend of new possibilities for biodegradable food packaging, capable of increasing the shelf life and helping the consumer identify the conservation status of food, which contributes to reducing food waste and the consequent environmental impact.

Packaging research constantly evolves with new raw materials, new technology applications, and smart packaging. Natural compounds are materials with ideal potential for the development of biodegradable packaging and with good film-forming properties [147].

Table 2 Application of bio-based materials in food packaging

Materials	Properties	Food	References
Calcium alginate film with zinc oxide nanoparticles	Antimicrobial activity	Poultry meat	[134]
Polylactic acid (PLA) with ZnO:Cu and Ag bio-nanocomposites	Mechanical, thermal, water vapor, oxygen, and carbon dioxide barrier; antimicrobial, and low migration of nanocomposites	Food simulants	[135]
Chitosan, nano zinc oxide, and neem essential oil	Tensile strength improved, and antimicrobial activity against *E. coli*	Carrot	[136]
Chitosan, poly-vinyl alcohol, and titanium nanocomposites	Effective antimicrobial activity against Gram-positive and Gram-negative microorganisms and fungi; extension in shelf life	Soft white cheese	[137]
Alginate/nano-Ag coating	Microbial spoilage reduction, and improved sensory attributions	Shiitake mushroom	[138]
Tilapia skin gelatin with coconut husk ethanolic extract and nanoclay	Lower lipid oxidation products; improvement in moisture barrier; extension in shelf life	Meat powder	[139]
Chitosan and nanoemulsion of mandarin essential oil	Antimicrobial activity in *L. monocytogenes*	Green beans	[140]
Cassava starch and grape skin anthocyanins	pH sensor showing color alteration with pH change in environment	Beef and fish	[141]
Starch, gelatin, and red radish anthocyanins	Color response in pH ranging from 2 to 12	Prawn and poultry meat	[142]
Whey protein isolate (WPI) and oregano oil	Antimicrobial activity for spoilage microorganisms	Fresh beef	[143]
Gelatin, chitosan, and clove essential oil	Antimicrobial activity	Fish	[144]

There are many possibilities for various raw materials for forming the polymeric matrix for biodegradable food packaging. Polysaccharides, proteins, lipids, plasticizers, and active compounds enable multiple functionalities for food preservation. These different types of substances help the polymer structure through improvements in physical–chemical characteristics, such as increased flexibility and reduced moisture transfer [69].

The development of active films has been identified as one of the main focus areas in packaging research in the past decades because, besides being effective in food preservation, they promote increasing food shelf life by adding antimicrobial and antioxidant components in the polymeric matrix [19].

The valorization of organic waste through green routes promoting the circular economy can considerably reduce greenhouse gas emissions and the energy used to produce food packaging material through industrial waste compared to petroleum-based production. The studies and knowledge about biodegradable packaging occur in parallel with the valorization of industrial by-products because they have low cost and help to develop the incorporation of this type of packaging in consumers' daily lives [23] economically.

Although biodegradable packaging has numerous advantages and applications, especially with recently developed new technologies, biofilms still face some commercial challenges. Ways to improve the properties of the polymer matrix to be equivalent to petroleum-based packaging are still under development. Water and gas resistance, mechanical properties, and physical strength are the main challenges to overcome when it comes to biodegradable packaging. The prerequisites for sustainable packaging are good gas barrier properties to keep food safe and resistance against water [148].

Nanotechnology is a trending area for biodegradable packaging. Multifunctional nanocomposites from biomass have been developed for application in active food packaging. With minimal environmental risks compared to petroleum-based packaging, they exhibit interesting functionalities and physicochemical properties [47].

Further studies on essential aspects regarding improving biodegradable packaging properties, investigation of new economic sources and commercial expansion, continuous production, and implementation of regulations and safety are crucial for successfully adopting biodegradable food packaging [47].

Consumer acceptance is a decisive factor in the concept of biodegradable packaging, and to achieve its commercial success, further studies and documentation are needed to prove characteristics such as biodegradability, organoleptic and sensory aspects, safety, and legal confirmation. To overcome these challenges, considerable research efforts should focus on the following key elements: promoting consumer awareness and advertising the importance of biodegradable packaging can increase its acceptability.

Acknowledgements The authors thank Fundação de Amparo à Pesquisa e Inovação do Estado de Santa Catarina (FAPESC) for the financial support (grant number 2022TR000162), Coordenação de Aperfeiçoamento de Pessoal de Nível Superior—CAPES [Financial Code: 001], Embrapa Florestas (Colombo), and the Food Preservation and Innovation Laboratory (FURB).

References

1. Coll A, Kleineidam J (2020) Improving low-loss food supply chains in developing countries based on a case study of Tanzania. https://www.researchgate.net/publication/337103004
2. Chi K, Wang H, Catchmark JM (2020) Sustainable starch-based barrier coatings for packaging applications. Food Hydrocoll 103:105696
3. Mangaraj S, Yadav A, Bal LM, Dash SK, Mahanti NK (2019) Application of biodegradable polymers in food packaging industry: a comprehensive review. J Packag Technol Res 3(1):77–96
4. Han JW, Ruiz-Garcia L, Qian JP, Yang XT (2018) Food packaging: a comprehensive review and future trends. Compr Rev Food Sci Food Saf 17(4):860–877
5. Russell DAM (2014) Sustainable (food) packaging—an overview. Food Addit Contam: Part A 31(3):396–401
6. de Léis CM, Nogueira AR, Kulay L, Tadini CC (2017) Environmental and energy analysis of biopolymer film based on cassava starch in Brazil. J Clean Prod 143:76–89
7. Sid S, Mor RS, Kishore A, Sharanagat VS (2021) Bio-sourced polymers as alternatives to conventional food packaging materials: a review. Trends Food Sci Technol [Internet] 115(June):87–104. https://doi.org/10.1016/j.tifs.2021.06.026
8. Trinetta V (2018) Definition and function of food packaging. In: Reference module in food science. Elsevier
9. Ivonkovic A, Zeljko K, Talic S, Lasic M (2017) Biodegradable packaging in the food industry. J Food Saf Food Qual 68:26–38
10. Song JH, Murphy RJ, Narayan R, Davies GBH (2009) Biodegradable and compostable alternatives to conventional plastics. Philos Trans R Soc B: Biol Sci 364(1526):2127–2139
11. Mendes AC, Pedersen GA (2021) Perspectives on sustainable food packaging: is bio-based plastics a solution? Trends Food Sci Technol 112:839–846
12. Yuvaraj D, Iyyappan J, Gnanasekaran R, Ishwarya G, Harshini RP, Dhithya V et al (2021) Advances in bio food packaging—an overview. Heliyon 7(9):e07998
13. Shaikh S, Yaqoob M, Aggarwal P (2021) An overview of biodegradable packaging in food industry. Curr Res Food Sci 4:503–520
14. Zhong Y, Godwin P, Jin Y, Xiao H (2020) Biodegradable polymers and green-based antimicrobial packaging materials: a mini-review. Adv Ind Eng Polym Res 3(1):27–35
15. Evyan YCY, Liew MS, Patricia J, Chong MY, Zairul ZA (2022) Biodegradable food packaging and film: a short review. Food Res [Internet] 6(S1):1–12. https://www.myfoodresearch.com/uploads/8/4/8/5/84855864/_1__fr-kliafp-007_evyan.pdf
16. Priyadarshi R, Roy S, Purohit SD, Ghosh T (2022) Biopolymers for food packaging and biomedical applications: options or obligations? Coatings 12(9):1261
17. Gupta V, Biswas D, Roy S (2022) A comprehensive review of biodegradable polymer-based films and coatings and their food packaging applications. Materials (MDPI) 15
18. Hoffmann TG, Peters DA, Angioletti BL, Bertoli SL, Péres LV, Ratto Reiter MG et al (2019) Potentials nanocomposites in food packaging. Chem Eng Trans 75:253–258
19. Pergentino Dos Santos S, Angioletti BL, Hoffmann TG, Gonçalves MJ, Bertoli SL, Hlebová M et al (2022) Whey based biopolymeric coating as an alternative to improve quality of fresh fruits (Malpighia emarginata D.C.) from Southern Brazil. J Microbiol Biotechnol Food Sci 11(5):e5433
20. Angioletti BL, dos Santos SP, Hoffmann TG, Bertoli SL, de Souza CK (2020) Aloe vera gel as natural additive to improve oxidative stability in refrigerated beef burger stored in aerobic and vacuum packaging. In: AIChE annual meeting
21. Ghosh T, Priyadarshi R, Krebs de Souza C, Angioletti BL, Rhim JW (2022) Advances in pullulan utilization for sustainable applications in food packaging and preservation: a mini-review. Trends Food Sci Technol 125:43–53
22. Mendes JF, Paschoalin RT, Carmona VB, Sena Neto AR, Marques ACP, Marconcini JM et al (2016) Biodegradable polymer blends based on corn starch and thermoplastic chitosan processed by extrusion. Carbohydr Polym 137:452–458

23. Hoffmann TG, Angioletti BL, Bertoli SL, de Souza CK (2021) Intelligent pH-sensing film based on jaboticaba peels extract incorporated on a biopolymeric matrix. J Food Sci Technol
24. Roy S, Ezati P, Biswas D, Rhim JW (2022) Shikonin functionalized packaging film for monitoring the freshness of shrimp. Materials 15(19):6615
25. Laorenza Y, Chonhenchob V, Bumbudsanpharoke N, Jittanit W, Sae-tan S, Rachtanapun C et al (2022) Polymeric packaging applications for seafood products: packaging-deterioration relevance, technology and trends. Polymers (Basel) 14(18):3706
26. Storz H, Vorlop KD (2013) Applied agricultural and forestry research. In: Landbauforsch, pp 321–32
27. Yuvaraj D, Iyyappan J, Gnanasekaran R, Ishwarya G, Harshini RP, Dhithya V et al (2021) Advances in bio food packaging—an overview. Heliyon 7(9)
28. Finardi S, Gabriela Hoffmann T, Louise Angioletti B, Mueller E, Souza Lazzaris R, Leandro Bertoli S et al (2022) Development and application of antioxidant coating on Fragaria spp. stored under isothermal conditions. J Microbiol Biotechnol Food Sci. https://doi.org/10.15414/jmbfs.5432
29. van den Oever M, Molenveld K, van der Zee M, Bos H (2017) Bio-based and biodegradable plastics: facts and figures: focus on food packaging in the Netherlands
30. Singh R, Sharma R, Shaqib M, Sarkar A, Chauhan KD (2021) Biodegradable polymers as packaging materials. In: Biopolymers and their industrial applications. Elsevier, pp 245–259
31. Chen L, Qiang T, Chen X, Ren W, Zhang HJ (2022) Gelatin from leather waste to tough biodegradable packaging film: one valuable recycling solution for waste gelatin from leather industry. Waste Manag 145:10–19
32. Jamróz E, Tkaczewska J, Kopeć M, Cholewa-Wójcik A (2022) Shelf-life extension of salmon using active total biodegradable packaging with tea ground waste and furcellaran-CMC double-layered films. Food Chem 383:132425
33. Priyadarshi R, Negi YS (2017) Effect of varying filler concentration on zinc oxide nanoparticle embedded chitosan films as potential food packaging material. J Polym Environ 25(4):1087–1098
34. Evyan YCY, Liew MS, Patricia J, Chong MY, Zairul ZA (2022) Biodegradable food packaging and film: a short review. Food Res 6(S1):1–12
35. Asgher M, Qamar SA, Bilal M, Iqbal HMN (2020) Bio-based active food packaging materials: sustainable alternative to conventional petrochemical-based packaging materials. Food Res Int 137:109625
36. Adam F, Othman NA, Yasin NHM, Cheng CK, Azman NAM (2022) Evaluation of reinforced and green bioplastic from carrageenan seaweed with nanocellulose. Fibers Polym 23(10):2885–2896
37. Goswami P, O'Haire T (2016) Developments in the use of green (biodegradable), recycled and biopolymer materials in technical nonwovens. In: Advances in technical nonwovens. Elsevier, pp 97–114
38. de Souza ML, de Freitas DAF, Gonçalves CS (2018) Biodegradability for the production of renewable energy. Divers@ Revista Eletrônica Interdisciplinar 11:26–38
39. Abbasi S, Ammar HS (2021) Biodegradable materials and their applications in sample preparation techniques—a review. Microchem J 171:106831
40. Elkington J (1994) Towards the sustainable corporation: win-win-win business strategies for sustainable development. Calif Manag Rev 36(2):90–100
41. Sartori S, Da Silva FL, de Campos LM S (2014) Sustainability and sustainable development: a taxonomy in the field of literature. Ambiente Sociedade 17:1–20
42. Olawumi TO, Chan DWM (2018) A scientometric review of global research on sustainability and sustainable development. J Clean Prod [Internet] 183:231–250. https://linkinghub.elsevier.com/retrieve/pii/S095965261830475X
43. Olawumi TO, Chan DWM (2018) A scientometric review of global research on sustainability and sustainable development. J Clean Prod 183:231–250
44. Angelstam P, Andersson K, Annerstedt M, Axelsson R, Elbakidze M, Garrido P et al (2013) Solving problems in social-ecological systems: definition, practice and barriers of transdisciplinary research. Ambio 42(2):254–265

45. Geissdoerfer M, Savaget P, Bocken NMP, Hultink EJ (2017) The circular economy—a new sustainability paradigm? J Clean Prod [Internet] 143:757–768. https://linkinghub.elsevier.com/retrieve/pii/S0959652616321023
46. de Vries H (2023) The role of food science and technology in the future partnership sustainable food systems. Trends Food Sci Technol 131:28–30
47. Shi J, Zhang R, Liu X, Zhang Y, Du Y, Dong H et al (2023) Advances in multifunctional biomass-derived nanocomposite films for active and sustainable food packaging. Carbohydr Polym 301:120323
48. Jambeck JR, Geyer R, Wilcox C, Siegler TR, Perryman M, Andrady A et al (2015) Plastic waste inputs from land into the ocean. Science 347(6223):768–771
49. Mendes AC, Pedersen GA (2021) Perspectives on sustainable food packaging: is bio-based plastics a solution? Trends Food Sci Technol [Internet] 112(March):839–846. https://doi.org/10.1016/j.tifs.2021.03.049
50. Fogt Jacobsen L, Pedersen S, Thøgersen J (2022) Drivers of and barriers to consumers' plastic packaging waste avoidance and recycling—a systematic literature review. Waste Manag 141:63–78
51. Lieder M, Rashid A (2016) Towards circular economy implementation: a comprehensive review in context of manufacturing industry. J Clean Prod 115:36–51
52. EC (2022) Sustainable food systems partnership for people, planet and climate
53. van Rossum T (2021) Marine biodegradability review of plastics. Water Cycle 2:38–43
54. Mahmoudi M, Parviziomran I (2020) Reusable packaging in supply chains: a review of environmental and economic impacts, logistics system designs, and operations management. Int J Prod Econ 228:107730
55. Li Y, Ren J, Zhang G, Li X (2021) High-barrier and antibacterial films based on PET/SiOx for food packaging applications. Food Sci Technol (Brazil) 41(3):763–767
56. Ahmad A, Qurashi A, Sheehan D (2023) Nano packaging—progress and future perspectives for food safety, and sustainability. Food Packag Shelf Life 35:100997
57. Food PathS (2022) The path towards a sustainable food system
58. Emadian SM, Onay TT, Demirel B (2017) Biodegradation of bioplastics in natural environments. Waste Manag 59:526–536
59. Qin M, Chen C, Song B, Shen M, Cao W, Yang H et al (2021) A review of biodegradable plastics to biodegradable microplastics: another ecological threat to soil environments? J Clean Prod 312:127816
60. Afrin H, Huda N, Abbasi R (2021) Study on end-of-life tires (ELTs) recycling strategy and applications. IOP Conf Ser Mater Sci Eng 1200(1):012009
61. Zink T, Geyer R (2017) Circular economy rebound. J Ind Ecol 21(3):593–602
62. Kumari SVG, Pakshirajan K, Pugazhenthi G (2022) Recent advances and future prospects of cellulose, starch, chitosan, polylactic acid and polyhydroxyalkanoates for sustainable food packaging applications. Int J Biol Macromol 221:163–182
63. Priyadarsini P, Biswal M, Gupta S, Mohanty S, Nayak SK (2022) Development and characterization of ester modified endospermic guar gum/polyvinyl alcohol (PVA) blown film: approach towards greener packaging. Ind Crops Prod 187:115319
64. Charlebois S, Walker TR, Music J (2022) Comment on the food industry's pandemic packaging dilemma. Front Sustain 24:3
65. Ellis LD, Rorrer NA, Sullivan KP, Otto M, McGeehan JE, Román-Leshkov Y et al (2021) Chemical and biological catalysis for plastics recycling and upcycling. Nat Catal 4(7):539–556
66. Ferreira A, Alves V, Coelhoso I (2016) Polysaccharide-based membranes in food packaging applications. Membranes (Basel) 6(2):22
67. Martins da Costa JC, Lima Miki KS, da Silva Ramos A, Teixeira-Costa BE (2020) Development of biodegradable films based on purple yam starch/chitosan for food application. Heliyon 6(4):e03718
68. García-Guzmán L, Cabrera-Barjas G, Soria-Hernández CG, Castaño J, Guadarrama-Lezama AY, Rodríguez LS (2022) Progress in starch-based materials for food packaging applications. Polysaccharides 3(1):136–177

69. Vlčko T, Golian J, Fikselová M, Rybnikár S (2022) Current overview in the field of application of edible coatings/films (meat products examples). J Microbiol Biotechnol Food Sci 3:e9281
70. Tang X, Alavi S, Herald TJ (2008) Effects of plasticizers on the structure and properties of starch–clay nanocomposite films. Carbohydr Polym 74(3):552–558
71. Silva RM, Ferreira GF, Shirai MA, Haas Â, Scherer ML, Franco CML et al (2008) Características físico-químicas de amidos modificados com permanganato de potássio/ácido lático e hipoclorito de sódio/ácido lático. Ciênc Tecnol Aliment 28(1):66–77
72. Leonel M (2007) Análise da forma e tamanho de grânulos de amidos de diferentes fontes botânicas. Ciênc Tecnol Aliment 27(3):579–588
73. Grujić R, Vujadinović D, Savanović D (2017) Biopolymers as food packaging materials. In: Advances in applications of industrial biomaterials. Springer International Publishing, Cham, pp 139–160
74. Bravin B, Peressini D, Sensidoni A (2006) Development and application of polysaccharide–lipid edible coating to extend shelf-life of dry bakery products. J Food Eng 76(3):280–290
75. Zhang X, Ismail BB, Cheng H, Jin TZ, Qian M, Arabi SA et al (2021) Emerging chitosan-essential oil films and coatings for food preservation—a review of advances and applications. Carbohydr Polym 273:118616
76. Galgano F, Condelli N, Favati F, di Bianco V, Perretti G, Caruso MC (2015) Biodegradable packaging and edible coating for fresh-cut fruits and vegetables. Ital J Food Sci 27(1):1–20
77. dos Santos PI, Galindo MV, de Medeiros JAS, Yamashita F, Alvim ID, Ferreira Grosso CR et al (2019) Comparative study of the properties of soy protein concentrate films containing free and encapsulated oregano essential oil. Food Packag Shelf Life 22:100419
78. Resta VG, Mali S (2019) Efeito de sacarose e glicerol como plastificantes em filmes orodispersíveis de amido e gelatina. Iniciação Científica Cesumar 21(1):15
79. de Oliveira Santos RP, Castro DO, Ruvolo-Filho AC, Frollini E (2014) Processing and thermal properties of composites based on recycled PET, sisal fibers, and renewable plasticizers. J Appl Polym Sci 131(12):n/a–n/a
80. Dhall RK, Alam MS (2020) Biodegradable packaging. In: Choudhury I, Hashmi S (eds) Encyclopedia of renewable and sustainable materials. Elsevier Inc., pp 26–43
81. Delvart A, Moreau C, Cathala B (2022) Dextrans and dextran derivatives as polyelectrolytes in layer-by-layer processing materials—a review. Carbohydr Polym 293:119700
82. de Souza CK, Ghosh T, Lukhmana N, Tahiliani S, Priyadarshi R, Hoffmann TG et al (2023) Pullulan as a sustainable biopolymer for versatile applications: a review. Mater Today Commun 36:106477
83. Du J, Li L, Zhou S (2019) Microbial production of cyanophycin: from enzymes to biopolymers. Biotechnol Adv [Internet] 37(7):107400. https://linkinghub.elsevier.com/retrieve/pii/S0734975019300813
84. Abang S, Wong F, Sarbatly R, Sariau J, Baini R, Besar NA (2023) Bioplastic classifications and innovations in antibacterial, antifungal, and antioxidant applications. J Bioresour Bioproducts [Internet]. https://linkinghub.elsevier.com/retrieve/pii/S2369969823000452
85. Kavitake D, Devi PB, Delattre C, Reddy GB, Shetty PH (2023) Exopolysaccharides produced by Enterococcus genus—an overview. Int J Biol Macromol [Internet] 226:111–120. https://linkinghub.elsevier.com/retrieve/pii/S0141813022029427
86. Garanger E, Lecommandoux S (2022) Emerging opportunities in bioconjugates of elastin-like polypeptides with synthetic or natural polymers. Adv Drug Deliv Rev [Internet] 191:114589. https://linkinghub.elsevier.com/retrieve/pii/S0169409X22004793
87. Hisham F, Maziati Akmal MH, Ahmad F, Ahmad K, Samat N (2023) Biopolymer chitosan: potential sources, extraction methods, and emerging applications. Ain Shams Eng J [Internet] 102424. https://linkinghub.elsevier.com/retrieve/pii/S2090447923003131
88. Gan I, Chow WS (2018) Antimicrobial poly(lactic acid)/cellulose bionanocomposite for food packaging application: a review. Food Packag Shelf Life 17:150–161
89. Vink ETH, Glassner DA, Kolstad JJ, Wooley RJ, O'Connor RP (2007) The eco-profiles for current and near-future NatureWorks® polylactide (PLA) production. Ind Biotechnol 3(1):58–81

90. Rydz J, Musioł M, Zawidlak-Węgrzyńska B, Sikorska W (2018) Present and future of biodegradable polymers for food packaging applications. In: Biopolymers for food design. Elsevier, pp 431–467
91. Mallick N, Pal D, Soni AB, Jhariya DC, Singh D (2020) Starch based antimicrobial food packaging film towards a sustainable environment. In: IOP conference series: earth and environmental science. IOP Publishing Ltd
92. Priyadarshi R, Purohit SD, Roy S, Ghosh T, Rhim JW, Han SS (2022) Antiviral biodegradable food packaging and edible coating materials in the COVID-19 era: a mini-review. Coatings 12(5):577
93. Kerry JP, O'Grady MN, Hogan SA (2006) Past, current and potential utilisation of active and intelligent packaging systems for meat and muscle-based products: a review. Meat Sci 74(1):113–130
94. Bazzano M, Barolo C, Buscaino R, D'Agostino G, Ferri A, Sangermano M et al (2016) Controlled atmosphere in food packaging using ethylene-α-cyclodextrin inclusion complexes dispersed in photocured acrylic films. Ind Eng Chem Res 55(3):579–585
95. Hoffmann TG, Ronzoni AF, da Silva DL, Bertoli SL, de Souza CK (2021) Cooling kinetics and mass transfer in postharvest preservation of fresh fruits and vegetables under refrigerated conditions. Chem Eng Trans 87:115–120
96. Gaikwad KK, Singh S, Ajji A (2019) Moisture absorbers for food packaging applications. Environ Chem Lett 17(2):609–628
97. Hoffmann TG, Ronzoni AF, da Silva DL, Bertoli SL, de Souza CK (2021) Impact of household refrigeration parameters on postharvest quality of fresh food produce. J Food Eng 1:306
98. Mousavi Khaneghah A, Hashemi SMB, Limbo S (2018) Antimicrobial agents and packaging systems in antimicrobial active food packaging: an overview of approaches and interactions. Food Bioprod Process 111:1–19
99. Gómez-Estaca J, López-de-Dicastillo C, Hernández-Muñoz P, Catalá R, Gavara R (2014) Advances in antioxidant active food packaging. Trends Food Sci Technol 35:42–51
100. Nemes SA, Szabo K, Vodnar DC (2020) Applicability of agro-industrial by-products in intelligent food packaging. Coatings (MDPI AG) 10
101. Gao Y, Jiao L, Jiao F, Dong D (2022) Non-intrusive prediction of fruit spoilage and storage time via detecting volatiles in sealed packaging using laser spectroscopy. LWT 155:112930
102. Shao P, Liu L, Yu J, Lin Y, Gao H, Chen H et al (2021) An overview of intelligent freshness indicator packaging for food quality and safety monitoring. Trends Food Sci Technol 118:285–296
103. Bhargava N, Sharanagat VS, Mor RS, Kumar K (2020) Active and intelligent biodegradable packaging films using food and food waste-derived bioactive compounds: a review. Trends Food Sci Technol 105:385–401
104. Pereira VA, de Arruda INQ, Stefani R (2015) Active chitosan/PVA films with anthocyanins from Brassica oleraceae (red cabbage) as time-temperature indicators for application in intelligent food packaging. Food Hydrocoll 43:180–188
105. Forghani S, Almasi H, Moradi M (2021) Electrospun nanofibers as food freshness and time-temperature indicators: a new approach in food intelligent packaging. Innov Food Sci Emerg Technol 73:102804
106. Vanderroost M, Ragaert P, Devlieghere F, de Meulenaer B (2014) Intelligent food packaging: the next generation. Trends Food Sci Technol 39(1):47–62
107. Schaefer D, Cheung WM (2018) Smart packaging: opportunities and challenges. Procedia CIRP. 72:1022–1027
108. Risch SJ (2009) Food packaging history and innovations. J Agric Food Chem 57(18):8089–8092
109. Teixeira-Costa BE, Andrade CT (2021) Natural polymers used in edible food packaging—history, function and application trends as a sustainable alternative to synthetic plastic. Polysaccharides. 3(1):32–58
110. Evyan YCY, Liew MS, Patricia J, Chong MY, Zairul ZA (2022) Biodegradable food packaging and film: a short review. Food Res 6:1–12

111. Asgher M, Qamar SA, Bilal M, Iqbal HMN (2020) Bio-based active food packaging materials: sustainable alternative to conventional petrochemical-based packaging materials. Food Res Int (Elsevier Ltd.) 137
112. Mendes AC, Pedersen GA (2021) Perspectives on sustainable food packaging: is bio-based plastics a solution? Trends Food Sci Technol (Elsevier Ltd.) 112:839–846
113. Reichert CL, Bugnicourt E, Coltelli MB, Cinelli P, Lazzeri A, Canesi I et al (2020) Bio-based packaging: materials, modifications, industrial applications and sustainability. Polymers (MDPI AG) 12
114. Melson CRR (1943) Food packaging for overseas use. Ind Eng Chem [Internet] 35(1):16–19. https://pubs.acs.org/sharingguidelines
115. Shlush E, Davidovich-Pinhas M (2022) Bioplastics for food packaging. Trends Food Sci Technol (Elsevier Ltd.) 125:66–80
116. Petersen K, Vñggemose Nielsen P, Bertelsen G, Lawther M, Olsen MB, Nilsson NH et al (1999) Potential of biobased materials for food packaging. Trends Food Sci Technol 10:52–68
117. Scannell AG, Hill C, Ross R, Marx S, Hartmeier W, Arendt EK (2000) Development of bioactive food packaging materials using ® immobilised bacteriocins Lacticin 3147 and Nisaplin [Internet]. Int J Food Microbiol. www.elsevier.nl/locate/ijfoodmicro
118. Bastioli C (2001) Global status of the production of biobased packaging materials. Starch/Staerke 53:351–355
119. Bucci DZ, Tavares LBB, Sell I (2005) PHB packaging for the storage of food products. Polym Test 24(5):564–571
120. Ban W, JianguoSong, Argyropoulos DS, Lucia LA (2006) Improving the physical and chemical functionality of starch-derived films with biopolymers. J Appl Polym Sci 100(3):2542–2548
121. Brody AL, Bugusu B, Han JH, Sand CK, McHugh TH (2008) Innovative food packaging solutions. J Food Sci 73
122. Farris S, Schaich KM, Liu LS, Piergiovanni L, Yam KL (2009) Development of polyion-complex hydrogels as an alternative approach for the production of bio-based polymers for food packaging applications: a review. Trends Food Sci Technol 20:316–332
123. Hermann BG, Blok K, Patel MK (2010) Twisting biomaterials around your little finger: environmental impacts of bio-based wrappings. Int J Life Cycle Assess 15(4):346–358
124. Rhim JW, Park HM, Ha CS (2013) Bio-nanocomposites for food packaging applications. Prog Polym Sci (Elsevier Ltd.) 38:1629–1652
125. Pathak S, Sneha C, Mathew BB (2014) Bioplastics: its timeline based scenario & challenges. J Polym Biopolym Phys Chem [Internet] 2(4):84–90. http://pubs.sciepub.com/jpbpc/2/4/5
126. Wang LF, Rhim JW (2015) Preparation and application of agar/alginate/collagen ternary blend functional food packaging films. Int J Biol Macromol 1(80):460–468
127. Caro N, Medina E, Díaz-Dosque M, López L, Abugoch L, Tapia C (2016) Novel active packaging based on films of chitosan and chitosan/quinoa protein printed with chitosan-tripolyphosphate-thymol nanoparticles via thermal ink-jet printing. Food Hydrocoll 1(52):520–532
128. Majid I, Ahmad Nayik G, Mohammad Dar S, Nanda V (2018) Novel food packaging technologies: innovations and future prospective. J Saudi Soc Agric Sci King Saud Univ 17:454–462
129. Sancheti I, Katkar P, Thorat S, Radhakrishnan S, Kulkarni MB (2019) PBAT/PLA/HNT filled blends for active food packaging. Mater Res Express 6(12)
130. Sid S, Mor RS, Kishore A, Sharanagat VS (2021) Bio-sourced polymers as alternatives to conventional food packaging materials: a review. Trends Food Sci Technol (Elsevier Ltd.) 115:87–104
131. Song T, Qian S, Lan T, Wu Y, Liu J, Zhang H (2022) Recent advances in bio-based smart active packaging materials. Foods (MDPI) 11
132. Mujtaba M, Lipponen J, Ojanen M, Puttonen S, Vaittinen H (2022) Trends and challenges in the development of bio-based barrier coating materials for paper/cardboard food packaging; a review. Sci Total Environ 10:851

133. Tan C, Han F, Zhang S, Li P, Shang N, di Pierro P et al (2021) Novel bio-based materials and applications in antimicrobial food packaging: recent advances and future trends. Int J Mol Sci [Internet] 22:9663. https://doi.org/10.3390/ijms
134. Akbar A, Anal AK (2014) Zinc oxide nanoparticles loaded active packaging, a challenge study against Salmonella typhimurium and Staphylococcus aureus in ready-to-eat poultry meat. Food Control 38(1):88–95
135. Vasile C, Râpă M, Ştefan M, Stan M, Macavei S, Darie-Niţă RN et al (2017) New PLA/ZnO:Cu/Ag bionanocomposites for food packaging. Express Polym Lett 11(7):531–544
136. Sanuja S, Agalya A, Umapathy MJ (2015) Synthesis and characterization of zinc oxide-neem oil-chitosan bionanocomposite for food packaging application. Int J Biol Macromol 1(74):76–84
137. Youssef AM, El-Sayed SM, Salama HH, El-Sayed HS, Dufresne A (2015) Evaluation of bionanocomposites as packaging material on properties of soft white cheese during storage period. Carbohydr Polym 6(132):274–285
138. Jiang T, Feng L, Wang Y (2013) Effect of alginate/nano-Ag coating on microbial and physicochemical characteristics of shiitake mushroom (Lentinus edodes) during cold storage. Food Chem 141(2):954–960
139. Nagarajan M, Benjakul S, Prodpran T, Songtipya P (2015) Properties and characteristics of nanocomposite films from tilapia skin gelatin incorporated with ethanolic extract from coconut husk. J Food Sci Technol 52(12):7669–7682
140. Donsì F, Marchese E, Maresca P, Pataro G, Vu KD, Salmieri S et al (2015) Green beans preservation by combination of a modified chitosan based-coating containing nanoemulsion of mandarin essential oil with high pressure or pulsed light processing. Postharvest Biol Technol 1(106):21–32
141. Vedove TMARD, Maniglia BC, Tadini CC (2021) Production of sustainable smart packaging based on cassava starch and anthocyanin by an extrusion process. J Food Eng 289
142. Chayavanich K, Thiraphibundet P, Imyim A (2020) Biocompatible film sensors containing red radish extract for meat spoilage observation. Spectrochim Acta A Mol Biomol Spectrosc 5:226
143. Zinoviadou KG, Koutsoumanis KP, Biliaderis CG (2009) Physico-chemical properties of whey protein isolate films containing oregano oil and their antimicrobial action against spoilage flora of fresh beef. Meat Sci 82(3):338–345
144. Gómez-Estaca J, López de Lacey A, López-Caballero ME, Gómez-Guillén MC, Montero P (2010) Biodegradable gelatin-chitosan films incorporated with essential oils as antimicrobial agents for fish preservation. Food Microbiol 27(7):889–896
145. Rodrigues C, Souza VGL, Coelhoso I, Fernando AL (2021) Bio-based sensors for smart food packaging—current applications and future trends. Sensors 21(6):1–24
146. Zubair M, Pradhan RA, Arshad M, Ullah A (2021) Recent advances in lipid derived bio-based materials for food packaging applications. Macromol Mater Eng 306(7)
147. Angioletti BL, Dos Santos SP, Hoffmann TG, Bertoli SL, de Souza CK (2020) Influence of whey protein edible film and refrigeration temperature on quality of acerola in natura during postharvest storage. In: AIChE annual meeting
148. Kumar VA, Hasan M, Mangaraj S, Verma DK, Srivastav PP (2022) Trends in edible packaging films and its prospective future in food: a review. Appl Food Res 2(1):100118

Plastics and Microplastics—Environmental and Health Hazards

Sai Kumar Tammina, Saraschandra Naraginti, Ajahar Khan, and L. Jyothi

Abstract Long-lasting use of plastic products and their exposure to high temperature and pressure conditions results in the release of hazardous chemicals into the soil, food, and water. Also, the disposal of plastics via landfilling or burning them in the air results in environmental pollution. Also, the plastics can wedge the urban drainage systems and cause floods during the rainy season by hindering the water passage. At the same time, the diversified physiological conditions of the ecosystem result in the gradual disintegration of plastic to release smaller plastic particles, known as microplastics, into the ecosystem. Microplastic accumulation alters the pH of the soil, which further influences soil enzymes with probable impact on carbon, nitrogen, and phosphate cycles. Even longer exposure to microplastics causes changes in soil respiration, enzymatic functionalities, and activities of the soil microbes. Microplastics also enter the food supply chain majorly through contaminated water bodies, especially seas and rivers. In the next step, these plastic particles ultimately reach and start accumulating in the human body through food and water, which may alter human chemical, physiological, and enzymatic conditions and pose a severe threat to human health. Therefore, there is an urgent need to control the plastics and microplastic accumulation in the ecosystem to save the soil, water, air, and human health for the current and upcoming generations. This chapter is important to understand the impact of microplastics on the environmental balance and human health, which calls

---

S. K. Tammina (✉)
Department of Chemical and Petroleum Engineering, Khalifa University of Science and Technology (KUST), SAN Campus, P.O. Box: 127788, Abu Dhabi, UAE
e-mail: saitammina@gmail.com

S. Naraginti
School of Chemical and Environmental Engineering, Anhui Polytechnic University, Wuhu 241000, China

A. Khan
Department of Food and Nutrition, BioNanocomposite Research Center, Kyung Hee University, Seoul 02447, Republic of Korea

L. Jyothi
School of Physics, University of Hyderabad, Hyderabad 500046, India

M. Alle et al. (eds.), *Functional Nanomaterials and Nanocomposites for Biodegradable Food Packaging*, Smart Nanomaterials Technology,
https://doi.org/10.1007/978-981-97-9084-5_2

for greener and eco-friendly alternatives such as biopolymer nanocomposites for future packaging applications.

Keywords Plastics · Microplastics · Ecosystem · Health hazard

1 Introduction

Non-renewable fossil resources, such as natural gas, oil, and coal, have been transformed by humans into long-chain polymers known as plastics, that comprise carbon, oxygen, hydrogen, silicon, and chloride [1]. Plastics are used in almost every industry, including packaging and wrapping, automotive, aquaculture, fisheries, biomedicine, transport, farming, construction and building, telecommunications, plumbing, personal care goods, textiles, and clothes [2]. In 2018, 359 million tons of plastic were produced worldwide, up 3.2% from 2017 and 46.5% from 2008 [3]. China produces the largest amount of plastics, i.e., 30% of the total production worldwide, followed by Canada, Mexico, the United States, and Europe [4]. However, plastics release harmful chemicals that are enduring and pervasive toxins that could impair organisms' ability to grow and develop, cause oxidative stress, impair their immune systems, shorten life expectancy, and affect their fertility [5].

Plastics degrade or break down into tiny fractions over time due to mechanical, chemical, physical, and biological activities to generate "microplastics" [6]. Because they are produced continually, are not biodegradable, are persistent, and have an extremely lengthy environmental lifespan, these microscopic and widespread plastic particles have steadily accumulated in the natural ecosystem over the years [7]. As a result, they were included as one of the ten emerging pollutants in the United Nations Environmental Programme (UNEP) Annual Book 2014 that might endanger the health of people and other creatures across all ecosystems [8]. Therefore, the buildup of microplastics in ecosystems is receiving more attention and growing in importance among academicians and scientists worldwide. Highly populated places or regions with a lot of anthropogenic activity have been shown to have a high concentration of plastic particles within water bodies, wetlands, seas, and coastlines [9]. Despite their small size, microplastics may impact human health by entering the food chain through the ingestion of seafood and other terrestrial foods [10–14]. In addition, the dumping of plastic trash in municipal waste systems produces toxic leachate that may pollute both soil and water [15]. Plastic waste keeps growing due to the unprecedented use of plastic goods and inadequate waste disposal methods [16]. Thus, the world has become a "plastic planet" due to the irresponsible use, improper disposal, inadequate management practices, and associated negative effects of plastics. The impact of microplastic pollution on human health has recently come into focus, even though there have been several studies on the subject. Fish, shellfish, oysters, shrimp, mussels, and other marine mollusks are all consumed by people. Microplastic contamination of these marine creatures has an impact on human health. However, some researchers found no concrete evidence that ingesting microplastics risks human health [17]. Also, the knowledge of the detrimental effects of plastic particles on the well-being of people, and the methods via which people are exposed to these particles is still in its infant stages, and not enough research is available on the subject. Additionally, plastics emit harmful organic pollutants, such as dioxins and bisphenol A, which can harm the reproductive system and cause cancer and other neurological disorders [15, 18]. Therefore, environmental contamination with microplastics is a serious hazard, and this component must be properly understood.

2 Plastic Types and Its Dimensions

The most common types of non-degradable petroleum-based plastics include polyethylene terephthalate (PET), polystyrene (PS), polyester (PES), polyethylene (PE), polypropylene (PP), polystyrene (PS), polyvinyl chloride (PVC), and polyurethanes [19]. To produce various plastic products, polymers like PET, PP, PE, PES, PVC, PS, Polyamide (PA), and nylon are frequently utilized [20].

According to Plastics Europe (2018), the most widely used plastic polymers in packaging are PP, PET, and low- and high-density polyethylene (LDPE and HDPE). Despite its propensity to cause human cancer, PET is widely used to make packaging materials, drinking water bottles, pipelines, insulating molding, plastic films, and other products [21]. PET has the tendency to enter the bloodstream and interfere with processes that induce reproductive disruption in marine filter feeders.

Further, polystyrene, which is largely utilized in producing containers and caps, boxes, and protective packaging, has been discovered to cause chronic toxicity [22]. Also, the endocrine-disrupting substances (EDCs) including phthalates and polybrominated diphenyl ethers are used to increase the fire resistance and flexibility of plastics. Due to the bioaccumulation of these substances, they have also been discovered in human bodies [22]. Also, these polymer particles have been found to exist in various sizes that may harm aquatic environments, livestock, and human beings [23]. According to their size, plastics can be categorized as huge (more than one meter), large (up to one meter), medium (2.5 cm), micro (up to five millimeters), and nano (less than one millimeter) [24]. In general, microplastics are characterized as plastic fragments with a minimum of a single dimension of 5 mm [25] or as any kind of polymer with the biggest dimension lesser compared to 5 mm [6, 26]. They are produced using polymerization or polycondensation techniques using organic and natural materials like coal, natural gas, and crude oil [27].

2.1 Fate and Transport of Microplastics in the Environment

Wind and water carry microplastics across great distances because of their small size, buoyancy, and light weight. Along with this, the effluent from wastewater treatment (WWT) plants acts as a significant pathway for introducing microplastics directly to aquatic and soil habitats [21]. During the initial stages of the WWT process, large-size microplastics and additional particles can be retained within the motorized screening and gritty chamber. Microplastic fibers settle by gravity during the initial treatment phase using the coagulation/flocculation mechanism [28]. Another potential pathway for microplastic transmission between WWT plants to the soil environment is sludge produced by WWT plants [29]. According to Obbard et al., plastic debris contributes to secondary plastics and carries plastic particles in soils and aquatic environments [30]. Likewise, microplastics are created when the plastic waste gradually degrades and builds up on the top layer of the soil. Agricultural practices, precipitation, fissures, and soil-related activities can all result in microplastics entering the soil subsurface [31]. These activities could enable microplastics to percolate in moist, deep soils, eventually contaminating waterways [25]. The ingestion and excretion processes of soil organisms, such as earthworms, bugs, nematodes, bacteria, fungi, and algae also impact microplastic transit and fate [32].

According to the study, fecal pellets may include microplastics that are a secondary source of microplastics in the aquatic environment [33]. Streams with freshwater and marine environments, runoff from the surface, and wind movement transport soil-borne microplastic debris [34]. Different morphologies of microplastics are constantly carried and deposited in sediments or soil after being discharged into the atmosphere, eventually endangering the higher organisms including humans. Microplastics have been found in marine organisms, including seabirds, fish, bivalves, mammals, crabs, and others [35].

Few studies have documented the existence of microplastic in remote locations and on the earth's continent [36]. Both ocean circulation and atmospheric fall can transfer microplastics. Microplastics are widely distributed throughout all ecosystems due to their tiny dimensions and low density. So, marine sediments and polar ice have developed into a worldwide sink for microplastics [30]. According to predictions, of the 269 million metric tons of particles discovered worldwide, 92% are microplastics. This hypothesis confirmed that most microplastics are found in marine sediments, acting as a sink, while significantly fewer microplastics are on the surface [29].

2.2 *Plastic Particles in Freshwater*

Freshwater is the major drinking source for human beings and there is a probability for human exposure to microplastics through it [37]. Industrial and secondary microplastics are the main types of microplastics that are produced majorly when the bigger sized plastic waste breaks down in freshwater bodies [38]. The shape of the microplastics can also indicate where they are found in a river; for instance, fragment-shaped microplastics easily float and are mostly found on the water surface, whereas fiber- and pellet-shaped microplastics settle at the bottom and have been discovered within sedimentary material [18].

Recently, several researchers found microplastics in sediments and freshwater bodies such as lakes, ponds, rivers, streams, and rivers. Microplastics in surface water have been documented across Asia, particularly in China and India. Freshwater research in China has primarily concentrated on the eastern parts of the country, which includes the Yangtze Estuary [39], Minjiang, Jiaojiang, and Oujiang Estuaries [40], Futuanhe River Estuary, Sha River Estuary [41], and Pearl River Estuary [42]. The mean quantity and estimated particle count were 0.2034 g/L and 226 particles/L, respectively [43].

Further, microplastics have been discovered in waterways and sediments of the USA and Canada in North American regions. According to Anderson et al., the surface waters of Winnipeg Lake (Canada) had 1.74 g/L of microplastics, with an estimated 1933 particle/L [44]. Likewise, another study reports that 0.50 g/L of plastic particles were found in the lakes and rivers of Canada [26]. According to some other studies, microplastics are also present in rivers and lakes in some European countries. The French river Seine was contaminated with microplastic [45]. Similar to this, microplastics have been discovered in surface water and sediments of rivers and lakes in other nations, including Lake Chiusi and Lake Bolsena in Italy [46], Flemish river surface water in Belgium [47], and Kelvin river sediments in the United Kingdom [48].

Compared to their terrestrial equivalents, aquatic systems have received significantly more scientific attention, as microplastic contamination in the oceans is several times greater than on land [49]. Furthermore, by absorbing or adsorbing chemicals or pollutants, microplastics act as vectors or carriers in the form of toxic sponges that transfer chemicals or pollutants [50]. Due to their chemical makeup, microplastics can concentrate and absorb hydrophobic persistent organic pollutants (POPs) in water, such as polychlorinated biphenyls (PCBs) and polycyclic aromatic hydrocarbons (PAHs), and inorganic pollutants, such as heavy metals [51]. According to reports, metals and microbes can also adhere to microplastics [10].

2.3 *Plastic Particles in the Atmosphere*

Synthetic textiles, tire wear from synthetic rubber, and urban dust are the principal contributors to microplastics in the atmosphere [52]. Other potential sources include building materials, industrial emissions, plastic fragments from furniture in homes, garbage dumps, traffic particles, waste incineration, tumble dryer emissions, synthetic particles used in horticultural soils, and sewage sludge used as fertilizer [53]. The many microplastics emitted into the atmosphere are constantly transported and settle in sediment or soil. According to Klein and Fischer, there is a chance that these microplastics will be translocated, bioaccumulated, and accumulated in trophic levels from the soil [54].

Airborne microplastics in both dry and wet atmospheric deposits make up the majority of microplastics in the environment. Dry deposition is defined as gravitationally settled microplastic deposits that are floating in gases and the atmosphere. They might be spread across any level surface or a canopy of forest that has had its canopy removed by rain. Rainout and washout-separated suspended, particulate, and dispersed particles are all included in wet deposition and are disseminated across an environmental matrix [20].

2.4 Plastic Particles in the Soil

The second-largest repository of microplastics and a major pathway for their transfer to diverse surfaces and subterranean substrates in the environment is soil [55]. The soil ecology is directly and indirectly contaminated by various sources such as industrial products, household waste made of plastic, sewage effluent, airborne depositions, and wastewater irrigation [56]. Once collected in soil, microplastics may naturally degrade and bioaccumulate in plants, soil creatures, and biodiversity [13]. According to He et al., microplastics also serve as a vehicle for transferring numerous other soil contaminants into soil organisms, aquatic ecosystems, and other toxins to living things [31]. A substantial amount of microplastics in the environment or rivers (via runoff) may be exposed to agricultural soil surfaces. Heavy rains and flooding in floodplains will likely introduce microplastics into aquatic systems [57]. Depending on the land application type, soil contains microplastic fragments [32]. The low-density polymer has been applied to Chinese fields as mulch recently, and this practice may be an important contributor to microplastics within the soil [58]. Many of these mulches have recyclable plastic residues that emit toxic chemicals, such as 50–120 mg phthalates/kg, causing phthalate concentrations in Chinese soil to be 74–208 times greater than in soil without mulch [59]. In Europe, between 63 and 430 thousand tons of microplastics are transferred annually from cropland to the ocean or surface water, while in North America, 44 and 300 thousand tons are transported annually [60]. Numerous studies have demonstrated that the gradual decomposition of plastic trash is the cause of plastic particles in soils from municipal garbage dumps [60]. According to Fuller and Gautam, soils in industrial areas in Australia were contaminated with microplastics in concentrations ranging from 300 to 67,500 mg/kg (0.03–6.7%) [61].

Furthermore, microplastics were found in nearly 90% of the soil in Swiss floodplains at depths ranging from 0 to 5 cm [62]. Researchers have discovered the highest value of 55.5 mg/kg and an average microplastic content of 5 mg/kg. According to research done on street dust samples from Chennai, India, there were, on average, 227.94 particles of microplastics per 100 g of street dust [63].

2.5 Plastic Particles in the Marine Environment

According to Shim and Thompson, microplastics are pervasive, emerging marine contaminants that pose a hazard to the environment [64]. The top 20 rivers that carry the microplastic pollutants to the oceans account for 67% of the total polluting rivers. Besides, most of the rivers are in Asia and carry between 1.12 and 2.41 million tons of plastic debris annually, according to estimations [65]. Plastic particles make their way to the rivers through home and commercial drainage systems, beauty products, granules, and air-blasting mediums [12]. Also, wastewater treatment facilities (WWTPs) contribute to the ocean's plastic debris by releasing wastewater straight into rivers, which is then transferred to the sea [22].

Huge amounts of plastics and microplastics are released into the sea in coastal regions by human activities like travel, leisure, industrial fishing, and transport, including the marine industry [34]. Plastics are difficult to biodegrade and may survive a very long time in the marine environment. An estimated 5 trillion pieces of plastic are floating in the waters around the world, from the Arctic to the Antarctic [66]. Since microplastics are difficult to physically separate after they reach the environment, further research is still needed to determine their distribution and destiny in the ocean. Studies have shown that microplastics are present in sediments [66], freshwater systems [67], and oceans [68].

Marine air and the ocean's surface are covered in microplastics suspended in the atmosphere. Plastic debris from terrestrial ecosystems may be transported to the sea and ocean–atmosphere via wind and air dynamics. For the first time in 2019, Liu et al. looked into suspended environmental microplastics (SAMPs) in the western part of the Pacific Ocean [69]. SAMP abundance was determined to be 0.01 particles per cubic meter in the maritime area and 0.13 particles per cubic meter in the coastal zone. Also, the SAMP abundance was 0.22 at night and 0.46 throughout the day. Most of the plastic particles, i.e. around 90%, were found as fibers and fragments. Additionally, 1.0×10^4 particles per square kilometer of microplastics were discovered in remote locations (Northwest Pacific) [70].

3 Effect of Plastic Particles on Human Health

Microplastic exposure to people has grown to be a major issue on a global scale. Microplastics are pervasive in the environment and mostly enter the body through two channels, namely ingestion and inhalation. Humans may experience negative health effects on exposure to microplastics [71]. The use of packaged foods is a prominent route by which people are exposed to microplastic contamination. The exact effects of microplastic exposure on human health are currently unknown. Aquatic food items have been recognized as the major reason humans are exposed to microplastics [72]. To evaluate the potential medical risk of microplastic exposure, it is necessary to quantify microplastic fragments in the food we eat and their passage to humans via food and food packaging [73–76].

3.1 Human Body Exposure Routes to Microplastics

Microplastics are pervasive in most ecosystems and pose potential risks to human health regarding their chemical, microbial, and particle composition [77]. Current awareness of the extent of microplastic contamination in humans is inadequate, and a complete study regarding exposure levels, physicochemical qualities, pollutants of concern, implicated tissues, and individual vulnerability to microplastics is required [78, 79]. There are three major ways by which people can be exposed to microplastics, i.e., skin contact, exposure through food, and exposure by inhalation.

It is becoming increasingly clear that various wildlife species can consume microplastics, which results in their entry into the food chain, ultimately reaching humans through diet [80]. Being substantial sources of microplastics and related hazardous pollutants, food, drinking water, and packaging made of plastic are all responsible for human exposure to microplastics [81]. Microplastics in the diet are becoming a serious problem for food safety. Numerous types of seafood have been shown to contain microplastic contamination, which could be quite dangerous for humans. They are frequently consumed entirely, including the gills and digestive systems, which are known to contain microplastics [82]. Since cultivated mussels and oysters frequently include microplastics, eating contaminated bivalves poses one of the greatest dietary dangers of microplastic exposure [83]. Research by Li et al. showed that mussels from shops and coastal waterways included microplastics. They also estimated that 100 g of mussels would contain around 70 microplastic particles [84]. In addition, microplastics were found in the liver and stomach of the fish *Squalius cephalus* and the cans of sardines and sprats [85]. No discernible variation was discovered in the amount of ingested microplastics among various fish species and sources. However, there was spatial variance. Fish near cities are more likely to consume microplastics [86].

Also, numerous nations have reported finding microplastics in table salt made from rock salt, wells, lakes, and oceans [87]. The most prevalent polymer types in commercial table salt were polyethylene terephthalate (PET), polypropylene (PP), and polyethylene (PE) [88]. Therefore, human salt consumption may result in the annual ingestion of several hundred microplastic particles because salt is a staple in the human diet and a media for long-term exposure [89].

Microplastics have been found in several investigations in bottled water, tap water, and groundwater sources [90, 91]. For instance, water from various containers, including beverage cartons, plastic bottles, and glass bottles exhibited microplastic contamination [92]. Traditional treatment plants can keep microplastics from drinking water [37]. However, plastic equipment's abrasion could contaminate the water filtration and distribution procedures [91].

Microplastics can also contaminate food during food processing and packaging. A takeout cup contains a PE liner that can only be recycled at a certain facility [93]. Food trays can produce microplastics because they are often constructed of extruded polystyrene [94]. Microplastic contamination in plastic mineral water bottles was substantially associated with the bottle's plastic pH, density, and thickness [95]. One study showed how frequently opening and shutting bottles with plastic tops caused microplastics to be released [96].

Airborne microplastic particles provide an underrated risk since they can overcome the respiratory system's defenses and travel deep inside human lungs. Microplastics comprise one of the major components of particulates that are rapidly being recognized as a serious public health issue [7]. Worst of all, commonplace indoor dust also contains microplastics [97]. According to Vianello et al., a minor portion of indoor airborne particles in household dust was found to contain microplastics [98]. The most common type of fiber in the air is microfiber. Industrial workers face health problems because they can enter the human respiratory system [5]. Tires, road-wear particles, sea salt aerosolization, the wind-driven discharge of wastewater sludge, the deterioration of building supplies, cloth drying, and synthetic textiles are all potential significant sources of airborne microplastics [80, 99]. According to Liu et al., Shanghai experiences an annual transfer of 120.72 kg of dispersed air microplastics [100]. An adult can breathe in about 21 microplastic particles every day. Furthermore, biofilms created by airborne microplastics may allow viruses to infect humans by transferring them to their lungs [52].

Microplastics may reach humans through skin contact after air fallout of synthetic fibers and tiny beads in cosmetic items. The physicochemical characteristics of microplastics play a significant influence in affecting human health. Microfibers are comparatively more harmful microplastic kinds for humans. Microplastics pose three main health risks: (1) release of hazardous chemical ingredients, including constituents and additives (both organic and inorganic), (2) chemical or biological vectors with microplastics adhering harmful unrelated chemicals, and (3) direct physical harm from plastic debris, including blockage in various organs because of ingested particles.

It is crucial to recognize the potential implications across all biological levels because human health concerns from microplastic pollution are related to the length of exposure, the implication to human organs, and the microplastic-specific exposure [17]. The effects of microplastic exposure are therefore investigated concerning the levels of the biological structure of the human body, starting with chemical levels, and moving up to cellular, connective tissue, organs, and system levels. There haven't been many studies on this subject, specifically including humans. As a result, we incorporate associated studies on other creatures to create a more comprehensive picture from an environmental health standpoint.

3.2 *Chemical Level*

The dosage, size, polymer type, shape, surface chemistry, and hydrophobicity of microplastics are potential factors in the chemical and physical poisons of microplastic pollution and can affect the bioavailability of microplastics [82]. As a part of the production process for plastic, many other chemicals may be used as additives, and even in extremely low concentrations, the additives can be hazardous to humans. Phthalates and bisphenol A (BPA), two chemical additives, are of special concern due to their biological consequences. Both substances are called "endocrine disrupters" because they have been proven to impact hormone system functions, inducing alterations in the body's energy and fat metabolism, mimicking natural hormones, and causing health issues like diabetes [101]. The diethylhexyl phthalate (DEHP), frequently used to soften PVC, may make up to 50% of the weight of the plastics and is reported to be malignant [34]. There has also been evidence of an etiological link between toxicant exposure (such as BPA and DEHP) and the emergence of autistic symptoms in children [102]. Additionally, flame retardants are frequently utilized in polymers. According to one study, co-exposure to microplastics and chemical flame retardants may increase the neurotoxicity in mice caused by oxidative stress [103, 104].

The gene expression associated with cellular stress response, immunological functions, energy metabolism, and development may be altered by exposure to microplastics in various species at the subcellular level [105]. A rise in reactive oxygen species (ROS) could eventually cause DNA damage.

An in-vitro investigation revealed that nano plastics could significantly upregulate interleukin genes in human gastric adenocarcinoma cells. These genes serve as cytokines in the pathology of gastric disorders by influencing cell survival, inflammatory response, and cell shape [106]. When mussels consume PS microbeads, they experience a range of oxidative stress reactions, such as an increase in hemocyte mortality, unbalanced cellular oxidative effects, upregulation of ROS generation in hemocytes, and antioxidant effects [107]. The defensive mechanisms triggered by exposure to microplastics were identified by Jeong et al. using in vitro testing. They demonstrated a size-dependent substantial activation of the associated enzymes and mitogen-activated protein kinases (MAPKs) [108].

3.3 Cellular Level

The particles' size, surface chemistry, quantity, charge, and interactions with nearby proteins, phospholipids, carbohydrates, and certain cell types substantially impact the particles' cellular uptake and translocation [80]. Additionally, tiny particles can pass through membranes like the placenta and blood–brain barrier (BBB) [109]. Endocytosis, passive diffusion, sticky interactions, and protein-mediated uptake, are a few of the various major pathways responsible for these processes [110].

In an uptake investigation in A549 cells, Salvati et al. showed that PS nanoparticles are endocytosed via an intracellular endocytotic route and accumulate in lysosomes [111]. In the lysosome, nanoparticle buildup appears permanent. Furthermore, normal cell lines from the human digestive tract have demonstrated the potential toxicity of microplastics [112].

Additionally, human brain and epithelial cells showed signs of oxidative stress when ROS impact and cell viability were assessed [113]. Additionally, it has been discovered that blue mussels exhibit substantial antioxidant responses to microplastic exposure. These included catalase (CAT), glutathione peroxidases, SOD activity, and total glutathione levels in the gills and digestive glands [114]. Cellular and subcellular levels of organisms are affected by the combined effects of microplastics. For instance, the p-JNK, p-p38, and MAPKs signaling pathways—all related to antioxidants—and inflammation in cells and apoptosis are intimately intertwined [108].

3.4 *Tissue Level*

Protein-coated nanoparticles typically exhibit a better interaction with tissue and organs than uncoated ones. Protein corona surrounding the particle surfaces, and their physical and chemical alterations may affect the nanoparticle's biological reactions, features, and properties [115].

It has been shown that microplastics accumulate at the tissue level, with size having a substantial impact on the rate of accumulation and distribution pattern. Additionally, the buildup of microplastics may have different consequences on metabolomics and biomarkers. Increased lactate dehydrogenase activity and lower levels of adenosine triphosphate (ATP), total cholesterol (T-CHO), and triglycerides (TG) are indicators of disturbances in energy and lipid metabolism, respectively. Upregulated glutathione peroxidase (GSH-Px), SOD, and downregulated CAT are indicators of induced oxidative stress. Blood indicators of neurotoxicity, such as acetylcholinesterase (AChE), were specifically elevated, indicating that exposure to microplastics resulted in significant health concerns [116]. After being exposed to microplastics in an experiment, blue mussels also had notable inflammatory reactions and tissue changes [117]. The absorption of polyethylene particles isolated from toothpaste caused structural alterations and necrosis in tissues in Mediterranean mussels [118].

3.5 *Organ Level*

Chronic microplastic exposure puts human organs in danger, and it's known that microplastics can move to other tissues and organs after the initial exposure. The exposure may happen in the GI tracts and lungs, except for endocytosis [11]. Rats have been used to study the movement of nanoparticles into secondary target organs through membranes [119]. According to a laboratory study, plastic nanoparticles have been shown to permeate fish's BBB and cause behavioral problems [120]. The liver of young fish, *Dicentrarchus labrax*, has also shown signs of oxidative stress and destruction exposure to microplastics [121]. Meng et al. showed that nanoparticles and microplastics can bioaccumulate in the kidney of mice through oral exposure, where they can then contribute to systemic circulation and impair renal function [122].

3.6 System Level

Eleven major organ systems make up the human body. A system comprises connected organs with a shared purpose, which operate in coherence to sustain a healthy human body. An organ can be a part of more than one organ systems. For instance, the pancreas is a component of the endocrine system as well as the digestive system [123]. According to one study, the toxicity of microplastics and nanoplastics differs in various systems. For instance, nanoplastics can potentially harm the reproductive system more severely than microplastics [124]. Consequently, a vast range exists in how microplastics interact with the body. The following highlights the potential effects of microplastics on the immunological, endocrine, respiratory, digestive, and respiratory systems.

3.6.1 Digestive System

Microplastics may threaten human health due to their chemical pollutants and biomagnification. Complex adsorptive reactions, such as the activation of macrophages and the production of cytokines, can be induced by the lumen's large surface areas and charge. These reactions may affect the gut immune system and trigger digestive irritation and inflammation [125]. Although developing a plastic bezoar in the digestive tract is uncommon, the presence of microplastics in human feces has been proven [126].

Human feces have been shown to include nine different microplastic compositions, with PP making up the samples' bulk (62.8%). This indicates the unintentional consumption of microplastics from multiple sources [127]. Further evidence that microplastics are pervasive in the human digestive tract came from discovering microplastics in human colectomy specimens [128]. There is a relationship between the mucosal immune system, epithelial barrier, and microbiome, which link the gut microbiota and host health. In light of this, gut flora may significantly influence microplastic difficulties [129]. Additionally, microplastics may damage intestinal enterocytes directly and indirectly by decreasing intestinal Ca^{2+} and raising gst-4 expression in nematodes *Caenorhabditis elegans* and zebrafish *Danio rerio* [130].

3.6.2 Respiratory System

Phagocytosis and lymphatic transport can clear the majority of inhaled microplastics from human lungs. However, the remaining amount may accumulate in the lungs, causing inflammatory reactions such as oxidative stress, translocation, and gene mutations resulting in acute and long-term respiratory conditions [52]. Particularly those who have respiratory illnesses or are exposed to microplastics regularly from their jobs, such as those who work in textiles or at facilities that produce PE and PP, are more susceptible to the development of airway and interstitial lung disorders [131]. Furthermore, enhanced lung sensitivity from inhaled nanoparticles was identified in a mouse cystic fibrosis model [132]. The possible mechanism included increased CXCL5, upregulated G-CSF, and lowered 15-HETE levels, which promoted exposure-related polymorphonuclear cell (PMN) formation on the surface of the airways. One recent study demonstrates that Sprague–Dawley rats may experience health effects from subacute contact with polystyrene micro(nano)plastics (PSMPs). TGF-β and TNF-α fibrosis and inflammation-related factors are expressed as proteins after 14 days of inhalation exposure. This implied that PSMP inhalation was a factor in the observed molecular action and that continued exposure could eventually have effects at much greater levels [133].

3.6.3 Endocrine System

Some substances found in microplastics are thought to be endocrine-disrupting chemicals (EDCs), which may change how the endocrine system functions. Even if the majority of microplastic exposure can be considered minimal, mounting research indicates that the effects of chronic EDC exposure on individuals should not be discounted as unimportant [134]. BPA is a well-known example of EDC and has a high estrogenic activity, which might have negative effects [135]. According to Lang et al., higher Gamma-glutamyl transferase (GGT) and preventable morbidity from cardiovascular disease and Type 2 diabetes are also linked to high BPA levels in adults' urine [136].

3.6.4 Reproductive System

One recent study showed that the placenta of six uncomplicated pregnant patients had microplastics [137]. The increased risk of harmful consequences, such as endometriosis and breast cancer, is of particular concern due to elevated exposure to EDCs in children and women of childbearing age [138]. High levels of BPA in the urine have been linked to lower ovarian activity and blastocyst formation in women undertaking in vitro fertilization (IVF), suggesting that BPA may affect early reproductive health outcomes [139]. Additionally, studies have shown a dependent association between the lower birth weight of kids and the amount of exposure to BPA during pregnancy [140]. Numerous research reports have looked into how exposure to microplastics affects the ability of marine organisms to reproduce. Nanoplastics dramatically reduced conception rates and negatively affected gametes and embryos in one research on Pacific oysters [141]. Similar to this, copepods exposed to microplastics showed decreased reproduction, which resulted in lower amounts of carbon biomass being absorbed [142].

3.6.5 Immune System

The human body may become toxic and inflammatory due to microplastic exposure. In the case where the particles are initially resistant to removal by the innate immune system, the exposure to and buildup of microplastics may result in long-lasting inflammatory reactions, changes in homeostasis, and an elevated risk of cancer, immunological disorders, and neurological illnesses [143, 144]. Numerous studies have been conducted on modifying genes and proteins involved in the immune response. As an illustration, an in vitro investigation using human lung cells showed that PS nanomaterials may stimulate the natural immune system by enhancing autophagic flux [145]. Additionally, another study found that PS microplastics may activate complement system genes in zebrafish larvae, demonstrating that microparticles have been incorporated into the immunological recognition processes [146]. Additionally, it was noted that exposure to microplastics has an impact on the family of genes responsible for producing the immune system's pattern recognition receptors (PRRs) and antimicrobial peptides (AMPs) [143]. The activity of several immune-related factors was downregulated due to the stress that microplastics caused. In the meantime, highly-regulated expressions of caspase and MyD88 in hemolymphocytes suggest that microplastics cause various levels of immunological suppression in innate immunity [147].

4 Conclusions

Even though plastics have a bright future ahead of them, it is clear that our current methods of production, use, and disposal are not sustainable and pose risks to both human health and wildlife. Although we know a lot about many environmental risks and more is becoming known about how they affect human health, there are still many worries and unknowns. Microplastics produced by the decomposition of plastics pose harm to the ecosystem, as well as to people, animals, and plants; therefore, it is now a matter of concern. Another subject of inquiry could be an assessment of the potential impacts of advanced treatment on the amount of microplastic in treated water. According to research, wastewater treatment plants (WWTPs) release microplastics into the environment. To model and identify contaminants' origin and buildup in the river environment, including river toxicity and biodiversity, it is necessary to investigate the potential toxicity of contaminants further retained on microplastic in the environment. Minimizing consumption, manufacturing, and trash generation is the best approach to decreasing microplastic pollution. To create management plans for the future, microplastic monitoring is crucial in places where seafood is gathered. The companies that produce cosmetics and beauty goods have already started to phase out plastic particles and replace them with more environmentally friendly chemicals. Even the usage and sales of cosmetics, including microplastics, have been outlawed in several nations. It is strongly advised that people adjust their behavior for the better because this will provide the microplastic issue a durable answer.

Moreover, a majority of the microplastics are generated from the waste related to food packaging, which primarily comprises petrochemical-based non-degradable plastic waste. This highlights the need to shift toward more sustainable options for food packaging, such as biopolymers and biopolymer-based nanocomposites, which are sustainable, safe, and environmentally friendly options.

References

1. Shah AA, Hasan F, Hameed A, Ahmed S (2008) Biological degradation of plastics: a comprehensive review. Biotechnol Adv (Elsevier) 26(3):246–265
2. Ogunola OS, Onada OA, Falaye AE (2018) Mitigation measures to avert the impacts of plastics and microplastics in the marine environment (a review). Environ Sci Pollut Res [Internet] (Springer) 25(10):9293–9310 [cited 2022 Sep 2]. https://doi.org/10.1007/s11356-018-1499-z
3. Tammina SK, Khan A, Rhim JW (2023) Advances and prospects of carbon dots for microplastic analysis. Chemosphere (Pergamon) 313:137433
4. Tiwari N, Santhiya D, Sharma JG (2020) Microbial remediation of micro-nano plastics: current knowledge and future trends. Environ Pollut (Elsevier) 265:115044
5. Chen G, Feng Q, Wang J (2020) Mini-review of microplastics in the atmosphere and their risks to humans. Sci Total Environ (Elsevier) 703:135504
6. Lestari P, Trihadiningrum Y, Wijaya BA, Yunus KA, Firdaus M (2020) Distribution of microplastics in Surabaya River, Indonesia. Sci Total Environ (Elsevier) 726:138560
7. Chamas A, Moon H, Zheng J, Qiu Y, Tabassum T, Jang JH et al (2020) Degradation rates of plastics in the environment. ACS Sustain Chem Eng [Internet] (American Chemical Society) 8(9):3494–3511 [cited 2023 Aug 7]. https://doi.org/10.1021/acssuschemeng.9b06635
8. Constant M, Ludwig W, Kerhervé P, Sola J, Charrière B, Sanchez-Vidal A et al (2020) Microplastic fluxes in a large and a small Mediterranean river catchments: the Têt and the Rhône, Northwestern Mediterranean Sea. Sci Total Environ (Elsevier) 716:136984
9. He B, Wijesiri B, Ayoko GA, Egodawatta P, Rintoul L, Goonetilleke A (2020) Influential factors on microplastics occurrence in river sediments. Sci Total Environ (Elsevier) 738:139901
10. Bondelind M, Sokolova E, Nguyen A, Karlsson D, Karlsson A, Björklund K (2020) Hydrodynamic modelling of traffic-related microplastics discharged with stormwater into the Göta River in Sweden. Environ Sci Pollut Res [Internet] (Springer) 27(19):24218–24230 [cited 2023 Aug 6]. https://doi.org/10.1007/s11356-020-08637-z
11. Rist S, Carney Almroth B, Hartmann NB, Karlsson TM (2018) A critical perspective on early communications concerning human health aspects of microplastics. Sci Total Environ (Elsevier) 626:720–726
12. Sharma S, Chatterjee S (2017) Microplastic pollution, a threat to marine ecosystem and human health: a short review. Environ Sci Pollut Res [Internet] (Springer) 24(27):21530–21547 [cited 2023 Aug 7]. https://doi.org/10.1007/s11356-017-9910-8
13. Chae Y, An YJ (2018) Current research trends on plastic pollution and ecological impacts on the soil ecosystem: a review. Environ Pollut (Elsevier) 240:387–395
14. Barboza LGA, Gimenez BCG (2015) Microplastics in the marine environment: current trends and future perspectives. Mar Pollut Bull (Pergamon) 97(1–2):5–12
15. Kataria N, Bhushan D, Gupta R, Rajendran S, Teo MYM, Khoo KS (2022) Current progress in treatment technologies for plastic waste (bisphenol A) in aquatic environment: occurrence, toxicity and remediation mechanisms. Environ Pollut (Elsevier) 315:120319
16. Geyer R, Jambeck JR, Law KL (2017) Production, use, and fate of all plastics ever made. Sci Adv [Internet] (American Association for the Advancement of Science) 3(7):1700782 [cited 2022 Aug 10]. https://doi.org/10.1126/sciadv.1700782

17. Rahman A, Sarkar A, Yadav OP, Achari G, Slobodnik J (2021) Potential human health risks due to environmental exposure to nano- and microplastics and knowledge gaps: a scoping review. Sci Total Environ (Elsevier) 143872
18. Rajmohan KVS, Ramya C, Raja Viswanathan M, Varjani S (2019) Plastic pollutants: effective waste management for pollution control and abatement. Curr Opin Environ Sci Heal (Elsevier) 12:72–84
19. Toxicology: new aspects to this scientific conundrum—Google Books [Internet] [cited 2023 Aug 7]. https://books.google.co.kr/books?hl=en&lr=&id=6cGQDwAAQBAJ&oi=fnd&pg=PA113&ots=nA-m_-n2yM&sig=gLELn5GyMUJpGMIasKQk471Kp9U&redir_esc=y#v=onepage&q&f=false
20. Rose PK, Jain M, Kataria N, Sahoo PK, Garg VK, Yadav A (2023) Microplastics in multimedia environment: a systematic review on its fate, transport, quantification, health risk, and remedial measures. Groundw Sustain Dev (Elsevier) 20:100889
21. Karbalaei S, Hanachi P, Walker TR, Cole M (2018) Occurrence, sources, human health impacts and mitigation of microplastic pollution. Environ Sci Pollut Res [Internet] (Springer) 25(36):36046–36063 [cited 2023 Aug 6]. https://doi.org/10.1007/s11356-018-3508-7
22. Sun J, Dai X, Wang Q, van Loosdrecht MCM, Ni BJ (2019) Microplastics in wastewater treatment plants: detection, occurrence and removal. Water Res (Pergamon) 152:21–37
23. Razeghi N, Hamidian AH, Wu C, Zhang Y, Yang M (2021) Scientific studies on microplastics pollution in Iran: an in-depth review of the published articles. Mar Pollut Bull (Pergamon) 162:111901
24. Elgarahy AM, Akhdhar A, Elwakeel KZ (2021) Microplastics prevalence, interactions, and remediation in the aquatic environment: a critical review. J Environ Chem Eng (Elsevier) 9(5):106224
25. Rillig MC, Ziersch L, Hempel S (2017) Microplastic transport in soil by earthworms. Sci Rep [Internet] (Nature Publishing Group) 7(1):1–6 [cited 2023 Aug 6]. https://www.nature.com/articles/s41598-017-01594-7
26. Anderson JC, Park BJ, Palace VP (2016) Microplastics in aquatic environments: implications for Canadian ecosystems. Environ Pollut (Elsevier) 218:269–280
27. Phuong NN, Zalouk-Vergnoux A, Poirier L, Kamari A, Châtel A, Mouneyrac C et al (2016) Is there any consistency between the microplastics found in the field and those used in laboratory experiments? Environ Pollut (Elsevier) 1(211):111–123
28. Zhou G, Wang Q, Li J, Li Q, Xu H, Ye Q et al (2021) Removal of polystyrene and polyethylene microplastics using PAC and FeCl3 coagulation: performance and mechanism. Sci Total Environ (Elsevier) 752:141837
29. Auta HS, Emenike CU, Fauziah SH (2017) Distribution and importance of microplastics in the marine environment: a review of the sources, fate, effects, and potential solutions. Environ Int (Pergamon) 102:165–176.
30. Obbard RW, Sadri S, Wong YQ, Khitun AA, Baker I, Thompson RC (2014) Global warming releases microplastic legacy frozen in Arctic Sea ice. Earth's Futur [Internet] (Wiley) 2(6):315–320 [cited 2023 Aug 6]. https://doi.org/10.1002/2014EF000240
31. He D, Luo Y, Lu S, Liu M, Song Y, Lei L (2018) Microplastics in soils: analytical methods, pollution characteristics and ecological risks. TrAC Trends Anal Chem (Elsevier) 109:163–172
32. Wang W, Ge J, Yu X, Li H (2020) Environmental fate and impacts of microplastics in soil ecosystems: progress and perspective. Sci Total Environ (Elsevier) 708:134841
33. Duis K, Coors A (2016) Microplastics in the aquatic and terrestrial environment: sources (with a specific focus on personal care products), fate and effects. Environ Sci Eur 2016 281 [Internet] (SpringerOpen) 28(1):1–25 [cited 2022 Aug 10]. https://doi.org/10.1186/s12302-015-0069-y
34. Cole M, Lindeque P, Halsband C, Galloway TS (2011) Microplastics as contaminants in the marine environment: a review. Mar Pollut Bull (Pergamon) 62(12):2588–2597
35. Hou L, Kumar D, Yoo CG, Gitsov I, Majumder ELW (2021) Conversion and removal strategies for microplastics in wastewater treatment plants and landfills. Chem Eng J (Elsevier) 406:126715

36. Guo X, Wang J (2019) The chemical behaviors of microplastics in marine environment: a review. Mar Pollut Bull (Pergamon) 142:1–14
37. Novotna K, Cermakova L, Pivokonska L, Cajthaml T, Pivokonsky M (2019) Microplastics in drinking water treatment—current knowledge and research needs. Sci Total Environ (Elsevier) 667:730–740
38. Horton AA, Walton A, Spurgeon DJ, Lahive E, Svendsen C (2017) Microplastics in freshwater and terrestrial environments: evaluating the current understanding to identify the knowledge gaps and future research priorities. Sci Total Environ (Elsevier) 586:127–141
39. Xu P, Peng G, Su L, Gao Y, Gao L, Li D (2018) Microplastic risk assessment in surface waters: a case study in the Changjiang Estuary, China. Mar Pollut Bull (Pergamon) 133:647–654
40. Zhao S, Zhu L, Li D (2015) Microplastic in three urban estuaries, China. Environ Pollut (Elsevier) 206:597–604
41. Zhou Q, Zhang H, Zhou Y, Dai Z, Li Y, Fu C et al (2018) Surface weathering and changes in components of microplastics from estuarine beaches. Chin Sci Bull [Internet] (Science China Press) 63(2):214–223 [cited 2023 Aug 6]. https://doi.org/10.1360/N972017-00911
42. Fok L, Cheung PK (2015) Hong Kong at the Pearl River Estuary: a hotspot of microplastic pollution. Mar Pollut Bull (Pergamon) 99(1–2):112–118
43. Yuan W, Liu X, Wang W, Di M, Wang J (2019) Microplastic abundance, distribution and composition in water, sediments, and wild fish from Poyang Lake, China. Ecotoxicol Environ Saf (Academic Press) 170:180–187
44. Anderson PJ, Warrack S, Langen V, Challis JK, Hanson ML, Rennie MD (2017) Microplastic contamination in Lake Winnipeg, Canada. Environ Pollut (Elsevier) 225:223–231
45. Dris R, Gasperi J, Rocher V, Saad M, Renault N, Tassin B et al (2015) Microplastic contamination in an urban area: a case study in Greater Paris. Environ Chem [Internet] (CSIRO Publishing) 12(5):592–599 [cited 2023 Aug 6]. https://www.publish.csiro.au/en/EN14167
46. Fischer EK, Paglialonga L, Czech E, Tamminga M (2016) Microplastic pollution in lakes and lake shoreline sediments—a case study on Lake Bolsena and Lake Chiusi (central Italy). Environ Pollut (Elsevier) 213:648–657
47. Slootmaekers B, Catarci Carteny C, Belpaire C, Saverwyns S, Fremout W, Blust R et al. (2019) Microplastic contamination in gudgeons (Gobio gobio) from Flemish rivers (Belgium). Environ Pollut (Elsevier) 244:675–684
48. Blair RM, Waldron S, Phoenix VR, Gauchotte-Lindsay C (2019) Microscopy and elemental analysis characterisation of microplastics in sediment of a freshwater urban river in Scotland, UK. Environ Sci Pollut Res [Internet] (Springer) 26(12):12491–12504 [cited 2023 Aug 6]. https://doi.org/10.1007/s11356-019-04678-1
49. de Souza Machado AA, Kloas W, Zarfl C, Hempel S, Rillig MC (2018) Microplastics as an emerging threat to terrestrial ecosystems. Glob Chang Biol [Internet] (Wiley) 24(4):1405–1416 [cited 2023 Aug 6]. https://doi.org/10.1111/gcb.14020
50. Amrutha K, Warrier AK (2020) The first report on the source-to-sink characterization of microplastic pollution from a riverine environment in tropical India. Sci Total Environ (Elsevier) 739:140377
51. Talvitie J, Mikola A, Koistinen A, Setälä O (2017) Solutions to microplastic pollution—removal of microplastics from wastewater effluent with advanced wastewater treatment technologies. Water Res (Pergamon) 123:401–407
52. Prata JC (2018) Airborne microplastics: consequences to human health? Environ Pollut (Elsevier Ltd.) 234:115–126
53. Liebezeit G, Liebezeit E (2014) Synthetic particles as contaminants in German beers. [Internet] (Taylor & Francis) 31(9):1574–1578 [cited 2023 Aug 6]. https://doi.org/10.1080/19440049.2014.945099
54. Klein M, Fischer EK (2019) Microplastic abundance in atmospheric deposition within the Metropolitan area of Hamburg, Germany. Sci Total Environ (Elsevier) 685:96–103
55. Zhang GS, Liu YF (2018) The distribution of microplastics in soil aggregate fractions in southwestern China. Sci Total Environ (Elsevier) 642:12–20

56. Zhou B, Wang J, Zhang H, Shi H, Fei Y, Huang S et al (2020) Microplastics in agricultural soils on the coastal plain of Hangzhou Bay, east China: multiple sources other than plastic mulching film. J Hazard Mater (Elsevier) 388:121814
57. Gao J, Pan S, Li P, Wang L, Hou R, Wu WM et al (2021) Vertical migration of microplastics in porous media: multiple controlling factors under wet-dry cycling. J Hazard Mater (Elsevier) 419:126413
58. Sarker A, Deepo DM, Nandi R, Rana J, Islam S, Rahman S et al (2020) A review of microplastics pollution in the soil and terrestrial ecosystems: a global and Bangladesh perspective. Sci Total Environ (Elsevier) 733:139296
59. Kong S, Ji Y, Liu L, Chen L, Zhao X, Wang J et al (2012) Diversities of phthalate esters in suburban agricultural soils and wasteland soil appeared with urbanization in China. Environ Pollut (Elsevier) 170:161–168
60. Guo JJ, Huang XP, Xiang L, Wang YZ, Li YW, Li H et al (2020) Source, migration and toxicology of microplastics in soil. Environ Int (Pergamon) 137:105263
61. Fuller S, Gautam A (2016) A procedure for measuring microplastics using pressurized fluid extraction. Environ Sci Technol [Internet] (American Chemical Society) 50(11):5774–5780 [cited 2023 Aug 6]. https://doi.org/10.1021/acs.est.6b00816
62. Scheurer M, Bigalke M (2018) Microplastics in Swiss floodplain soils. Environ Sci Technol [Internet] (American Chemical Society) 52(6):3591–3598 [cited 2023 Aug 6]. https://doi.org/10.1021/acs.est.7b06003
63. Patchaiyappan A, Dowarah K, Zaki Ahmed S, Prabakaran M, Jayakumar S, Thirunavukkarasu C et al (2021) Prevalence and characteristics of microplastics present in the street dust collected from Chennai metropolitan city, India. Chemosphere (Pergamon) 269:128757
64. Shim WJ, Thomposon RC (2015) Microplastics in the ocean. Arch Environ Contam Toxicol [Internet] (Springer Science and Business Media, LLC) 69(3):265–268 [cited 2023 Aug 7]. https://doi.org/10.1007/s00244-015-0216-x
65. Lebreton LCM, Van Der Zwet J, Damsteeg JW, Slat B, Andrady A, Reisser J (2017) River plastic emissions to the world's oceans. Nat Commun [Internet] (Nature Publishing Group) 8(1):1–10 [cited 2023 Aug 7]. https://www.nature.com/articles/ncomms15611
66. Matsuguma Y, Takada H, Kumata H, Kanke H, Sakurai S, Suzuki T et al (2017) Microplastics in sediment cores from Asia and Africa as indicators of temporal trends in plastic pollution. Arch Environ Contam Toxicol [Internet] (Springer Science and Business Media, LLC) 73(2):230–239 [cited 2023 Aug 7]. https://doi.org/10.1007/s00244-017-0414-9
67. Sarkar DJ, Das Sarkar S, Das BK, Manna RK, Behera BK, Samanta S (2019) Spatial distribution of meso and microplastics in the sediments of river Ganga at eastern India. Sci Total Environ (Elsevier) 694:133712
68. Desforges JPW, Galbraith M, Dangerfield N, Ross PS (2014) Widespread distribution of microplastics in subsurface seawater in the NE Pacific Ocean. Mar Pollut Bull (Pergamon) 79(1–2):94–99
69. Liu K, Wu T, Wang X, Song Z, Zong C, Wei N et al (2019) Consistent transport of terrestrial microplastics to the ocean through atmosphere. Environ Sci Technol [Internet] (American Chemical Society) 53(18):10612–10619 [cited 2023 Aug 7]. https://doi.org/10.1021/acs.est.9b03427
70. Pan Z, Guo H, Chen H, Wang S, Sun X, Zou Q et al (2019) Microplastics in the Northwestern Pacific: abundance, distribution, and characteristics. Sci Total Environ (Elsevier) 650:1913–1922
71. Wang C, Zhao J, Xing B (2021) Environmental source, fate, and toxicity of microplastics. J Hazard Mater (Elsevier) 407:124357
72. Huang W, Song B, Liang J, Niu Q, Zeng G, Shen M et al (2021) Microplastics and associated contaminants in the aquatic environment: a review on their ecotoxicological effects, trophic transfer, and potential impacts to human health. J Hazard Mater (Elsevier) 405:124187
73. Prata JC, Paço A, Reis V, da Costa JP, Fernandes AJS, da Costa FM et al (2020) Identification of microplastics in white wines capped with polyethylene stoppers using micro-Raman spectroscopy. Food Chem (Elsevier) 331:127323

74. Kosuth M, Mason SA, Wattenberg EV (2018) Anthropogenic contamination of tap water, beer, and sea salt. PLoS ONE [Internet] (Public Library of Science) 13(4):e0194970 [cited 2023 Aug 7]. https://doi.org/10.1371/journal.pone.0194970
75. Karami A, Golieskardi A, Keong Choo C, Larat V, Galloway TS, Salamatinia B (2017) The presence of microplastics in commercial salts from different countries. Sci Rep [Internet] (Nature Publishing Group) 7(1):1–11 [cited 2023 Aug 7]. https://www.nature.com/articles/srep46173
76. Mason SA, Welch VG, Neratko J (2018) Synthetic polymer contamination in bottled water. Front Chem (Frontiers Media S.A.) 6:389699
77. Kontrick AV (2018) Microplastics and human health: our great future to think about now. J Med Toxicol [Internet] (Springer New York LLC) 14(2):117–119 [cited 2023 Jul 28]. https://doi.org/10.1007/S13181-018-0661-9
78. Koelmans AA, Besseling E, Foekema E, Kooi M, Mintenig S, Ossendorp BC et al (2017) Risks of plastic debris: unravelling fact, opinion, perception, and belief. Environ Sci Technol [Internet] (American Chemical Society) 51(20):11513–11519 [cited 2023 Jul 28]. https://doi.org/10.1021/acs.est.7b02219
79. Revel M, Châtel A, Mouneyrac C (2018) Micro(nano)plastics: a threat to human health? Curr Opin Environ Sci Heal (Elsevier) 1(1):17–23
80. Lehner R, Weder C, Petri-Fink A, Rothen-Rutishauser B (2019) Emergence of nanoplastic in the environment and possible impact on human health. Environ Sci Technol [Internet] (American Chemical Society) [cited 2023 Jul 27]. https://doi.org/10.1021/acs.est.8b05512
81. Qu C, Li B, Wu H, Wang S, Giesy JP (2015) Multi-pathway assessment of human health risk posed by polycyclic aromatic hydrocarbons. Environ Geochem Health [Internet] (Kluwer Academic Publishers) 37(3):587–601 [cited 2023 Jul 28]. https://doi.org/10.1007/s10653-014-9675-7
82. Smith M, Love DC, Rochman CM, Neff RA (2018) Microplastics in seafood and the implications for human health. Curr Environ Heal reports [Internet] (Springer) 5(3):375–386 [cited 2023 Jul 27]. https://doi.org/10.1007/s40572-018-0206-z
83. Zhu J, Zhang Q, Li Y, Tan S, Kang Z, Yu X et al (2019) Microplastic pollution in the Maowei Sea, a typical mariculture bay of China. Sci Total Environ (Elsevier) 658:62–68
84. Li J, Green C, Reynolds A, Shi H, Rotchell JM (2018) Microplastics in mussels sampled from coastal waters and supermarkets in the United Kingdom. Environ Pollut (Elsevier) 241:35–44
85. Karami A, Golieskardi A, Choo CK, Larat V, Karbalaei S, Salamatinia B (2018) Microplastic and mesoplastic contamination in canned sardines and sprats. Sci Total Environ (Elsevier) 612:1380–1386
86. Chan HSH, Dingle C, Not C (2019) Evidence for non-selective ingestion of microplastic in demersal fish. Mar Pollut Bull (Pergamon) 149:110523
87. Gündoğdu S (2018) Contamination of table salts from Turkey with microplastics. [Internet] (Taylor & Francis) 35(5):1006–1014 [cited 2023 Jul 28]. https://doi.org/10.1080/19440049.2018.1447694
88. Lee H, Kunz A, Shim WJ, Walther BA (2019) Microplastic contamination of table salts from Taiwan, including a global review. Sci Rep [Internet] (Nature Publishing Group) 9(1):1–9 [cited 2023 Jul 28]. https://www.nature.com/articles/s41598-019-46417-z
89. Peixoto D, Pinheiro C, Amorim J, Oliva-Teles L, Guilhermino L, Vieira MN (2019) Microplastic pollution in commercial salt for human consumption: a review. Estuar Coast Shelf Sci (Academic Press) 219:161–168
90. Oßmann BE, Sarau G, Holtmannspötter H, Pischetsrieder M, Christiansen SH, Dicke W (2018) Small-sized microplastics and pigmented particles in bottled mineral water. Water Res (Pergamon) 141:307–316
91. Mintenig SM, Löder MGJ, Primpke S, Gerdts G. Low numbers of microplastics detected in drinking water from ground water sources. Sci Total Environ (Elsevier) 648:631–635
92. Schymanski D, Goldbeck C, Humpf HU, Fürst P (2018) Analysis of microplastics in water by micro-Raman spectroscopy: release of plastic particles from different packaging into mineral water. Water Res (Pergamon) 129:154–162

93. Embrandiri A, Quaik S, Emmanuel MI, Rahma M, Rupani PF, Jamaludin MH et al. "Microplastics": the next threat to mankind? https://services.igi-global.com/resolvedoi/resolve.aspx?doi=104018/978-1-7998-0369-0.ch006 [Internet]. IGI Global 106–22 [cited 2023 Jul 28]. https://www.igi-global.com/chapter/microplastics/242013
94. Kedzierski M, Lechat B, Sire O, Le Maguer G, Le Tilly V, Bruzaud S (2020) Microplastic contamination of packaged meat: Occurrence and associated risks. Food Packag Shelf Life (Elsevier) 24:100489
95. Zuccarello P, Ferrante M, Cristaldi A, Copat C, Grasso A, Sangregorio D et al (2019) Exposure to microplastics (<10 μm) associated to plastic bottles mineral water consumption: the first quantitative study. Water Res (Pergamon) 157:365–371
96. Winkler A, Santo N, Ortenzi MA, Bolzoni E, Bacchetta R, Tremolada P (2019) Does mechanical stress cause microplastic release from plastic water bottles? Water Res (Pergamon) 1(166):115082
97. Enyoh CE, Verla AW, Verla EN, Ibe FC, Amaobi CE (2019) Airborne microplastics: a review study on method for analysis, occurrence, movement and risks. Environ Monit Assess [Internet] (Springer International Publishing) 191(11):1–17 [cited 2023 Jul 28]. https://doi.org/10.1007/s10661-019-7842-0
98. Vianello A, Jensen RL, Liu L, Vollertsen J (2019) Simulating human exposure to indoor airborne microplastics using a breathing thermal manikin. Sci Rep [Internet] (Nature Publishing Group) 9(1):1–11 [cited 2023 Jul 28]. https://www.nature.com/articles/s41598-019-45054-w
99. Jan Kole P, Löhr AJ, Van Belleghem FGAJ, Ragas AMJ (2017) Wear and tear of tyres: a stealthy source of microplastics in the environment. Int J Environ Res Public Heal 2017 [Internet] (Multidisciplinary Digital Publishing Institute) 14(10):1265 [cited 2023 Jul 28]. https://www.mdpi.com/1660-4601/14/10/1265/htm
100. Liu K, Wang X, Fang T, Xu P, Zhu L, Li D (2019) Source and potential risk assessment of suspended atmospheric microplastics in Shanghai. Sci Total Environ (Elsevier) 675:462–471
101. Teuten EL, Saquing JM, Knappe DRU, Barlaz MA, Jonsson S, Björn A et al (2009) Transport and release of chemicals from plastics to the environment and to wildlife. Philos Trans R Soc B Biol Sci [Internet] (The Royal Society) 364(1526):2027–2045 [cited 2023 Jul 27]. https://doi.org/10.1098/rstb.2008.0284
102. Ye BS, Leung AOW, Wong MH (2017) The association of environmental toxicants and autism spectrum disorders in children. Environ Pollut (Elsevier) 227:234–242
103. Deng Y, Zhang Y, Qiao R, Bonilla MM, Yang X, Ren H et al (2018) Evidence that microplastics aggravate the toxicity of organophosphorus flame retardants in mice (Mus musculus). J Hazard Mater (Elsevier) 357:348–354
104. Yang J, Zhao Y, Li M, Du M, Li X, Li Y (2019) A review of a class of emerging contaminants: the classification, distribution, intensity of consumption, synthesis routes, environmental effects and expectation of pollution abatement to organophosphate flame retardants (OPFRs). Int J Mol Sci [Internet] (Multidisciplinary Digital Publishing Institute) 20(12):2874 [cited 2023 Jul 27]. https://www.mdpi.com/1422-0067/20/12/2874/htm
105. Liu Z, Yu P, Cai M, Wu D, Zhang M, Huang Y et al (2019) Polystyrene nanoplastic exposure induces immobilization, reproduction, and stress defense in the freshwater cladoceran Daphnia pulex. Chemosphere (Pergamon) 215:74–81
106. Forte M, Iachetta G, Tussellino M, Carotenuto R, Prisco M, De Falco M et al (2016) Polystyrene nanoparticles internalization in human gastric adenocarcinoma cells. Toxicol Vitr (Pergamon) 31:126–136
107. Paul-Pont I, Lacroix C, González Fernández C, Hégaret H, Lambert C, Le Goïc N et al (2016) Exposure of marine mussels Mytilus spp. to polystyrene microplastics: toxicity and influence on fluoranthene bioaccumulation. Environ Pollut (Elsevier) 216:724–737
108. Jeong CB, Won EJ, Kang HM, Lee MC, Hwang DS, Hwang UK et al (2016) Microplastic size-dependent toxicity, oxidative stress induction, and p-JNK and p-p38 activation in the monogonont rotifer (Brachionus koreanus). Environ Sci Technol [Internet] (American Chemical Society) 50(16):8849–8857 [cited 2023 Jul 27]. https://doi.org/10.1021/acs.est.6b01441

109. Vethaak AD, Leslie HA (2016) Plastic debris is a human health issue. Environ Sci Technol [Internet] (American Chemical Society) 50(13):6825–6826 [cited 2023 Jul 27]. https://doi.org/10.1021/acs.est.6b02569
110. Monti DM, Guarnieri D, Napolitano G, Piccoli R, Netti P, Fusco S et al (2015) Biocompatibility, uptake and endocytosis pathways of polystyrene nanoparticles in primary human renal epithelial cells. J Biotechnol (Elsevier) 193:3–10
111. Salvati A, Åberg C, dos Santos T, Varela J, Pinto P, Lynch I et al (2011) Experimental and theoretical comparison of intracellular import of polymeric nanoparticles and small molecules: toward models of uptake kinetics. Nanomed Nanotechnol Biol Med (Elsevier) 7(6):818–826
112. Zhang Y, Wang S, Olga V, Xue Y, Lv S, Diao X et al (2022) The potential effects of microplastic pollution on human digestive tract cells. Chemosphere (Pergamon) 291:132714
113. Schirinzi GF, Pérez-Pomeda I, Sanchís J, Rossini C, Farré M, Barceló D (2017) Cytotoxic effects of commonly used nanomaterials and microplastics on cerebral and epithelial human cells. Environ Res [Internet] 159:579–587 [cited 2022 Aug 10]. https://pubmed.ncbi.nlm.nih.gov/28898803/
114. Magara G, Elia AC, Syberg K, Khan FR (2018) Single contaminant and combined exposures of polyethylene microplastics and fluoranthene: accumulation and oxidative stress response in the blue mussel, Mytilus edulis. [Internet] (Taylor & Francis) 81(16):761–773 [cited 2023 Jul 27]. https://doi.org/10.1080/15287394.2018.1488639
115. Treuel L, Brandholt S, Maffre P, Wiegele S, Shang L, Nienhaus GU (2014) Impact of protein modification on the protein corona on nanoparticles and nanoparticle-cell interactions. ACS Nano [Internet] (American Chemical Society) 8(1):503–513 [cited 2023 Jul 27]. https://doi.org/10.1021/nn405019v
116. Deng Y, Zhang Y, Lemos B, Ren H (2017) Tissue accumulation of microplastics in mice and biomarker responses suggest widespread health risks of exposure. Sci Rep [Internet] (Nature Publishing Group) 7(1):1–10 [cited 2022 Aug 10]. https://www.nature.com/articles/srep46687
117. Von Moos N, Burkhardt-Holm P, Köhler A (2016) Uptake and effects of microplastics on cells and tissue of the blue mussel Mytilus edulis L. after an experimental exposure. Environ Sci Technol [Internet] (American Chemical Society) 46(20):11327–11335 [cited 2023 Jul 27]. https://doi.org/10.1021/es302332w
118. Bråte ILN, Blázquez M, Brooks SJ, Thomas KV (2018) Weathering impacts the uptake of polyethylene microparticles from toothpaste in Mediterranean mussels (M. galloprovincialis). Sci Total Environ (Elsevier) 626:1310–1318
119. Kreyling WG, Semmler-Behnke M, Seitz J, Scymczak W, Wenk A, Mayer P et al (2009) Size dependence of the translocation of inhaled iridium and carbon nanoparticle aggregates from the lung of rats to the blood and secondary target organs. [Internet] (Taylor & Francis) 21(Suppl. 1):55–60 [cited 2023 Jul 28]. https://doi.org/10.1080/08958370902942517
120. Mattsson K, Johnson EV, Malmendal A, Linse S, Hansson LA, Cedervall T (2017) Brain damage and behavioural disorders in fish induced by plastic nanoparticles delivered through the food chain. Sci Rep [Internet] (Nature Publishing Group) 7(1):1–7 [cited 2023 Jul 28]. https://www.nature.com/articles/s41598-017-10813-0
121. Barboza LGA, Vieira LR, Branco V, Carvalho C, Guilhermino L (2018) Microplastics increase mercury bioconcentration in gills and bioaccumulation in the liver, and cause oxidative stress and damage in Dicentrarchus labrax juveniles. Sci Rep 81 [Internet] (Nature Publishing Group) 8(1):1–9 [cited 2023 Jul 28]. https://www.nature.com/articles/s41598-018-34125-z
122. Meng X, Zhang J, Wang W, Gonzalez-Gil G, Vrouwenvelder JS, Li Z (2022) Effects of nano- and microplastics on kidney: physicochemical properties, bioaccumulation, oxidative stress and immunoreaction. Chemosphere (Pergamon) 288:132631
123. Tortora GJ, Derrickson BH (2018) Principles of anatomy and physiology. Google Books [Internet] (Wiley) 659–683 [cited 2023 Jul 28]. https://books.google.co.kr/books?hl=en&lr=&id=aSaVDwAAQBAJ&oi=fnd&pg=PR4&ots=le7j_I0BwO&sig=82cikHsc5TXwAlFvyMDM_ti5oEE&redir_esc=y#v=onepage&q&f=false

124. Yin K, Wang Y, Zhao H, Wang D, Guo M, Mu M et al (2021) A comparative review of microplastics and nanoplastics: toxicity hazards on digestive, reproductive and nervous system. Sci Total Environ (Elsevier) 774:145758
125. Bouwmeester H, Hollman PCH, Peters RJB (2015) Potential health impact of environmentally released micro- and nanoplastics in the human food production chain: experiences from nanotoxicology. Environ Sci Technol [Internet] (American Chemical Society) 49(15):8932–8947 [cited 2023 Jul 28]. https://doi.org/10.1021/acs.est.5b01090
126. Yaka M, Ehirchiou A, Alkandry TTS, Sair K (2015) Huge plastic bezoar: a rare cause of gastrointestinal obstruction. Pan Afr Med J (African Field Epidemiology Network) 21
127. Schwabl P, Koppel S, Konigshofer P, Bucsics T, Trauner M, Reiberger T et al (2019) Detection of various microplastics in human stool. Ann Intern Med [Internet] (American College of Physicians) 171(7):453–457 [cited 2022 Aug 10]. https://doi.org/10.7326/M19-0618
128. Ibrahim YS, Tuan Anuar S, Azmi AA, Wan Mohd Khalik WMA, Lehata S, Hamzah SR et al (2021) Detection of microplastics in human colectomy specimens. JGH Open [Internet] (Wiley) 5(1):116–121. https://doi.org/10.1002/jgh3.12457
129. Li B, Ding Y, Cheng X, Sheng D, Xu Z, Rong Q et al (2020) Polyethylene microplastics affect the distribution of gut microbiota and inflammation development in mice. Chemosphere (Pergamon) 244:125492
130. Lei L, Wu S, Lu S, Liu M, Song Y, Fu Z et al (2018) Microplastic particles cause intestinal damage and other adverse effects in zebrafish Danio rerio and nematode Caenorhabditis elegans. Sci Total Environ [Internet] 619–620:1–8 [cited 2022 Aug 10]. https://pubmed.ncbi.nlm.nih.gov/29136530/
131. Wright SL, Kelly FJ (2017) Threat to human health from environmental plastics. BMJ [Internet] (British Medical Journal Publishing Group) 358:j4334 [cited 2023 Jul 28]. https://www.bmj.com/content/358/bmj.j4334
132. Geiser M, Stoeger T, Casaulta M, Chen S, Semmler-Behnke M, Bolle I et al (2014) Biokinetics of nanoparticles and susceptibility to particulate exposure in a murine model of cystic fibrosis. Part Fibre Toxicol [Internet] (BioMed Central Ltd.) 11(1):1–15 [cited 2023 Jul 28]. https://doi.org/10.1186/1743-8977-11-19
133. Lim D, Jeong J, Song KS, Sung JH, Oh SM, Choi J (2021) Inhalation toxicity of polystyrene micro(nano)plastics using modified OECD TG 412. Chemosphere (Pergamon) 262:128330
134. Vandenberg LN, Luthi D, Quinerly DA (2017) Plastic bodies in a plastic world: multi-disciplinary approaches to study endocrine disrupting chemicals. J Clean Prod (Elsevier) 140:373–385
135. Hengstler JG, Foth H, Gebel T, Kramer PJ, Lilienblum W, Schweinfurth H et al (2011) Critical evaluation of key evidence on the human health hazards of exposure to bisphenol A. [Internet] (Taylor & Francis) 41(4):263–291 [cited 2023 Jul 28]. https://doi.org/10.3109/10408444.2011.558487
136. Lang IA, Galloway TS, Scarlett A, Henley WE, Depledge M, Wallace RB et al (2008) Association of urinary bisphenol A concentration with medical disorders and laboratory abnormalities in adults. JAMA [Internet] (American Medical Association) 300(11):1303–1310 [cited 2023 Jul 28]. https://jamanetwork.com/journals/jama/fullarticle/182571
137. Ragusa A, Svelato A, Santacroce C, Catalano P, Notarstefano V, Carnevali O et al (2021) Plasticenta: first evidence of microplastics in human placenta. Environ Int (Pergamon) 146:106274
138. Vandenberg LN, Maffini MV, Sonnenschein C, Rubin BS, Soto AM (2009) Bisphenol-A and the great divide: a review of controversies in the field of endocrine disruption. Endocr Rev [Internet] (Oxford Academic) 30(1):75–95 [cited 2023 Jul 28]. https://doi.org/10.1210/er.2008-0021
139. Ehrlich S, Williams PL, Missmer SA, Flaws JA, Ye X, Calafat AM et al (2012) Urinary bisphenol A concentrations and early reproductive health outcomes among women undergoing IVF. Hum Reprod [Internet] (Oxford Academic) 27(12):3583–3592 [cited 2023 Jul 28]. https://doi.org/10.1093/humrep/des328

140. Miao M, Yuan W, Zhu G, He X, Li DK (2011) In utero exposure to bisphenol-A and its effect on birth weight of offspring. Reprod Toxicol (Pergamon) 32(1):64–68
141. Tallec K, Huvet A, Di Poi C, González-Fernández C, Lambert C, Petton B et al (2018) Nanoplastics impaired oyster free living stages, gametes and embryos. Environ Pollut (Elsevier) 242:1226–1235
142. Cole M, Lindeque P, Fileman E, Halsband C, Galloway TS (2015) The impact of polystyrene microplastics on feeding, function and fecundity in the marine copepod Calanus helgolandicus. Environ Sci Technol [Internet] (American Chemical Society) 49(2):1130–1137 [cited 2023 Jul 28]. https://doi.org/10.1021/es504525u
143. Détrée C, Gallardo-Escárate C (2018) Single and repetitive microplastics exposures induce immune system modulation and homeostasis alteration in the edible mussel Mytilus galloprovincialis. Fish Shellfish Immunol (Academic Press) 83:52–60
144. Prata JC, da Costa JP, Lopes I, Duarte AC, Rocha-Santos T (2020) Environmental exposure to microplastics: an overview on possible human health effects. Sci Total Environ [Internet] 702 [cited 2022 Aug 10]. https://pubmed.ncbi.nlm.nih.gov/31733547/
145. Chiu HW, Xia T, Lee YH, Chen CW, Tsai JC, Wang YJ (2014) Cationic polystyrene nanospheres induce autophagic cell death through the induction of endoplasmic reticulum stress. Nanoscale [Internet] (Royal Society of Chemistry) 7(2):736–746 [cited 2023 Jul 28]. https://pubs.rsc.org/en/content/articlehtml/2015/nr/c4nr05509h
146. Veneman WJ, Spaink HP, Brun NR, Bosker T, Vijver MG (2017) Pathway analysis of systemic transcriptome responses to injected polystyrene particles in zebrafish larvae. Aquat Toxicol [Internet] Aquat Toxicol 190:112–120. https://pubmed.ncbi.nlm.nih.gov/28704660/. Accessed 10 Aug 2022
147. Liu Z, Yu P, Cai M, Wu D, Zhang M, Chen M et al (2019) Effects of microplastics on the innate immunity and intestinal microflora of juvenile Eriocheir sinensis. Sci Total Environ (Elsevier) 685:836–846

Biopolymers—Isolation and Properties from a Food Packaging Perspective

Wafa Taktak, Hajer Ben Hlima, Slim Abdelkafi, and Slim Smaoui

Abstract Currently, food industries are confronting numerous issues in guaranteeing and preserving food quality as well as extending its shelf life. Biopolymer-based biodegradable and sustainable food packaging material, produced from renewable resources, has assisted food industries in fulfilling these challenges. As food contact materials, these eco-friendly biopolymers are progressively being utilized as substitutes for conventional plastics. Mainly produced by microorganisms through fermentative processes, microbial polymers have achieved great attention. In addition, plant-based and animal-based biopolymers, including polysaccharides, proteins, and lipids were widely applied in food packaging. This chapter primarily covers the recent advancements relating to the extraction, isolation of biopolymers originating from biomass products and microorganisms. The most common biodegradable polymers, general isolation and extraction principles, their biological activities, and their potential for food packaging applications were discussed in this chapter. Additionally, the biodegradation mechanism and issues linked to safety concerns about the use of biopolymers have been detailed.

Keywords Biopolymers · Biomass · Food packaging

W. Taktak · H. Ben Hlima · S. Abdelkafi
Laboratoire de Génie Enzymatique et Microbiologie, Equipe de Biotechnologie des Algues, Ecole Nationale d'Ingénieurs de Sfax, Université de Sfax, 3038 Sfax, Tunisia
e-mail: wafa.taktak@gmail.com

H. Ben Hlima
e-mail: hajer.benhlima@enis.tn

S. Abdelkafi
e-mail: slim.abdelkafi@enis.tn

S. Smaoui (✉)
Laboratory of Microbial Biotechnology and Engineering Enzymes (LMBEE), Center of Biotechnology of Sfax (CBS), University of Sfax, Road of Sidi Mansour Km 6, 3018 Sfax, Tunisia
e-mail: slim.smaoui@cbs.rnrt.tn

M. Alle et al. (eds.), *Functional Nanomaterials and Nanocomposites for Biodegradable Food Packaging*, Smart Nanomaterials Technology,
https://doi.org/10.1007/978-981-97-9084-5_3

1 Introduction

The worldwide population is incessantly increasing which presses for the distribution of freshwater, renewable energy, and vital bioproducts to meet the rising demands [1]. Recently, with the rising global energy demands, the consumption of all fuel types has risen by 2.3% after 2018. Fossil fuel alone accounts for 70% of the total fuel consumption. According to the International Energy Agency (IEA), the plastic industry is one of the primary drivers of the total oil demand as it utilizes petrochemical resources as a raw material [2]. Furthermore, petroleum-based plastics are among the major ingredients employed for the production of packaging materials in most industries, from electronics to food [3]. Thus, the production of petroleum-based plastic and synthetic polymers reached 380 MT in 2015 alone and is estimated to reach 12 billion tons by 2050 if there are no developments in the commercialization of biodegradable alternatives realized through scientific innovations and other involvements [4]. Indeed, only about 9% of the produced plastic waste is recycled, 12% is incinerated, and ~79% is either landfilled or ends up in the oceans or other water bodies, with half-lives extending nearly 1200 years [4]. The mishandled plastic waste, along with the harmful emissions produced due to the incineration process are proven to accumulate in the environment. This waste as well as emergent contaminants (micro- and nano-plastics) affects the environmental quality, ecological balance, biodiversity, and human health [5]. Thus, several studies are raising consciousness about the requirement to protect the environment [6].

Consequently, the research on alternative renewable resources is essential for controlling polymer production and utilization [7]. Nowadays, the development of polymer analogues from biological resources is well-investigated, to substitute artificial polymers produced from fossil resources which currently predominate the global marketplace [8]. Thus, the use of biological resources and processes could be considered as a possible solution to the problem regarding the generation of enormous quantities of plastic waste and its negative environmental and ecological impacts [9, 10]. Biopolymers like polysaccharides and proteins can be obtained from several renewable sources, such as microorganisms, animals, plants, and algae. They are safe, environmentally friendly, and biodegradable materials [11]. Several other types of biopolymers can be efficiently produced in pure form by various microorganisms using natural resources as substrates. For instance, polyhydroxybutyrate (PHB), a type of polyhydroxyalkanoate (PHA), which is a biocompatible and non-toxic polymer, is produced by several specific bacteria that use carbohydrates as substrates. However, the industrial scale production of PHB faces some constraints such as higher cost due to high-value substrates [12]. Subsequently, sustainable feedstocks, such as agricultural or food industry wastes, are alternative resources that can be used as substrates for PHB production on a commercial scale with lower cost [13]. Certain other biopolymers are obtained from bio-derived monomers using a synthetic process [14]. For instance, polylactic acid (PLA), is produced by the polymerization of bio-derived lactic acid monomers produced by the fermentation process [15, 16].

For the past several years, biorenewable polymers have continued to undergo substantial development before attracting a significant market share in the near future. This chapter mainly covers the recent advancements regarding the extraction and isolation of biopolymers from biomass products and microorganisms, and their development as alternatives to plastic-based packaging.

2 Biopolymers and Their Sources

The production of biopolymers has recently attracted the interest of researchers to obtain efficient, ecological, green, and biodegradable materials [17]. The biodegradability of biopolymers could be affected by the source materials, temperature, and the production process [18]. The main biological sources of biopolymers include proteins, polysaccharides, lipids, polyesters, and secondary metabolites. Furthermore, aliphatic polymers based on other non-biobased sources, such as poly(ε-caprolactone) (PCL) and polybutylene succinate (PBS), could be degraded by microorganisms, and hence, are considered as biodegradable polymers [19]. The major benefit of biopolymers is related to their macrostructure, chemical durability, configuration of functional groups, and their bioavailability in comparison to chemically assimilated biopolymers [20]. Due to their high biocompatibility, biopolymers find wide-ranging applications in various fields, such as food packaging, medicine, pharmaceuticals, and cosmetics. They are commonly classified into (1) natural biopolymers and (2) semi-synthetic biopolymers, each having their advantages and limitations (Table 3.1). Natural biopolymers can be further classified into plant-based, animal-based, and microorganism-based [21].

Table 3.1 Advantages and limitations of natural and semi-synthetic biopolymers

Biopolymers	Advantages	Limitations
Natural polymers	High biodegradability and biocompatibility [22]	Poor thermal and mechanical properties [23] hereafter their instinctive structures are simply destroyed through high-temperature processing thus restricting their application potential
Semi-synthetic polymers	Biopolymers modified by blending/grafting/crosslinking with synthetic polymers or produced by polymerizing bio-derived monomers. Hence, covers properties of both synthetic polymers and biopolymers, thus owning good mechanical and thermal properties and higher biocompatibility [23]	The semi-synthetic polymers depend on the specific combination of synthetic and biopolymers. Not degradable in the environment and requires specific degradation conditions

2.1 Plant-Based Biopolymers

A large expanse of food is consumed by the world population each year, which arises directly or indirectly from fruits and vegetables [24]. Additionally, an important quantity of biological waste or crop residue (e.g., wood dust, sugar cane bagasse, rice husks) and other portions of fruits and vegetable crops is left behind, when food is harvested [25]. Nevertheless, agricultural wastes can be exploited as raw materials to produce biopolymers, such as starch and cellulose, that could be formed after depolymerization or conversion of the key constituents of the raw materials [26] with a production of approximately 7.5×10^{10} tons per year [27]. These plant-based ingredients, including proteins (e.g., zein), polysaccharides (e.g., starch and cellulose), and lipids are used in the production of film/coating especially in the field of food packaging and edible coatings [28]. These biomaterials provide superior qualities and could substitute numerous synthetic polymers in the fields of medicine, clothing, packaging, and other areas [29–31].

Among these biopolymers found in plant-based biomass (cereals, tubers, and legumes), polysaccharides derived from biological resources have unceasingly been used in various applications [32]. For example, zein is a class of prolamin protein obtained from corn that displays remarkable potential in food, food packaging, and food coating applications [33]. Starch constitutes a widespread polysaccharide in nature. There are two forms of starch, the first one is a linear polymer named amylose, and the second one is a branched polymer known as amylopectin [34]. Starch has been applied for numerous purposes, such as in the preparation of industrial paper, which is actually one of the most widespread applications. Besides, the textile industry exploits substantial quantities of starch to enhance the fiber strength in the weaving development [35].

Agar is a polysaccharide derived from the sulfated galactan of red algae [36]. This polymer is known for its gel-forming capacity with interesting mechanical properties. Thus, agar could be combined with other natural polymers to improve the blended gel properties [37]. It is also applied in diverse food and medicinal fields [38].

Alginate is another marine polysaccharide with high molecular weight [39] and heterogeneous configuration that constitutes the main element of the cell walls of brown algae. Alginate is extracted from marine microalgae as a salt of alginic acid [40]. α-L-gluconic and β-(1,4)-D-mannuronic acids are the major monomeric units that characterize alginate [41]. The functional groups (CO and OH) of alginates improve their aqueous solubility. While, this property is influenced by other factors including pH, gelling ions of the solvent, and total ionic strength of the solute [42]. Alginates have special characteristics such as gelation, viscosity, adsorption capacity, non-toxicity, biocompatibility, biodegradability, and cell proliferation ability [43–45] that extend their range of applications. Thus, alginate is used in several applications like medicine (tissue engineering and drug delivery), food industry [43], textiles, paper coating, wound treatment, etc. [42].

Carrageenan is another linear sulfated polysaccharide obtained from red marine macroalgae [46]. The source type, the atmosphere used for growing, and the isolation procedures will affect the morphological and biological properties of carrageenan [32]. Due to the interesting functional properties of carrageenan especially gelling, thickening, and stabilizing abilities, it has been used in diverse applications, such as an encapsulating material in food additives. Furthermore, it has been applied in medicine and pharmaceutical fields as an antioxidant, anticoagulant, antitumor, and antiviral agent [32].

Cellulose and lignin are other important and widely used commercial polysaccharides obtained primarily from higher plants. Lignin is a polyphenolic macromolecule with exceptional properties that is obtained in abundance as plant-based basic material and as a byproduct of cellulosic ethanol production, paper manufacturing, and other industries utilizing plant-based raw materials [47]. Moreover, the development of nano-fibrillated cellulose (NFC) has unfastened new prospects in the field of food packaging using cellulose in place of plastics [48]. Furthermore, some other potential crops of biomass pretreatment are converted to polymers with lower molecular weight and high value, such as glucose, xylose, arabinose, mannose, phenols, furan, vanillin, or organic acids [49], and have a significant demand in various applications areas [50].

2.2 *Biopolymers from Microorganisms*

Biopolymers could be produced via numerous routes using the bacteria-mediated fermentation process. Examples of such biopolymers are PLA and PHA. Although PLA has found various commercial applications, PHA has recently been the one with more research focus. PHAs possess remarkable biocompatibility and significant mechanical attributes. The versatility in their properties arises due to the ability of different bacteria to link more than 150 different monomers in various combinations, resulting in PHAs with various functionalities [51]. Thus, PHAs could also be used to develop composite materials through chemical modifications besides physical combination with nanoparticles and nanofillers. The modification of PHAs improves their endurance and durability and makes them economical, allowing them to compete with conventional plastics regarding both physicochemical properties and manufacturing expenses [52]. Due to these properties, PHAs are exploited in various fields such as packaging, containers, and films, such as coatings, pharmaceuticals, and medical applications [35]. Based on monomers, the common PHAs include polyhydroxybutyrate (PHB), polyhydroxyvalerate (PHV), and their copolymers. Among these biopolymers, PHB is considered the most promising microbial bioprocessed polyester as a biodegradable non-toxic polymer [53]. This biopolymer has a high potential to substitute conventional plastics. Nevertheless, the industrial-scale production of PHB has several limitations. In fact, the expensive production process is a main challenge in the commercialization of PHBs arising principally from the high prices of the feedstock for the bacteria [13]. Nevertheless, agricultural or food industrial waste has been increasingly considered as cheap and maintainable feedstock resources for PHB production [13]. On the other hand, marine bacteria or microalgae are considered prospective sources for PHB production [54]. Owing to their fast evolution rate, high productivity with little dietary needs, microalgae are regarded as a promising feedstock [9].

Moreover, recent works showed biopolymers obtained from the microalgae are characterized by strong mechanical properties in comparison with conventional oil-based polymers [55, 56] and can bind with inorganic or carbon-based particles via intermolecular interactions, forming strong composites [56].

2.3 *Animal-Based Biopolymers*

Animal sources have also been widely explored as biopolymer sources. Most biopolymers obtained from animal sources are proteins. Keratin, a proteinaceous biopolymer derived from chicken feather sources, is applied as reinforcements due to their good dispersion properties that enhance the compatibility between polysaccharides and proteins [57]. Albumen is another protein-based polymer mostly obtained from blood serum and egg white that exhibits interesting properties such as viscoelasticity, tensile strength, and optical properties, as well as higher compatibility with other biopolymers in creating novel materials [58]. Collagen is one of the important structural proteins originating in the extracellular matrices of many connective tissues in mammals and it includes different types (type I, II, and III) [59]. Due to its predominance in native tendons and ligaments, collagen type I plays an important role in tissue regeneration [60]. Collagen possesses versatile properties, including surface activity, biodegradability, nontoxicity, and biocompatibility [61]. Gelatin is also a protein-based biopolymer which is mainly obtained from collagen. The physicochemical properties such as strong gas barrier properties and appropriate mechanical potentials are manipulated through its crosslinking and its ability for modification, to obtain excellent materials with several conceivable applications in biomedicine and food packaging [62].

Besides protein-based polymers, a common animal-based polysaccharide is chitosan, which is composed of N-acetyl-D-glucosamine and D-glucosamine units (Fig. 1) obtained in various grades depending upon the degree of acetylated moieties [63]. It is obtained through the deacetylation of chitin, which is the second most abundant polysaccharide on earth after cellulose. Chitin is present as a structural component in the exoskeleton of crustaceans, insects, and certain microorganisms [64]. Chitosan is renowned for its non-toxicity, biocompatibility, biodegradability, and gel-forming ability [65]. This bio-based polymer with active functional groups, as well as the NH_2 at the C-2 position and OH groups at the C-3 and C-6 positions, can exhibit interesting properties [42]. Thus, chitosan contributes to the development of modern science and technology especially in drug delivery, tissue engineering, green chemistry, and other industrial application areas [66, 67].

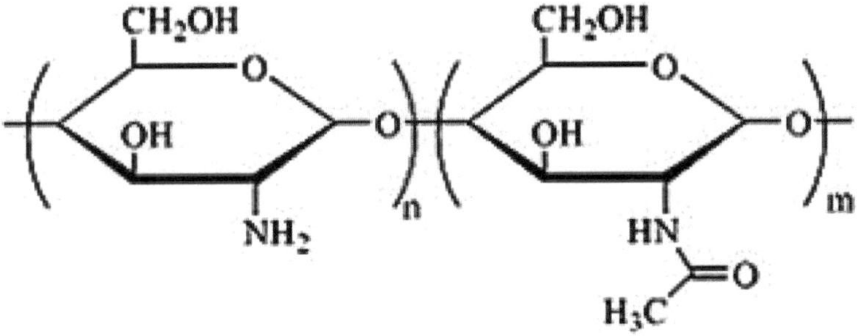

Fig. 1 Illustration of the chitosan structure [63]

2.4 *Semi-synthetic Biopolymers*

Attention has increased significantly in the development of materials that combine both synthetic and natural polymers to obtain novel hybrid polymer structures [68]. The semi-synthetic biopolymers could be prepared by modification of biopolymers through blending, crosslinking, or grafting with synthetic polymers. The flexibility and adaptability of synthetic polymers could be combined with the functionality and the biocompatibility of biopolymers creating a distinct polymer blend with exceptional properties [69]. The combination of the properties of both polymers gives rise to certain advantages of the hybrid material in several applications, including biomolecule stabilizers, therapeutics, biosensors, drug-delivery vehicles, biomedical adhesives, anti-fouling materials, and biomimetic scaffolds. The potentiality of the novel semi-synthetic biopolymers in the drug delivery systems as drug carriers has overcome the limitations related to both synthetic and natural polymers, such as the lower solubility of the drug particles [70]. Furthermore, the conjugation of certain natural biopolymers, such as protein, with synthetic polymer results in a blend with synergistic characteristics and can benefit from the advantages of both constituents. For example, Sithole et al. [70] illustrated (Fig. 2) the blending process of biopolymer collagen with synthetic polymers polyvinylpyrrolidone and PVP, resulting in enhanced properties of the blends. Additionally, the novel biohybrid materials are appropriate for numerous applications, especially in the range of polymeric therapeutics [71]. In fact, the polymer-protein interaction could overcome the problems encountered by pure proteins as precise therapeutic agents, such as the low stability in vivo, and the rapid clearance from the body [72].

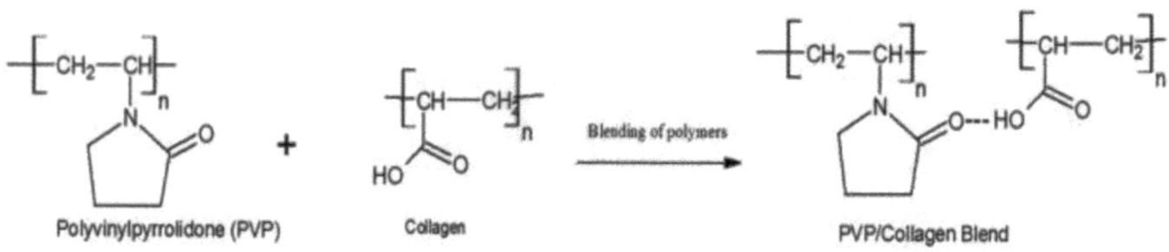

Fig. 2 Semi-synthetic collagen/PVP complex formed through hydrogen bonding blending

3 Biopolymers for Food Packaging Application

Plastic packaging was revealed to have numerous properties. For example, they are lightweight, strong, versatile, shatter-resistant, and inexpensive that makes them innovative for conserving, protecting, and transporting various commodities and offer robust surfaces to print merchandise data. However, plastics have shaped a new habitat named the plastisphere. Plastics stay in the environment for centuries and take around 400–500 years to degrade. In the past few years, bioplastics have emerged as potential substitutes for plastic as they present a crucial solution to plastic waste [73].

The ingredients suitable for use in food packaging are designated depending on the definite provisions of the packaged foodstuff, especially its shelf life, safety, freshness, the atmosphere wherein it could be stored, and the relevant favorable properties of the package [74]. The employment of environmentally friendly packaging materials (Fig. 3) is essential from the manufacturer to the end consumer while being attractive to customers. Therefore, advancements in biopolymer-based packaging materials are required to counter conventional synthetic plastics [75]. Biopolymer-based packaging can also integrate active and smart elements as release systems, to increase the shelf life and nutritional quality of food products [76].

Among the biodegradable and sustainable biological sources, plant and animal proteins (i.e. whey protein, casein, zein, gluten, etc.) are being focused [77]. Various physical and chemical modifications could enhance the functional properties of proteins to expand their application regime as coatings and films [78]. Recently, agro-industrial residues may be exploited for biopolymers extraction and applied as economically viable alternatives for the films/coatings (Fig. 4) development in the food chain [28].

3.1 Techniques for Producing Food Packaging Films

Nowadays, the expansion of a diverse variety of packaging materials has been investigated to avoid food spoilage. Glass containers and jars have been frequently used for packaging liquids (e.g., juice, sauce), and metal sacks and boxes are generally employed for packaging perishable food products (e.g., soups, vegetables, and fruits

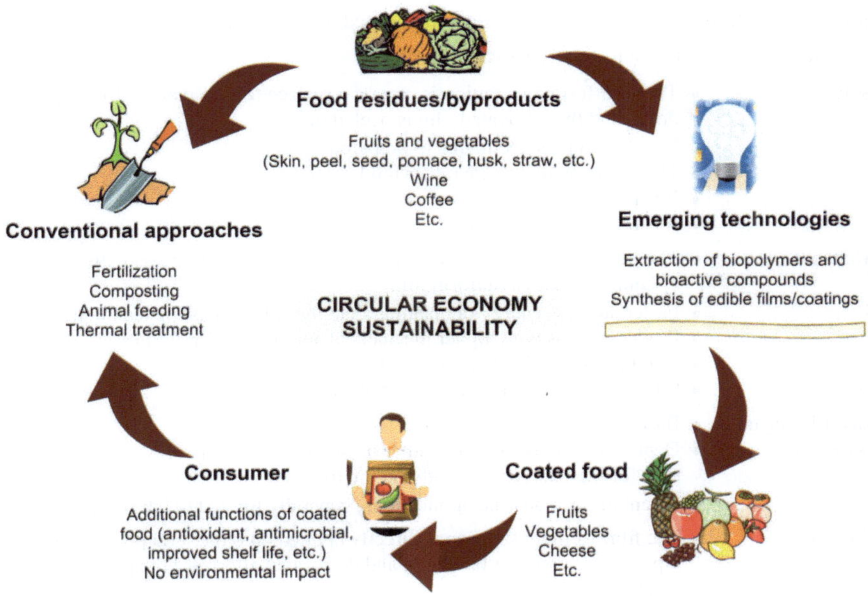

Fig. 3 Schematic illustration of the food chain of biodegradable edible films/coatings based on food residues [28]

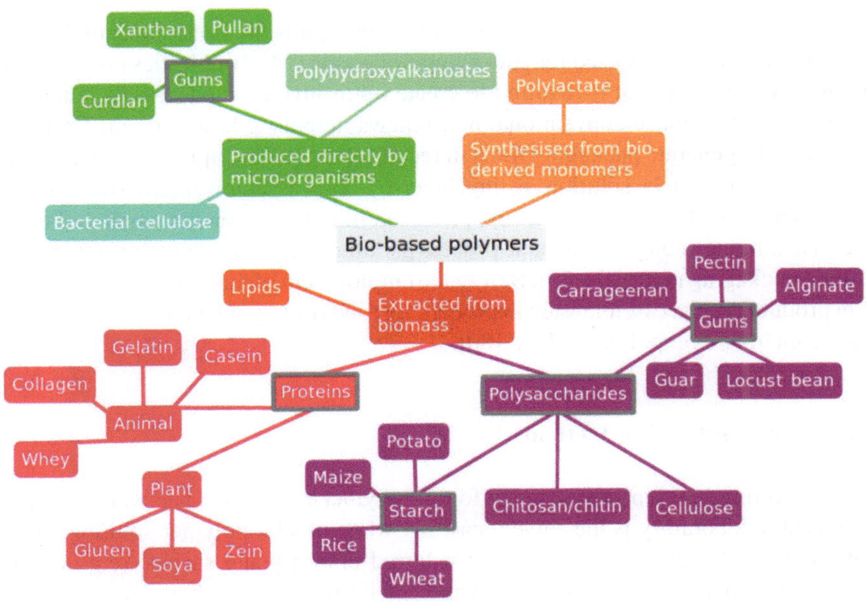

Fig. 4 Examples of biopolymers used in food packaging applications

Table 2 Main technique used to prepare biopolymer-based films [81, 82]

Technique	Important characteristics
Solvent casting	• The film-forming solution is cast on a surface (e.g., a petri dish), properly dried, and the fabricated film is peeled off • It is the simplest technique to prepare film • The used conditions are rather mild • It is time-consuming • It is limited to a lab scale
Melt extrusion	• The mixture containing the biopolymer is transferred into an extruder system, to obtain a uniform film • Faster and less energy-demanding than the solvent casting technique • Produces films with higher mechanical and thermal properties • Conditions may be deteriorative for biopolymers • Can be scaled up for industrial applications
Layer-by-layer assembly	• Based on the deposition of alternating layers • Deposition can be realized either by submersion in or spraying the film-forming solvents directly on the food • Potential for industrial applications, generally limited to lab scale
Coating	• The film-forming solution is directly applied onto the food product via dipping, spraying, or brushing, and desiccated afterward in appropriate conditions • Materials must be food grade if the coating is meant to be eaten • Some applications in the food industry (mostly wax coatings)

etc.), plastic flasks containers used for packaging liquids, powders, and solid foods, and paperboard packets used for packaging dry goods [74]. Bioplastics constitute an innovative approach that is required to create innovative food packaging. Their use can potentially replace conventional non-renewable plastic-based packaging materials [77]. To generate packaging films in recent years, several approaches have been used (Table 2), namely, solution casting, layer-by-layer assembly, melt extrusion, and electrospinning. The properties of the produced film depend on the employed fabrication technique, along with diverse process parameters, and affect their performance as food packaging materials [79]. Before the employment of the desired technique for film production, the essential step in processing food packaging includes solubilizing and/or melting a biopolymer blend [80].

3.1.1 Solvent Casting Method

The first commonly reported strategy for biopolymer-based food packaging production used in laboratory is the solvent casting, wherein a biopolymer solution, along with any functional components, is transferred onto a plane surface (e.g., a petri dish or glass plate), conferring to the anticipated thickness and homogeneity of the films [82], formed by solvent evaporation. The solution can be simply dried either at room temperature or in a high-temperature oven with or without an additional air system. Following this, the film is peeled off from the surface [83]. For example, this technique has been used to produce bioactive nanocomposite films containing PVA, tragacanth (TG), ascorbic acid (AA), and ZnO nanoparticles [84].

3.1.2 Melt Extrusion Method

The melt extrusion is another useful method to produce food packaging films. During the process, polymers remain homogenized and extruded at elevated pressures and temperatures. The equipment contains an extruder that has a hopper for loading the raw materials into the system. The raw materials pass via a cylinder containing one or two rotating screws, which are set at a temperature that melts the polymer. Here, as the polymer melts, the screws help in uniformly mixing the constituents, and push the mixture to a die. The die is the final component of the extrusion setup. At the die, the molten polymer mixture leaves the system taking an anticipated form which is governed by the shape of the die. This technique is represented in the development of the composite-based film containing low-density polyethylene (LDPE) and curcumin, and in the polypropylene-based films combined with synthetic antioxidant substances [85, 86].

3.1.3 Layer-By-Layer Assembly Method

The layer-by-layer (LBL) assembly method is based on the deposition of alternating layers of differing polymer materials like polysaccharides, nucleic acids, proteins, carbohydrates, synthetic polymers, and other substances. Different methods can be used for the development of the multilayer film. For example, casting various layers of a solvent onto a surface, directly spray-coating layers against the surface of a material, dipping the substrate in various solutions, etc. [82, 87].

3.1.4 Electrospinning Method

The electrospinning technique is one of the modern methods for fabricating biopolymer-based packaging materials. The production of fibrous materials is carried out by the deposition of a polymer, with the help of an electrically charged single jet, onto a grounded collector. The electrospun fibers keep depositing onto the substrate under the influence of the electric field to form a film with highly effective surface area and porosity. This method is widely employed to fabricate smart packaging materials, which perform various functions, such as sensing, recording, and communicating information about the food quality throughout the transportation and storage process, while interacting with the food to enhance its shelf life. The information about the quality and shelf life of packaged food products, along with information on any potential problems that consumers or food producers may face, are provided by intelligent packaging materials [88]. On the other hand, the active materials respond to the variations occurring in the immediate vicinity of the food by releasing active components, such as antioxidants and antimicrobials, to enhance the shelf life of the packaged food [89]. This method is fundamental, cost-effective, inimitable, and appropriate for extensive industrial nanofiber development [90]. The preparation of active food packaging based on the electrospinning method has been approved by several studies. The production of lentil flour/poly (ethylene oxide) (PEO) nanofibers loaded with gallic acid is an example of an active food packaging material prepared by this method [91].

3.2 Practical Application of Biopolymers in Food Packaging

Recently, the fabrication of food packaging materials that are made from biopolymers, as alternatives to plastic films is investigated in several studies to obtain the qualities that are required in packaging films. These novel packaging systems can incorporate smart active compounds which are released into the packaging interior, thereby directly interacting with the food to extend the shelf life and nutritional quality of the packaged food products [76].

The biopolymer-based food packaging can be produced by using proteins, lipids, or polysaccharides. The main advantages of using polysaccharides, such as carrageenan, pectin, and alginates, along with xanthan gum, comprise their profusion, availability, affordability, non-toxicity, and biocompatibility. Moreover, these polymers provide selective permeability to oxygen and carbon dioxide when fabricated into film [92]. According to the study of Farhan et al. [93], the semi-purified k-carrageenan is exploited to develop active packaging films added with water extract germinated fenugreek seeds for the preservation of chicken breast. However, due to their hydrophilic nature, the polysaccharides-based film shows poor water barrier properties. Lipid-based films such as beeswax and oils have shown an interesting barrier property of films due to the high value of hydrophobicity of lipids, but they are inflexible. Protein-based films such as gelatin, casein, zein, and whey protein present a high nutritional value, interesting biodegradability, advanced gas barrier, and better mechanical properties compared to polysaccharides and lipids-based film. The release of bioactive molecules, such as antioxidants, antimicrobials, probiotics, and nanomaterials, from protein-based packaging is facilitated by the physicochemical variations in the material surfaces, resulting in enhanced freshness, shelf life, and quality of food products [94]. The functionalization of protein-based edible films can enhance their properties to be applied in several applications (Fig. 5). Among the used approaches, the crosslinking process which allows the binding of adjacent protein molecules chemically, may strengthen the mechanical characteristics of the films. Integrating plasticizers or other polymers to the protein films also improves their mechanical properties [92, 95]. The incorporation of bioactive molecules, such as vitamin E as an antioxidant, and peptide as natural antimicrobial agents, in the film-based protein can expand their stability, shelf life and defend packaged foods from bacterial spoilage [96]. Other approaches are applied to enhance mechanical,

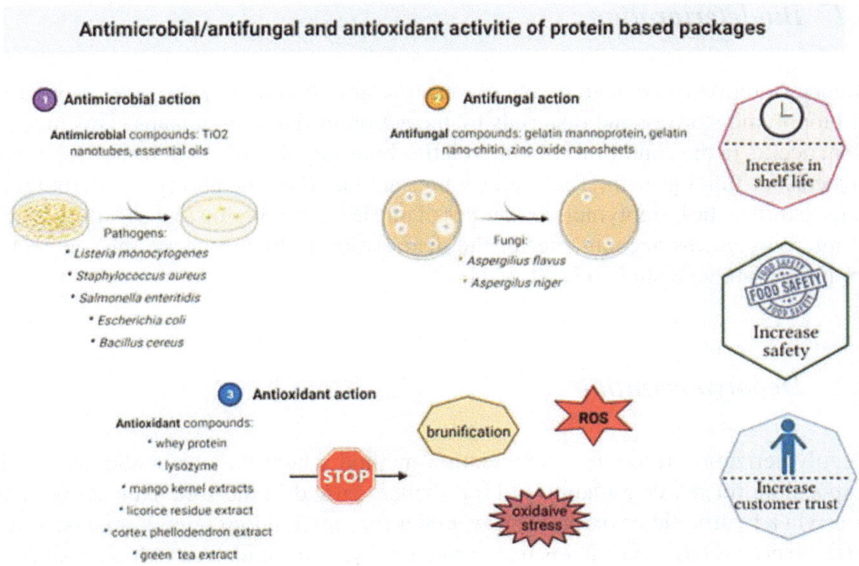

Fig. 5 Protein-based packages and their additions of antimicrobial/antifungal and antioxidant activities [98]

barrier properties and reduce their cost, such as the adhesion to other surfaces of protein films [97], blending protein with another polymer (e.g., chitosan, cellulose, and alginate), or by modifying the film surface properties by adding diverse molecules (e.g., lipids, polysaccharides).

4 Biopolymers Biodegradability

Biodegradation is among one of the various factors based on which biopolymers are categorized. The biodegradation procedure is influenced by environmental factors, microorganisms, related enzymes, and biopolymer properties, resulting in significant variations in their biodegradation periods and rates. The environmental factors that affect the biodegradation of biopolymers include temperature, moisture, and acidic or basic conditions. On the other hand, the intrinsic properties of the biopolymers, for example, chemical structure and physical characteristics, also play an important role. The main advantage of biodegradable polymers formed from bio-based resources is that they quickly degrade into biomass, unlike fossil-based polymers [99].

4.1 Biodeterioration

Biodeterioration is the degradation of the structural, functional, and other properties of biopolymers or organic materials by the action of living organisms. This mechanism occurs in the conditions of high relative humidity that result in bacteria, actinomycetes, or fungi growth. These procedures are based on the ability of microorganisms to utilize the biopolymers as sources of nutrition. Belonging to the leading group of microorganisms accountable for the degradation of biopolymers, micromycetes (or molds) are well studied [100, 101].

4.2 Depolymerization

Depolymerization refers to a degradation method where the enzymatic activity is responsible for the degradation of biopolymers buried in the soil. This action may arise via a hydrolytic or oxidative way, either through random or end-chain scission [102]. The oxidative mechanism is termed enzymatic oxidation, which results in hydrolysis of the polymer. However, when the polymer is non-hydrolyzable, owing to the absence of hydrolyzable groups, redox reactions are considered the predominant way to disrupt the C–C chains. Nevertheless, to allow electron abstraction from non-reactive C–H or C–C chains, the extracellular enzymes must have high enough redox potentials that could be a significant impediment to the eventual biodegradability of polymer [103].

4.3 Bio-assimilation

Bio-assimilation is associated with the procurement or uptake of constituents by the microorganisms for their metabolic development. After the depolymerization phase, compounds are sufficiently small to pass the semi-permeable membrane and can be theoretically processed by the metabolism of the microorganism and lastly mineralized (dissimilation) or exploited through metabolic ways (assimilation) in the biosynthesis of innovative products. Overall, the periplasmic space—the cell membrane—is where the cleavage occurs and oligomers can be delivered across the cytoplasmic membrane for more oxidation in the cycle of β-oxidation. With the aid of surfactants produced by microorganisms during biofilm formation, oligomers may be assumed and be used as carbon and energy sources through the action of intracellular enzymes. In the bio-assimilation stage, the existence of water for the transport of components is an important factor [104].

4.4 *Mineralization*

Mineralization, or eventual biodegradation, refers to the degradation of residual polymers by the mineralization mechanisms, with the liberation of CO_2 and H_2O in aerobic conditions. Other compounds can be released, depending on the polymer composition, including sulfide, phosphate or phosphite, ammonia, nitrite or nitrate, sulfate or sulfite, chloride, and fluoride. Biodegradation rates and mineralization percentage can be measured by determining the mineralization stages (i.e., CO_2 released or evolved). All through this phase, organic compounds, such as carbohydrates and proteins, are utilized by aerobic microorganisms as metabolites of the tricarboxylic (TCA) cycle to generate energy [105, 106].

5 Conclusions

This study has inspected the actual benefits of using biopolymers and presented them as possible eco-friendly alternatives to plastic packaging. Among the widely used biopolymers polysaccharides, proteins, and aliphatic polyesters have been used either as matric or as substrates in fabricating bio-based plastics. Biopolymers are susceptible to various challenges considering their inferior physicochemical properties compared to fossil-based plastics. Thus, the industrial application of biopolymers requires more innovations to enhance their mechanical properties, barrier, and moisture resistance properties. Additionally, the use of natural bioactive biomolecules was found to enhance the functional properties of active bio-based packaging. Alongside the biopolymers, these composite materials enhance the development of packaging materials that are not only biodegradable and sustainable but also possess active and intelligent properties such as biological activities, physicochemical properties, and ultraviolet barrier ability.

6 Future Challenges and Opportunities

Many people are not able to distinguish between the diverse types of plastic and between bio-related plastic labels. Some of them trust that all bioplastics are biodegradable [3]. Thus, it is crucial for scientists to recognize all the physicochemical, mechanical, thermal, and chemical properties of the biopolymers-based packaging [73]. Additionally, researchers have to understand the mechanism of interaction between biopolymers and various food substances. Presently, natural polymers are mixed with numerous harmful chemicals or fossil-based polymers to generate plastics that are not completely biodegradable and have a harmful impact on the environment. On other hand, efforts must be focused on improving the mechanical properties of the biopolymers facilitating them to compete with synthetic plastics, resulting in their commercialization and production [73]. Another main challenge with using biopolymers is to enhance the shelf life of the food product [75]. Furthermore, the appropriate selection of the constituents, especially the active agents, the manufacturing conditions for the films, and the film properties are fundamental while considering the impact of the packaging on the quality preservation of the food product [79].

References

1. Vyas S, Prajapati P, Shah AV, Varjani S (2022) Municipal solid waste management: dynamics, risk assessment, ecological influence, advancements, constraints and perspectives. Sci Total Environ 814:152802
2. International Energy Agency. IEA World Energy Statistics and Balances (2018) https://www.oecd-ilibrary.org/energy/data/iea-world-energy-statisticsand-balances_enestats-data-en
3. Babaremu K, Oladijo OP, Akinlabi E (2023) Biopolymers: a suitable replacement for plastics in product packaging. Adv Ind Eng Polym Res. ISSN 2542-5048 (In Press)
4. Geyer R, Jambeck JR, Law KL (2017) Production, use, and fate of all plastics ever made. Sci Adv 3:e1700782
5. Lau WW, Shiran Y, Bailey RM, Cook E, Stuchtey MR, Koskella J, Palardy JE et al (2020) Evaluating scenarios toward zero plastic pollution. Science 369(6510):1455–1461
6. Kumar R, Verma A, Shome A, Sinha R, Sinha S, Jha PK, Vara Prasad PV et al (2021) Impacts of plastic pollution on ecosystem services, sustainable development goals, and need to focus on circular economy and policy interventions. Sustainability 13 (17):9963
7. Smaoui S, Chérif I, Ben Hlima H, Khan MU, Rebezov M, Thiruvengadam M, Sarkar T, Shariati MA, Lorenzo JM (2023) Zinc oxide nanoparticles in meat packaging: a systematic review of recent literature. Food Packag Shelf Life 36:101045
8. Ahmed A, Rahman MZ, Ou Y, Liu S, Mobasher B, Guo S, Zhu D (2021) A review on the tensile behavior of fiber-reinforced polymer composites under varying strain rates and temperatures. Constr Build Mater 294:123565
9. Mehariya S, Plöhn M, Jablonski P, Stagge S, Jönsson LJ, Funk C (2023) Biopolymer production from biomass produced by Nordic microalgae grown in wastewater. Bioresour Technol 376:128901
10. Lange L, Connor KO, Arason S, Bundgård-Jørgensen U, Canalis A, Carrez D, Vieira H et al (2021) Developing a sustainable and circular bio-based economy in EU: by partnering across sectors, upscaling and using new knowledge faster, and for the benefit of climate, environment and biodiversity, and people and business. Front Bioeng Biotech 1456
11. Nilani P, Raveesha P, Kasthuribai BRNN, Duraisamy B, Dhamodaran P, Elango K (2010) Formulation and evaluation of polysaccharide based biopolymer-an ecofriendly alternative for synthetic polymer. J Pharm Sci Res 2(3):178
12. Palmeiro-Sánchez T, O'Flaherty V, Lens PN (2022) Polyhydroxyalkanoate bio-production and its rise as biomaterial of the future. J Biotech 348:10–25
13. Talan A, Pokhrel S, Tyagi RD, Drogui P (2022) Biorefinery strategies for microbial bioplastics production: sustainable pathway towards circular bioeconomy. Bioresour Technol Rep 17:100875
14. Davachi SM, Heidari BS, Hejazi I, Seyfi J, Oliaei E, Farzaneh A, Rashedi H (2017) Interface modified polylactic acid/starch/poly ε-caprolactone antibacterial nanocomposite blends for medical applications. Carbohydr polym 155:336–344
15. Shen H, Li Y, Yao W, Yang S, Yang L, Pan F, Yin X et al (2021) Solvent-free cellulose nanocrystal fluids for simultaneous enhancement of mechanical properties, thermal conductivity, moisture permeability and antibacterial properties of polylactic acid fibrous membrane. Compos B Eng 222:109042

16. Yang F, Xu C, Zhang W, Sun L, Feng G, Ning T, Fan Y et al (2023) Biodegradable magnesium incorporated microspheres enable immunomodulation and spatiotemporal drug release for the treatment of osteonecrosis of the femoral head. Compos B Eng 250:110430
17. Yaashikaa PR, Kumar PS, Karishma S (2022) Review on biopolymers and composites–evolving material as adsorbents in removal of environmental pollutants. Environ Res 212:113114
18. Rai P, Mehrotra S, Priya S, Gnansounou E, Sharma SK (2021) Recent advances in the sustainable design and applications of biodegradable polymers. Bioresour technol 325:124739
19. Muthuraj R, Misra M, Mohanty AK (2018) Biodegradable compatibilized polymer blends for packaging applications: a literature review. J Appl Polym Sci 135(24):45726
20. Khanra A, Vasistha S, Rai, MP, Cheah WY, Khoo KS, Chew KW, Show PL et al (2022) Green bioprocessing and applications of microalgae-derived biopolymers as a renewable feedstock: circular bioeconomy approach. Environ Technol Innov:102872
21. Bernaerts TM, Gheysen L, Foubert I, Hendrickx ME, Van Loey AM (2019) The potential of microalgae and their biopolymers as structuring ingredients in food: a review. Biotechnol Adv 37(8):107419
22. Liu G, Franssen E, Fitch MI, Warner E (1997) Patient preferences for oral versus intravenous palliative chemotherapy. J Clin Oncol 15:110–115
23. Banna GL, Collova E, Gebbia V, Lipari H, Giuffrida P, Cavallaro S, Ferraù F et al (2010) Anticancer oral therapy: emerging related issues. Cancer Treat Rev 36:595–605
24. Faizan M, Bhat JA, Chen C, Alyemeni MN, Wijaya L, Ahmad P, Yu F (2021) Zinc oxide nanoparticles (ZnO-NPs) induce salt tolerance by improving the antioxidant system and photosynthetic machinery in tomato. Plant Physiol Biochem 161:122–130
25. Lin M, Begho T (2022) Crop residue burning in South Asia: a review of the scale, effect, and solutions with a focus on reducing reactive nitrogen losses. J Environ Manage 314:115104
26. Vu HP, Nguyen LN, Vu MT, Johir MAH, McLaughlan R, Nghiem LD (2020) A comprehensive review on the framework to valorise lignocellulosic biomass as biorefinery feedstocks. Sci Total Environ 743:140630
27. Habibi Y (2014) Key advances in the chemical modification of nanocelluloses. Chem Soc Rev 43(5):1519–1542
28. Gaspar MC, Braga ME (2023) Edible films and coatings based on agrifood residues: a new trend in the food packaging research. Curr Opin Food Sci 50:101006
29. Klemm D, Cranston ED, Fischer D, Gama M, Kedzior SA, Kralisch D, Rauchfuß F et al (2018) Nanocellulose as a natural source for groundbreaking applications in materials science: today's state. Mater Today Chem 21(7):720–748
30. Mahmud S, Hasan KF, Jahid MA, Mohiuddin K, Zhang R, Zhu J (2021) Comprehensive review on plant fiber-reinforced polymeric biocomposites. J Mater Sci 56:7231–7264
31. Sugiarto S, Leow Y, Tan CL, Wang G, Kai D (2022) How far is Lignin from being a biomedical material? Bioact Mater 8:71–94
32. Gereniu CRN, Saravana PS, Chun BS (2018) Recovery of carrageenan from solomon islands red seaweed using ionic liquid-assisted subcritical water extraction. Sep Purif Technol 196:309–317
33. Torres-Giner S, Gimenez E, Lagaron JM (2008) Characterization of the morphology and thermal properties of zein prolamine nanostructures obtained by electrospinning. Food Hydrocoll 22(4):601–614
34. Kalita D, Netravali AN (2017) Thermoset resin-based fiber reinforced biocomposites. Text Finish Recent Dev Future Trends:423–484
35. Phiri R, Sanjay MR, Siengchin S, Oladijo OP, Dhakal HN (2023) Development of sustainable biopolymer-based composites for lightweight applications from agricultural waste biomass: a Review. Adv Ind Eng Polym Res
36. Arvizu-Higuera DL, Rodríguez-Montesinos YE, Murillo-Álvarez JI, Muñoz-Ochoa M, Hernández-Carmona G (2008) Effect of alkali treatment time and extraction time on agar from *Gracilaria vermiculophylla*. J Appl Phycol 20(5):515–519

37. Rhim JW, Wang LF (2013) Mechanical and water barrier properties of agar/carrageenan/konjac glucomannan ternary blend biohydrogel films. Carbohydr Polym 96:71–81
38. Lisha VS, Kothale RS, Sidharth S, Kandasubramanian B (2022) A critical review on employing algae as a feed for polycarbohydrate synthesis. Carbohydr Polym Technol Appl:100242
39. Rastogi P, Kandasubramanian B (2019) Review of alginate-based hydrogel bioprinting for application in tissue engineering. Biofabrication 11(4):042001
40. Chudasama NA, Sequeira RA, Moradiya K, Prasad K (2021) Seaweed polysaccharide based products and materials: an assessment on their production from a sustainability point of view. Molecules 26(9):2608
41. Aprilliza M (2017) Characterization and properties of sodium alginate from brown algae used as an ecofriendly superabsorbent. International symposium on current progress in functional materials. In: IOP conference series: materials science and engineering, vol 188(1), p 012019. IOP Publishing
42. Amenorfe LP, Agorku ES, Sarpong F, Voegborlo RB (2022) Innovative exploration of additive incorporated biopolymer-based composites. Sci Afr:e01359
43. Draget KI, SmidsrØd O, Skjak-Braek G (2005) Alginates from algae in polysaccharides and polyamides in the food industry: properties, production and patents:1–30
44. Tong Z, Chen Y, Liu Y, Tong L, Chu J, Xiao K, Chu X et al (2017) Preparation, characterization and properties of alginate/poly (ɣ-glutamic acid) composite microparticles. Mar Drugs 15(4):91
45. Zhang C, Show PL, Ho SH (2019) Progress and perspective on algal plastics—a critical review. Bioresour Technol 289:121700
46. Pourjavadi A, Harzandi A, Hosseinzadeh H (2004) Modified carrageenan 3 synthesis of a novel polysaccharide-based superabsorbent hydrogel via graft copolymerization of acrylic acid onto kappa-carrageenan in air. Europ Polym J 40:1363–1370
47. Lu F, Rodriguez-Garcia J, Van Damme I, Westwood NJ, Shaw L, Robinson JS, Charalampopoulos D et al (2018) Valorisation strategies for cocoa pod husk and its fractions. Curr Opin Green Sustain Chem 14:80–88
48. Rol F, Belgacem MN, Gandini A, Bras J (2019) Recent advances in surface-modified cellulose nanofibrils. Prog Polym Sci 88:241–264
49. Madadi M, Tu Y, Abbas A (2017) Pretreatment of lignocelollusic biomass based on improving enzymatic hydrolysis. Int J Appl Sci and Biotechnol. 5(1):1–11
50. Deng F, Amarasekara AS (2021) Catalytic upgrading of biomass derived furans. Ind Crops Prod 159:113055
51. Pryadko A, Surmeneva MA, Surmenev RA (2021) Review of hybrid materials based on polyhydroxyalkanoates for tissue engineering applications. Polymers 13(11):1738
52. Visakh PM, Morlanes MJM (2015) Polyethylene-based blends, composites and nanocomposities: state-of-the-art, new challenges and opportunities, polyethylene-based blends. Compos Nanocompos:1–19
53. McAdam B, Brennan Fournet M, McDonald P, Mojicevic M (2020) Production of polyhydroxybutyrate (PHB) and factors impacting its chemical and mechanical characteristics. Polymers 12(12):2908
54. Onen Cinar S, Chong ZK, Kucuker MA, Wieczorek N, Cengiz U, Kuchta K (2020) Bioplastic production from microalgae: a review. IJERPH 17(11):3842
55. Devadas VV, Khoo KS, Chia WY, Chew KW, Munawaroh HSH, Lam MK, Show PL et al (2021) Algae biopolymer towards sustainable circular economy. Bioresour Technol 325:124702
56. Lutzu GA, Ciurli A, Chiellini C, Di Caprio F, Concas A, Dunford NT (2021) Latest developments in wastewater treatment and biopolymer production by microalgae. J Environ Chem Eng 9(1):104926
57. Flores-Hernández CG, Colin-Cruz A, Velasco-Santos C, Castaño VM, Almendarez-Camarillo A, Olivas-Armendariz I, Martínez-Hernández AL (2018) Chitosan–starch–keratin composites: improving thermo-mechanical and degradation properties through chemical modification. J Polym Environ 26:2182–2191

58. González-Gutiérrez J, Partal P, García-Morales M, Gallegos C (2011) Effect of processing on the viscoelastic, tensile and optical properties of albumen/starch-based bioplastics. Carbohydr polym 84(1):308–315
59. Rodrigues A, Emeje M (2012) Recent applications of starch derivatives in nanodrug delivery. Carbohydr polym 87(2):987–994
60. Chopra AS, Lordan R, Horbańczuk OK, Atanasov AG, Chopra I, Horbańczuk JO, Arkells N et al (2022) The current use and evolving landscape of nutraceuticals. Pharmacol Res 175:106001
61. Zargar V, Asghari M, Dashti A (2015) A review on chitin and chitosan polymers: structure, chemistry, solubility, derivatives, and applications. Chem Bio Eng Rev 2(3):204–226
62. Bigi A, Cojazzi G, Panzavolta S, Rubini K, Roveri N (2001) Mechanical and thermal properties of gelatin films at different degrees of glutaraldehyde crosslinking. Biomaterials 22(8):763–768
63. Hoppe-Seiler F (1994) Chitin and chitosan. Ber Dtsch Chem Ges 27:3329–3331
64. Elieh-Ali-Komi D, Hamblin MR (2016) Chitin and chitosan: production and application of versatile biomedical nanomaterials. Int J Adv Res 4(3):411
65. Mahmodi G, Zarrintaj P, Taghizadeh A, Taghizadeh M, Manouchehri S, Dangwal S, Saeb MR et al (2020) From microporous to mesoporous mineral frameworks: an alliance between zeolite and chitosan. Carbohydr Res 489:107930
66. Farhadihosseinabadi B, Zarebkohan A, Eftekhary M, Heiat M, Moghaddam MM, Gholipour-malekabadi M (2019) Crosstalk between chitosan and cell signaling pathways. Cell Mol Life Sci:1–22
67. Muzzarelli R, Muzzarelli C (2005) Chitosan chemistry: relevance to the biomedical sciences. Adv Polym Sci 186:151–209
68. Deming TJ (1997) Facile synthesis of block copolypeptides of defined architecture. Nature 390:386–389
69. Van Hest JCM (2007) Biosynthetic-synthetic polymer conjugates. J Macromol Sci C Polymer Rev 47(1):63–92
70. Sithole MN, Choonara YE, du Toit LC, Kumar P, Pillay V (2017) A review of semi-synthetic biopolymer complexes: modified polysaccharide nano-carriers for enhancement of oral drug bioavailability. Pharm Dev Technol 22(2):283–295
71. Liu S, Maheshwari R, Kiick KL (2009) Polymer-based therapeutics. Macromolecules 42:3–13
72. Jung B, Theato P (2013) Chemical strategies for the synthesis of protein–polymer conjugates. Bio-synthetic Polym Conjug:37–70
73. de Matos Fonseca J, Koop BL, Trevisol TC, Capello C, Monteiro AR, Valencia GA (2022) An overview of biopolymers in food packaging systems. Nanotechnol Enhanc Food Packag:19–53
74. Bhaskar R, Zo SM, Kanan BN, Purohit S, Gupta MK, Han SS (2023) Recent development of protein-based biopolymers in food packaging applications: a review. Polym Test:108097
75. Sid S, Mor RS, Kishore A, Sharanagat VS (2021) Bio-sourced polymers as alternatives to conventional food packaging materials: a review. Trends Food Sci Technol 115:87–104
76. Stanley J, John A, Pušnik Črešnar K, Fras Zemljič L, Lambropoulou DA, Bikiaris DN (2022) Active agents incorporated in polymeric substrates to enhance antibacterial and antioxidant properties in food packaging applications. Macromol 3(1):1–27
77. Jabeen N, Majid I, Nayik GA (2015) Bioplásticos y envasado de alimentos: una revisión. Aliment Agricul Convincent 1(1):2
78. Shlush E, Davidovich-Pinhas M (2022) Bioplastics for food packaging. Trends Food Sci Technol
79. Lai WF, Wong WT (2022) Design and practical considerations for active polymeric films in food packaging. Int J Mol Sci 23(11):6295
80. Priyadarshi R, Rhim JW (2020) Chitosan-based biodegradable functional films for food packaging applications. IFSET 62:102346
81. Kumar S, Mukherjee A, Dutta J (2020) Chitosan based nanocomposite films and coatings: emerging antimicrobial food packaging alternatives. Trends Food Sci Technol 97:196–209

82. Wang H, Qian J, Ding F (2018) Emerging chitosan-based films for food packaging applications. J Agric Food Chem 66(2):395–413
83. da Rocha M, de Souza MM, Prentice C (2018) Biodegradable films: an alternative food packaging. In: Food packaging and preservation. Academic Press, pp 307–342
84. Janani N, Zare EN, Salimi F, Makvandi P (2020) Antibacterial tragacanth gum-based nanocomposite films carrying ascorbic acid antioxidant for bioactive food packaging. Carbohydr polym 247:116678
85. Zia J, Paul UC, Heredia-Guerrero JA, Athanassiou A, Fragouli D (2019) Low-density polyethylene/curcumin melt extruded composites with enhanced water vapor barrier and antioxidant properties for active food packaging. Polymer 175:137–145
86. Fasihnia SH, Peighambardoust SH, Peighambardoust SJ, Oromiehie A, Soltanzadeh M, Peressini D (2020) Migration analysis, antioxidant, and mechanical characterization of polypropylene-based active food packaging films loaded with BHA, BHT, and TBHQ. J Food Sci 85(8):2317–2328
87. Taketa TB, Bataglioli RA, Neto JBMR, de Carvalho BG, de la Torre LG, Beppu MM (2020) Fundamentals and biomedical applications of biopolymer-based layer-by-layer films. In: Biopolymer membranes and films. Elsevier, pp 219–242
88. Kalpana S, Priyadarshini SR, Leena MM, Moses JA, Anandharamakrishnan C (2019) Intelligent packaging: trends and applications in food systems. Trends Food Sci Technol 93:145–157
89. Li L, Wang H, Chen M, Jiang S, Jiang S, Li X, Wang Q (2018) Butylated hydroxyanisole encapsulated in gelatin fiber mats: volatile release kinetics, functional effectiveness and application to strawberry preservation. Food Chem 269:142–149
90. Homocianu M, Pascariu P (2020) Electrospun polymer-inorganic nanostructured materials and their applications. Polym Rev 60(3):493–541
91. Aydogdu A, Yildiz E, Aydogdu Y, Sumnu G, Sahin S, Ayhan Z (2019) Enhancing oxidative stability of walnuts by using gallic acid loaded lentil flour based electrospun nanofibers as active packaging material. Food Hydrocol 95:245–255
92. Mohamed MG, Atayde EC Jr, Matsagar BM, Na J, Yamauchi Y, Wu KCW, Kuo SW (2020) Construction hierarchically mesoporous/microporous materials based on block copolymer and covalent organic framework. J Taiwan Inst Chem Eng 112:180–192
93. Farhan A, Hani NM (2022) Active edible films based on semi-refined κ-carrageenan: antioxidant and color properties and application in chicken breast packaging. Food Packag Shelf Life 24:100476
94. Teixeira-Costa BE, Andrade CT (2022) Natural polymers used in edible food packaging—history, function and application trends as a sustainable alternative to synthetic plastic. Polysaccharides 3(1):32–58
95. Pushp P, Bhaskar R, Kelkar S, Sharma N, Pathak D, Gupta MK (2021) Plasticized poly(vinylalcohol) and poly(vinylpyrrolidone) based patches with tunable mechanical properties for cardiac tissue engineering applications. Biotechnol Bioeng 118(6):2312–2325
96. Fahmy HM, Eldin RES, Serea ESA, Gomaa NM, AboElmagd GM, Salem SA, Shalan AE et al (2020) Advances in nanotechnology and antibacterial properties of biodegradable food packaging materials. RSC Adv 10(35):20467–20484
97. Wu J, Sun Q, Huang H, Duan Y, Xiao G, Le T (2019) Enhanced physico-mechanical, barrier and antifungal properties of soy protein isolate film by incorporating both plant-sourced cinnamaldehyde and facile synthesized zinc oxide nanosheets. Colloids Surf B 180:31–38
98. Mihalca V, Kerezsi AD, Weber A, Gruber-Traub C, Schmucker J, Vodnar DC, Pop OL et al (2021) Protein-based films and coatings for food industry applications. Polymers 13(5):769
99. Bher A, Mayekar PC, Auras RA, Schvezov CE (2022) Biodegradation of biodegradable polymers in mesophilic aerobic environments. Int J Molecul Sci 23(20):12165
100. Ni J, Li, SS, Geng XY (2022) Mechanical and biodeterioration behaviours of a clayey soil strengthened with combined carrageenan and casein. Acta Geotech:1–17
101. Orlita A (2004) Microbial biodeterioration of leather and its control: a review. Int Biodeterior Biodegrad 53(3):157–163

102. Devi RS, Kannan VR, Natarajan K, Nivas D, Kannan K, Chandru S, Antony AR (2016) The role of microbes in plastic degradation. Environ Waste Manage 341:341–370
103. Krueger MC, Harms H, Schlosser D (2015) Prospects for microbiological solutions to environmental pollution with plastics. Appl Microbiol Biotechnol 99:8857–8874
104. Nekkaa A, Benaissa A, Mutelet F, Canabady-Rochelle L (2021) Rhamnus alaternus plant: extraction of bioactive fractions and evaluation of their pharmacological and phytochemical properties. Antioxidants 10(02):300
105. Ghosh S, Qureshi A, Purohit HJ (2019) Microbial degradation of plastics: biofilms and degradation pathways. Contam Agric Environ Health Risks Rem 1:184–199
106. Flemming HC, Wingender J (2010) The biofilm matrix. Nat Rev Microbiol 8(9):623–633

Metal/Metal Oxide Nanomaterials in Biodegradable Food Packaging

Pınar Terzioğlu

Abstract Nanotechnology has important roles in almost every sector, as well as food packaging, in order to succeed in sustainable and cost-effective food protection. Nowadays, biodegradable polymer-based nanocomposites for sustainable food packaging applications have gained significant attention in both the scientific and industrial fields due to environmental concerns. There is a growing demand to develop new bionanocomposite packaging systems for food preservation to maintain high-quality, healthy foods with an extended shelf life. The incorporation of metal and metal oxide nanoscale fillers into the biodegradable polymer matrix is one of the promising alternatives that should enhance the various functional properties (barrier, bioactivity, mechanical, etc.) of the final food packaging. This chapter will give a comprehensive update on the recent advancements of mostly used metal (Ag, Au, and Cu) and metal oxide (CuO, Cu_2O, Fe_3O_4, Fe_2O_3, MgO, ZnO, and TiO_2) nanomaterials incorporated biodegradable food packaging systems. The role of nanoparticles on the properties and biodegradability of developed bionanocomposites and the migration of nanoparticles from polymer matrix are also summarized.

Keywords Biodegradable food packaging · Nanocomposites · Metal nanoparticles · Metal · Oxide nanoparticles

1 Introduction

Due to strict environmental regulations, the food packaging sector is investigating innovative commercially viable biodegradable packaging systems with desired features. Because annual plastics in manufacturing use approximately 40% of global oil production and approximately 50% of plastics are evaluated as packaging material [1]. Worldwide plastic bag usage is about 500 billion, that's close to one million bags

P. Terzioğlu (✉)
Faculty of Engineering and Natural Sciences, Department of Polymer Materials Engineering, Bursa Technical University, Bursa, Türkiye
e-mail: pinar.terzioglu@btu.edu.tr

M. Alle et al. (eds.), *Functional Nanomaterials and Nanocomposites for Biodegradable Food Packaging*, Smart Nanomaterials Technology,
https://doi.org/10.1007/978-981-97-9084-5_4

per minute [1]. This results in huge amounts of plastic waste flowing into the oceans from coastal areas which are expected to nearly triple by 2040. The chained negative effects observed in the ecosystem due to synthetic polymer-based non-degradable packaging materials. Because they are resistant to biological, chemical, and physical degradation [2]. Hence, the search for biodegradable polymer-based food packaging and coatings for sustainable packaging requirements to replace non-biodegradable plastics has become increasingly popular [3].

Frequently, biopolymers are used to develop biodegradable packaging materials. Biopolymer-based ones are produced from proteins, polysaccharides, and their derivatives. Also, biodegradable packaging materials can be obtained from fossil fuel sources based on biodegradable plastics such as polyvinyl alcohol (PVA), polybutylene succinate (PBS), and polybutylene adipate terephthalate (PBAT). Besides the environmentally friendly nature of the biopolymers, they do not create any intrinsic food contamination (e.g. phthalate leaching) [4]. However, the insufficient chemical, mechanical, physical, permeability, and thermal properties of biopolymers and biodegradable plastics is another challenge [5]. The limitations and drawbacks associated with these polymers can be overcome by using micro/nano reinforcing agents. Nanotechnology is a promising technology in food packaging that has great potential advantages. This technology can be defined as the application of materials that have a minimum of one dimension in 1–100 nm [6]. Nanoparticles (NPs) differ from the macroscale or traditional bulk materials of the same substance by displaying prominently different chemical and physical features [7]. Especially, the surface area to volume ratio rises owing to the greatly decreased particle size of NPs [8]. Hence, nanoparticles can be combined with biodegradable polymers to develop bionanocomposites with improved properties for the food sector. The obtained biodegradable nanocomposites can offer advanced properties in terms of bioactivity, mechanical performance, oxygen permeability, moisture permeability, and thermal stability than neat biopolymer-based products.

There are several NPs that can be classified as carbon-based NPs, organic NPs, and inorganic NPs in general [9]. As inorganic NPs, metal, and metal oxide NPs are focusing more attention at the academic and commercial level due to their importance in various fields (Fig. 1). These nanomaterials are also used in food packaging to produce less permeable, lighter, and stronger packages [10]. Especially, some of the metal and metal oxide NPs preferred to prepare active packaging materials due to being more stable at high temperatures with an intrinsic antimicrobial activity compared to organic counterparts (enzymes, organic acids, etc.) [11]. Moreover, metal and metal oxide NPs have potential ethylene and oxygen scavenging, as well as UV-light blocking ability, which make them ideal candidates for active packaging applications. However, the bionanocomposite packaging performance is based on the nanofiller's size and the homogeneity of the NPs distribution and dispersion in the polymer matrix [5].

Taken together, this chapter will focus on the recent trends in metal and metal oxide NPs incorporated biodegradable polymer-based food packaging materials. It will specifically include an overview of (i) the fundamental properties of metal and metal oxide NPs; (ii) the effect of metal and metal oxide NPs on the barrier, mechanical,

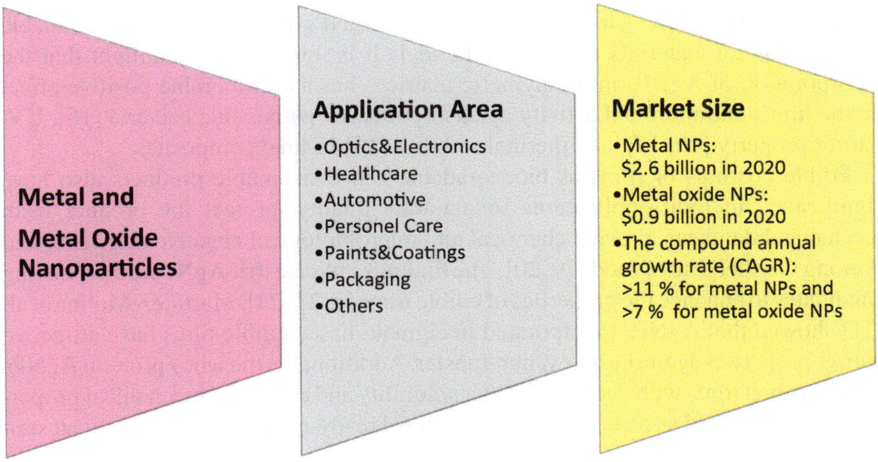

Fig. 1 Application areas of metal and metal oxide nanoparticles [12, 13]

and other functional properties of the biodegradable polymer-based food packaging (iii) recent trends of metal and metal oxide NPs incorporated bionanocomposites in active packaging applications; (iv) recent trends of metal and metal oxide NPs incorporated bionanocomposites in intelligent packaging applications; (v) metal and metal oxide NPs migration phenomenon from packaging.

2 Metal Nanomaterials in Biodegradable Food Packaging

Metal nanoparticles including silver (Ag), gold (Au), and copper (Cu) have been extensively investigated and frequently evaluated in food packaging applications due to their superior biological and physicochemical properties.

2.1 Silver

Among the metal nanoparticles, silver is the most accepted metal for food packaging. Silver is a precious metal that belongs to the transition metals group. Silver nanoparticles size is between 1–100 nm [14]. Silver nanoparticles (AgNPs) are potent antimicrobial and antifungal agents with low cytotoxicity. They have also anti-fungi, anti-filarial, anti-yeast, and anti-viral activities [11]. AgNPs exhibit better antimicrobial activities compared to metallic silver via providing better contact with the microorganism depending on their very large surface area [15]. Therefore, AgNPs incorporated biodegradable packaging materials generally developed for active packaging applications to find a solution for the growing threat of foodborne outbreaks.

A summary of the selected papers on the AgNPs incorporated biodegradable food packaging materials is given in Table 1. It is important to highlight that the incorporation of AgNPs into polymeric matrices has a considerable positive effect on the film's antimicrobial activity against foodborne pathogenic bacteria [16], UV-barrier property [16, 17], and thermal stability [18] of final composites.

Edible coatings or films as biodegradable and consumable products also have significance in food applications to maintain quality, protect the product from mechanical damages, prevent chemical and microbiological negative activities, and prolong the shelf life of food [19, 20]. The findings indicate that AgNPs are promising candidates to enhance the properties of edible films [7, 21–24]. Martínez-Molina et al. [21] showed that AgNPs incorporated in alginate-based edible films have attractive barrier properties against water vapor transfer. Additionally, the whey protein/ AgNPs antimicrobial films with decreased water solubility and enhanced mechanical properties were suggested as placement between sliced ready-to-eat cheeses or delicatessen [22].

Moreover, some recent studies focused on the synergistic effect of silver and essential oils in biodegradable composite packaging [25–30]. Positive results were obtained to extend the shelf-life of cheese [26], pork [28] and strawberries [25].

Intelligent packaging offers a convenient approach for real-time monitoring of the quality of packaged food. Hydrogen sulfide (H_2S) is a volatile gas that occurs during spoilage of meat. AgNPs are preferred in sensing systems because of their selective detection of H_2S and ideal optical properties [30]. The colorimetric sensor process depends on the ultra-strong binding ability of Ag with H_2S to form silver sulfide (Ag_2S) followed by a silver-silver sulfide ($Ag@Ag_2S$) core–shell structure formation [30, 31]. It has been confirmed that AgNPs are suitable metal nanoparticles owing to their unique local surface plasmon resonance property to develop biodegradable colorimetric freshness indicators [30–33]. The gellan gum-AgNPs were found to be suitable for monitoring chicken breast spoilage with yellow-to-colorless color changes under H_2S gas [31]. Kwon and Ko [30] reported a remarkable color change for cellulose nanocrystal- AgNPs composite from yellowish-dark wine red to colorless metallic silver based on the AgNPs amount when exposed to H_2S. The studies revealed the potential of AgNPs to develop user-friendly and highly effective platforms for intelligent packaging applications.

Time–temperature indicators (TTIs) are also a type of intelligent packaging that can give information on food quality during transportation, storage, distribution, and consumption with the time between disposals [34]. Metal NPs with thermochromic properties as Ag, and Au NPs can be used in physical TTIs [35]. These NP_S are thermal conductors that show visible color changes when absorbing heat [35]. A recent study revealed the potential use of AgNPs as an additive for a TTI film. The higher thermal conductivity of AgNPs make a positive effect on the color change behavior of polydiacetylene/AgNPs embedded carboxymethyl cellulose film. The developed film was suggested as a TTI for fruits and vegetables [34].

The biodegradability of the PVA/montmorillonite/AgNPs films via indoor soil burial degradation analysis was performed by Mathew et al. [36] by simulating real field conditions. To determine the biodegradation behavior, degradation degrees of

Table 1 Recent studies on silver NPs incorporated biodegradable packaging materials

Nanomaterial	Size (nm)	Polymer matrix	Category	Main findings	Tested food	References
AgNPs (spherical)	~64	Carrageenan	Active packaging	100% UV protection, Antioxidant (DPPH and ABTS radical scavenging) activity Antibacterial activity against *S. aureus* and *E. coli*	–	[17]
AgNPs	–	Soluble soybean polysaccharide	Active antimicrobial packaging	Improved antibacterial, thermal, and UV barrier properties, Lowered mechanical properties and wettability	Green grapes	[18]
AgNPs (spherical)	~10	Starch/agar	Active packaging	Antibacterial activity toward *E. coli* and *L. monocytogenes*	–	[37]
AgNPs	–	Cellulose nanocrystal	Intelligent packaging	Ability in detecting the spoilage gases	Chicken breast	[30]
AgNPs	–	Cellulose	Active packaging	Antibacterial (*E. coli and S. aureus*) and antifungal (*Rhizopus stolonifer*) activity	Strawberry storage time extended to 12 days	[38]
AgNPs	50–80	Starch/ polyvinyl alcohol / cardanol oil	Active packaging	Reduced strength Increased flexibility Antibacterial (*E. coli, P. aeruginosa, B. subtilis, S. aureus, and S. mutans*), and antifungal (*C. albicans*) activity	–	[39]

(continued)

Table 1 (continued)

Nanomaterial	Size (nm)	Polymer matrix	Category	Main findings	Tested food	References
AgNPs	~86	Gelatin	Active packaging	Antibacterial activity against *S. aureus*	–	[40]
AgNPs	–	Tapioca starch	Active packaging	Antibacterial activity against *S. aureus and E. coli*	–	[41]
AgNPs	–	Poly vinyl alcohol/rice hull powder	Active packaging	Antibacterial activity against *E. coli, P. aeruginosa, S. oralis, S. aruginosa*	–	[42]
AgNPs (spherical)	~13	Starch	Coating	–	Strawberry Extend the shelf-life from 2 to 6 days at room storage and from 8 to 16 days in cold storage	[43]
AgNPs (spherical)	10–30	Chitosan	Coating-active packaging	–	Grape Grapes showed reduced decay percentage and weight loss	[44]

films were recorded in terms of weight loss (%) on different time intervals as 15th, 30th, 60th, 90th, and 110th days. The results showed that AgNPs incorporated in PVA/montmorillonite film initially demonstrated a slower biodegradation rate when compared to the PVA/montmorillonite film. The complete disintegration of AgNPs incorporated film was 90 days. These results were related to the antibacterial potential of AgNPs that might have slowed down the growth of soil microorganisms and led to resist degradation.

2.2 Gold

Currently, gold nanoparticles (AuNPs) were also evaluated in biodegradable polymer-based composites. Bumbudsanpharoke and Ko [45] regarded the AuNPs as an antioxidant agent due to their catalytic activity to obtain AuNPs-lignocellulose fiber nanocomposite sheets for potential active packaging applications. Similarly, Namasivayam et al. [46] used spherical AuNPs (80–90 nm) to fabricate gum acacia-gold nanocomposite containing food preservative agent natamycin. The composite was coated on butter paper to be applied as packaging material for bread slices. It was suggested that fabricated bionanocomposite can be evaluated as biosafe efficient food preservative. Chowdhury et al. [47] studied the antimicrobial effect of AuNPs on PVA-based biocomposite film. Moreover, the success of the biocomposite film on banana preservation was investigated. The results showed that AuNPS contributed to the enhancement of the mechanical and antibacterial properties of the film. PVA-glyoxal-AuNPs extended the shelf-life of banana up to five days leading to the formation of minimum black spots as compared to pure PVA film. In another study, a composite film composed of AuNPs (ca. 4.0 nm), chitosan, and 3-aminopropyltrimethoxysilane (APTMS) was developed using the drop-casting method. The synergistic influence of the three components resulted in a greater antimicrobial activity against *Salmonella enterica serovar typhimurium* bacterium [48]. However, the price of gold may be an obstacle in the way of commercialization.

2.3 Copper

Copper nanoparticles (CuNPs) are also among the investigated reinforcements in biobased food preservation materials. The interests depend on the low cost and comparable antibacterial abilities of CuNPs compared to expensive antibacterial agents (gold, platinum, silver, etc.) [49]. It was shown that cellulose/ CuNPs-based composites are good candidates for active packaging applications [49, 50]. Kumar et al. [49] introduced the cellulose-based CuNPs-modified tamarind nut powder incorporated composite films. The results demonstrated that the hybrid nanocomposites exhibited better thermal and antibacterial activities against *E. coli, P. aeruginosa, S. aureus,* and *B. licheniformis*. The antibacterial mechanism of metal nanoparticles is well known (Fig. 2). Similar to other metal ions, the copper ions in the CuNPs hinder cell respiration. The formation of holes in the bacterial cell membrane occurs and leads to membrane disruption. Subsequently, reactive oxygen species formation happens due to the reaction with Cu^{2+} ions. The increased reactive oxygen species (ROS) production is the reason for bacterial proteins, DNA, RNA, and lipids damage and cell death [51]. Recently, CuNPs doped carboxymethyl cellulose/pectin composites were investigated by Sougandhi et al. [52] to explore the effect of copper concentration on the antibacterial, oxygen, and water barrier and thermal properties. The incorporation of CuNPs at 2% improved the barrier properties of the films by

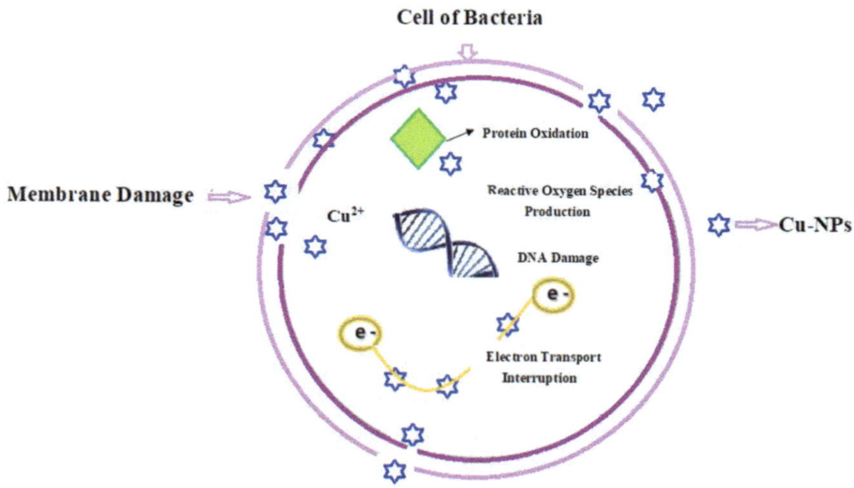

Fig. 2 Antibacterial mechanisms of metal nanoparticles

reducing the oxygen transmission rate (OTR) and water vapor transmission permeability (WVTR) values. The developed composites offered for active packaging applications due to showing distinctive antimicrobial activity against Gram-positive (*S. aureus*) and Gram-negative (*E. coli and P. aeruginosa*) foodborne pathogenic bacteria.

3 Metal Oxide Nanomaterials in Biodegradable Food Packaging

Metal oxide nanoparticles including copper oxides (CuO and Cu_2O), iron oxides (Fe_3O_4, Fe_2O_3), magnesium oxide (MgO), zinc oxide (ZnO), and titanium dioxide (TiO_2) are deeply investigated for food packaging materials. To enhance the probability of their use in the food packaging industry, researchers searching to find optimal approaches to develop metal oxide nanoparticles in biodegradable packaging.

3.1 Copper Oxides

Copper oxide nanoparticles (CuO-NPs) have prospective useful properties such as better stability, lower cost, long life compared to other nanoparticles, high surface-to-volume ratio, high-temperature superconductivity, non-toxicity, barrier to visible and UV light [53–55]. The potential use of CuO-NPs in biodegradable packaging to enhance the shelf life of foods by reducing the growth and development of coliforms, microorganisms, molds, and yeast has been proven [53]. The CuO-NPs were evaluated in agar [56], carboxymethyl chitosan [57], kefiran-carboxymethyl cellulose [55, 58], carboxymethyl cellulose [59], cellulose/sodium alginate [60], methylcellulose [53], whey protein isolate [54] and zein [61] based biodegradable films. The positive results for anti-bacterial properties against food-borne pathogens [55, 56, 59, 60] antioxidant activity [60], mechanical properties [55, 59], visible [55], and UV light barrier property [56], as well as water vapor barrier [55] property of the developed composite films were reported by the researchers. However, it is important to note that the variance of properties is based on the size, stability, and concentration of CuO-NPs in the polymer matrix, like the other nanoparticles.

3.2 Iron Oxides

Iron oxide nanoparticles are especially preferred as UV-blocking additives. Research is also underway to use iron oxide nanoparticles loaded with biodegradable packaging materials [62–69]. A recent study has reported the utilization of iron oxide nanoparticles with lysed yeast cells residue to develop whey protein concentrate-based biodegradable films [62]. The incorporation of iron oxide nanoparticles improved the tensile strength and water vapor permeability of films. Additionally, the iron oxide nanoparticles loaded film color became darker which indicated the photo-protective potential of the composite. In another study, Fe_2O_3-NPs (<50 nm) loaded poly(L-lactide) based films were developed to be evaluated in various applications including food and drug packaging. The developed films were brown which could be a UV-protection material [63]. PLA/zero-valent iron and PLA/Fe_2O_3-based coating films were developed by Ligaj et al. [64]. The neat PLA and iron oxide loaded PLA film did not reveal antimicrobial activity against gram-negative bacteria (*E. coli*), gram-positive bacteria (*Bacillus subtilis and Staphylococcus epidermidis*), yeast (*Rhodotorula rubra*), and molds (*Geotrichum candidum*) in contrast to PLA/zero-valent iron composite film. It is suggested that PLA/zero-valent iron films should find application as active packaging. Another research showed that the magnetic iron oxide NPs improved the antibacterial, barrier, and mechanical properties of gelatin films [65].

Iron oxide nanoparticles were also evaluated to develop ethyl cellulose based composite coating to extend the shelf life of Bell Pepper. The shelf-life extension was observed for α-Fe_2O_3/ethyl cellulose nanocomposite dip-coated bell peppers. The as-prepared nanocomposite paste was offered as an inexpensive and biodegradable packaging film for dynamic active vegetable packaging applications [66].

3.3 Magnesium Oxide (MgO)

Magnesium oxide is also a favorable naturally occurring inorganic metal oxide filler with its excellent low dielectric constant, low refractive index, simple stoichiometry, and thermodynamic stability [70–72]. It is a permitted food additive in the EU (E 530) with no quantity limitation except for chocolate and cocoa products [8]. The low price and easy availability provide MgO-NPs popularity in the field of packaging [71]. The good antibacterial properties, good biocompatibility, and UV-blocking ability are the other charming properties of MgO-NPs [71]. It has been asserted that the availability of numerous defects or oxygen vacancy on the MgO-NPs surface will take part in the formation of ROS, which can destroy cells of bacteria and protein peptides [73, 74].

Interesting investigations have been carried out with different composite films to show the potential evaluation of MgO nanoparticles in food packaging applications. Recent research highlighted that the incorporation of MgO-NPs into the PLA matrix leads to improvement in mechanical and gas barrier properties [72]. For instance, enhancement in thermal stability, UV shielding property, water resistance as well as antimicrobial activity of MgO-NPs loaded carboxymethyl chitosan-based biopolymer films was reported by Wang et al. (Table 2) [73]. Due to the obtained results the developed nanocomposite films were suggested as suitable candidates for UV-protecting coatings and water-rich food packaging applications [73].

There are several studies in the literature that reported the synergistic effect of MgO-NPs with other bioactive materials such as cellulose [75], essential oil [76], and polyphenol [77] on biodegradable polymer-based composite properties. The developed films showed good antibacterial activity [76, 77].

Metal oxides are suitable candidates to improve the gas barrier properties of biodegradable packaging materials by reducing the transportation of gas molecules. Swaroop and Shukla [78] reported a 65% reduction in oxygen permeability corresponding to 1% MgO-NPs reinforced PLA films in comparison to the neat PLA films. It was determined that the increased amount of nano-filler leads to an increase in oxygen permeability. This situation can be explained via the agglomeration of NPs at higher levels which causes the formation of free volumes at the interface of polymer and filler, and finally generation of a preferential path for permeating molecules diffusion. Moreover, the film production process is a key factor in the permeability feature. When MgO-NPs reinforced PLA films were developed by blowing [78] the films exhibited reduced oxygen and water vapor transmission compared to the solvent-cast films [72].

To show the environmental implications of biocomposites investigation of the biodegradation behavior is essential. The biodegradation of Polyvinyl Alcohol/Carboxymethyl Cellulose Composite Incorporated with L-Alanine Functionalized MgO Nanoplates under natural environmental circumstances was searched by Amaregouda et al. [79]. It was determined that MgO-L-alanine incorporation into the PVA/CMC matrix negatively affected the soil degradability of the composite. The degradability was approximately decreased from 40 to 38% with MgO-L-alanine

Table 2 Recent studies on magnesium oxide NPs incorporated biodegradable packaging materials

Nanomaterial	Size (nm)	Polymer matrix	Category	Main findings	Tested food	References
MgO-NPs	<60	PLA	Active packaging	Enhanced tensile strength, oxygen barrier, and antimicrobial activity	–	[72]
MgO-L-alanine NPs	–	PVA/CMC	Active packaging	Enhanced UV light screening property, antimicrobial activity against food-borne pathogens *E. coli* and *C. tropical*	–	[79]
MgO-NPs	20–30	Carboxymethyl chitosan	Active packaging-Water rich food packaging	Enhanced thermal stability, UV shielding performance, water-insolubility, antimicrobial activity against *Listeria monocytogenes* and *Shewanella baltica*	–	[73]
MgO-NPs	20	Modified starch		Decreased the solubility, moisture content, and water vapor permeability Enhanced antioxidant and antimicrobial activity against *E. coli* and *S. aureus*	–	[80]
Cetrimonium bromide (CTAB) modified MgO-NPs	40	Poly(butylene adipate-co-terephthalate)	Active packaging	Enhanced mechanical properties and antimicrobial activity against *E. coli* and *S. aureus*	–	[81]

addition. This was mainly because the microbial resistance capacity of MgO-L-alanine could withstand the microbe activation and retarded the disruption of the polymer chain.

3.4 Zinc Oxide (ZnO)

ZnO-NPs have a white appearance strong antimicrobial activity in a broad spectrum of microorganisms, antioxidant activity, photocatalytic, near-UV emission, and optical transparency properties [4, 82, 83]. Behind the above functional properties, ZnO-NPs can be considered as reinforcement for biodegradable packaging materials owing to their biocompatibility, being non-toxic, availability at commercial scale, and low price. ZnO is also among the GRAS (generally recognized as safe) materials according to the Food and Drug Administration (FDA) of the USA [83].

ZnO-NPs are deeply investigated as an active antimicrobial agent in various biodegradable polymer-based packaging (Table 3). The literature presents a couple of biodegradable films, with ZnO-NPs, tried as packaging for bread [84, 85], cheese [86], fish [87], fruits [88–90], meat [91–94] or vegetables [95]. The antimicrobial activity of ZnO-NPs mainly depends on Zn^{2+} ions release that interact with the anionic groups of the bacterial cell wall and thus damage the membrane [86, 96]. The other mechanism is the formation of reactive oxygen species (ROS) and low-concentration H_2O_2 through photocatalysis resulting in oxidative stress to the cell and inhibiting cell growth and/or killing bacteria by reacting with enzymes, DNA, lipids, and proteins [86, 97]. However, the antimicrobial effect of nanocomposite may differ against Gram-positive and Gram-negative cells according to the difference in their cell wall structural and chemical characteristics [86, 98]. In addition, a generally stronger antibacterial effect is provided with ZnO-NPs that have smaller particle sizes [97].

In addition, the synergistic effect of ZnO-NPs with different extracts including bamboo leaves [99], grape seed [100], Mentha pulegium [101], and essential oil [102–105] for the preservation of various foods such as high-fat meat [100], fruits [101, 102] and rainbow trout fillets [105] have been investigated to develop antimicrobial active packaging materials. Interestingly, the combined use of ZnO-NPs and extracts in the polymer matrix showed stronger antimicrobial activity than those containing either of these antimicrobial agents [104–106].

Recently, many studies have been published concerning ZnO-NPs as nanofillers to improve the antibacterial performance, color stability, mechanical properties, water resistance, and UV-blocking properties of the biodegradable films for application in intelligent packaging beside colorimetric indicators including anthocyanin [106–111], betanin nanoliposomes [93]. The developed pH-sensitive films were expected to be safe indicators for real-time monitoring of the freshness of foods during storage.

ZnO-NPs can be also used to carry antioxidants and then prepare active packaging. This approach was highlighted by Yadav et al. [4]. A biocomposite film was developed by mixing ZnO-NPs loaded gallic acid with chitosan. The tensile strength,

Table 3 Recent studies on zinc oxide NPs incorporated biodegradable packaging materials

Nanomaterial	Size (nm)	Polymer matrix	Category	Main findings	Tested food	References
ZnO-NPs	–	Poly(3-hydroxybutyrate) (PHB)	Active packaging	Bionanocomposite with ZnO (5.0 wt%) is a potential alternative to synthetic plastics, with an optimal balance of toughness, barrier properties, and antibacterial activity against *E. coli* and *S. aureus*	–	[98]
ZnO-NPs	–	PVA	Coating/UV protective packaging	Decreased water permeability UV blocking effect (blocked the rays below 380 nm)	Apple Storage time extended	[88]
ZnO-NPs (rod like)	50	PVA/starch	Active/UV protective packaging	Antimicrobial activity against both E. coli and S. aureus UV-shielding capability High optical transparency	Carrot slices İmproved freshness	[95]
ZnO-NPs	30 ± 10	Carboxymethyl chitosan/fish skin gelatin	Active packaging	Enhanced water insolubility and mechanical properties Antibacterial activity against E. coli and S. aureus	–	[113]

(continued)

Table 3 (continued)

Nanomaterial	Size (nm)	Polymer matrix	Category	Main findings	Tested food	References
ZnO-NPs	30	Chitosan/coix seed starch	Edible packaging	Enhanced tensile strength, barrier properties, and antibacterial activities	Pork Prolonged shelf life	[91]
ZnO-NPs	–	Starch/chitosan Starch	Coating	Easy degradation in soil and seawater	Bread Increased microbial shelf life from 10 to 30 days	[120]
ZnO-NPs (hexagonal)	36.5	Polylactic acid (PLA)	Active packaging	Antibacterial activity against *E. coli* and *S. aureus*	Grape Improved food quality and shelf life up to two weeks at 40 °C	[89]
ZnO-NPs	–	Poly (acrylamide-co-2-hydroxyethyl methacrylate)/chitosan	Active packaging	No oxygen permeation up to 1.5 kg/cm^2 feed pressure, Antibacterial activity against *E. coli* and *S. aureus*	-	[122]
ZnO-NPs	10–30	Salvia mirzayanii seed gum	Active packaging	Enhanced physicochemical and mechanical characteristics, reduced the light transmission in the biopolymers at 280 and 660 nm	Wheat bread Antifungal activity during storage for 12 days	[85]

(continued)

Table 3 (continued)

Nanomaterial	Size (nm)	Polymer matrix	Category	Main findings	Tested food	References
ZnO-NPs (nanorods)	500 (diameter)	PLA	Active packaging	Antibacterial activity against *E. coli* and *S. aureus*	–	[123]
ZnO-NPs (spherical)	30–40	Corn starch	Active packaging	Highly water resistant	–	[119]
ZnO-NPs	–	Cassava starch/chitosan/ pineapple leaf fiber	Coating-active packaging	Completely decomposed 21 days in ordinary soil and 18 days in seawater	–	[120]
ZnO-NPs	–	PLA (first layer) Carboxymethyl chitosan/ carboxymethyl cellulose sodium (second layer)	Coating for degradable fast food paperboard	Improved barrier and antibacterial properties	–	[124]
ZnO-NPs (nanorods)	30–60 (diameter)	Poly(ethylene oxide)/ Carboxymethyl cellulose	Active packaging	Antibacterial activity against *E. coli* and *S. aureus*	–	[125]
ZnO-NPs	<100	Tamarind kernel powder	Active packaging	Improved mechanical properties Antimicrobial efficacy against *Bacillus cereus* and *E. coli*	–	[112]
ZnO-NPs	20	Polybutylene adipate-co-terephthalate (PBAT)/ thermoplastic starch	Active packaging	Prevented microbial growth and lipid oxidation	Pork Meat Extended the shelf life by more than 3 days under refrigerated conditions	[126]

elongation at break values, and antibacterial and antioxidant properties of the as-prepared films were effected significantly by the incorporation of ZnO-NPs loaded gallic acid. The homogeneous dispersion of ZnO-NPs/gallic acid in the polymer matrix caused an improvement in the light barrier feature of the chitosan-based films by blocking light transmission and scattering light through the films. Unfortunately, barrier properties like oxygen permeability, as well as water vapor permeability, decreased compared to pure chitosan films.

The superior interfacial interaction of ZnO-NPs with biopolymers made them suitable nanofiller candidates for biocomposites [112]. Enhanced mechanical properties for ZnO-NPs incorporated carboxymethyl chitosan/fish skin gelatin [113], hydroxyethylcellulose/citric acid [114], starch/polyvinyl alcohol/ /phytochemicals [115], tamarind kernel powder [112] films were reported by researchers. The mechanical properties are used to verify the capability of food packaging materials to carry developed stresses throughout the processing, transporting, handling, and storage of foods [112]. Both the microstructural network and the physicochemical properties of the material play key roles in the mechanical properties of the biodegradable packaging. The enhancement of mechanical properties also strongly depends on the dispersion of the NPs in the polymer matrix which affects the efficient transfer of stress at the interface of the filler and matrix. It was observed that tensile strength generally improved with the addition of NPs at lower concentrations depending on the uniform dispersion of fillers resulting in better interfacial interaction between the NPs and polymer [112]. The incorporation of NPs amounts further a specific limit causes aggregation, leading to a decreased contact area between the NPs and the polymer matrix. Then, reduction occurs in the tensile strength value at higher loading concentrations due to the defects within the polymer films referring to agglomeration and non-uniform distribution of NPs [112, 114].

ZnO-NPs are also widely evaluated as ultraviolet (UV) light-blocking additives in biodegradable composites because of having strong barrier properties against UV-B (from 290 to 320 nm) and UV-A (from 320 to 400 nm) light [87]. Therefore, the performance of ZnO-NPs proved in the development of UV-screening biodegradable food packaging materials [116–118].

Since biodegradation is a charming alternative to valid practices for waste disposal, the biodegradability of ZnO-NPs incorporated polymer nanocomposites is important. The biodegradation rate of 7 wt % ZnO-NPs loaded corn flour-based films was investigated by calculating the amount of emitted CO_2 for 28 days at room temperature [119]. The results demonstrated that the biodegradation behavior of the film was nonresistant in the soil in the availability of compost and mineral salts. The composite film was determined to be highly biodegradable and showed high degradability in the first week. Another research group reported the biodegradation behavior of ZnO-NPs incorporated starch/chitosan/pineapple leaf fiber-based bioplastic in ordinary soil and seawater for up to 21 days. It was determined that the composite films were easily decomposed in soil and seawater for 21 days and 18 days, respectively. The researchers suggested that the developed antimicrobial bioplastics have high potential as food packaging with faster biodegradation performance [120]. ZnO-NPs-biodegradable polyester-based composites are also the focus of the research due to the adjustable biodegradation rate of the polymer matrix. Preliminary biodegradation of ZnO-NPs functionalized with ionic liquid (IL)/ polycaprolactone (PCL) composites in the presence of microorganisms (bacterium *Pseudomonas sp.* and fungus *Fusarium solani*) for 28 days was studied by Bujok et al. [121]. There was not any visible biodegradation for the pure PCL and the PCL with 1 wt % IL-ZnO-NPs (Fig. 3). Higher biodegradations were observed for 2 wt% and 3 wt% IL functionalized ZnO-NPs/PCL nanocomposites. This gave evidence that the presence of surface-bonded ionic liquid has a positive influence on the biodegradation rate of PCL-based films. These new ZnO-NPs-loaded nanocomposites can find great applications for biodegradable packaging systems.

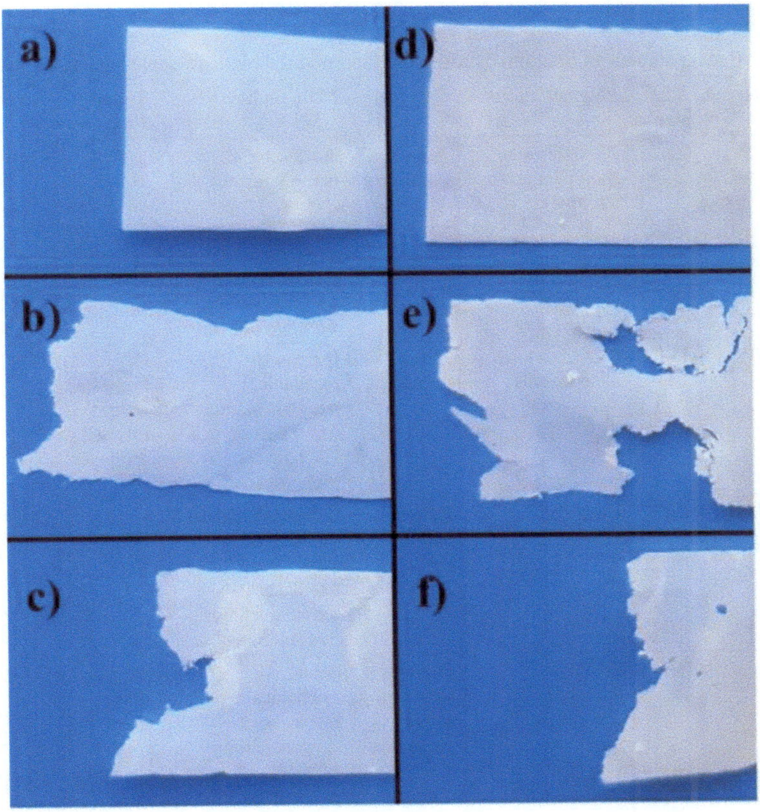

Fig. 3 Nanocomposite films after 28 days of biodegradation **a** pure PCL, **b** 1 wt% ZnO-NPs/PCL, **c** 2 wt% ZnO-NPs/PCL, **d** 1 wt% IL functionalized ZnO-NPs/PCL, **e** 2 wt% IL functionalized ZnO-NPs/PCL and **f** 3 wt% IL functionalized ZnO-NPs/PCL. Reprinted from Bujok et al. [121], Copyright (2022), with permission from Elsevier

3.5 Titanium Dioxide (TiO_2)

Titanium dioxide (TiO_2) also known as titania or titanium (IV) oxide is the naturally occurring form of oxidized titanium [127]. TiO_2 nanoparticles (TiO_2-NPs) are accepted as multifunctional and purchasable GRAS metal oxides according to their excellent features including bioactivity, biocompatibility, high refractive index, UV absorptivity, photocatalytic properties, low toxicity, high thermal and chemical stability [127–130]. Therefore, TiO_2-NPs can be used as excellent metal oxides for the development of biodegradable packaging films (Table 4).

The size, concentration, and distribution of TiO_2 had a significant effect on the final properties of biocomposites. Researchers have explored the 0.25% and 1% w/w

Table 4 Recent studies on titanium dioxide NPs incorporated biodegradable packaging materials

Nanomaterial	Size (nm)	Polymer matrix	Category	Main findings	Tested food	References
TiO_2_NPs	21	PVA/chitosan (polyphosphonate additive)	Active packaging	Antibacterial activity against *S. aureus, Methicillin-resistant S. aureus, E. coli, Pseudomonas aeruginosa, Candida albicans*	–	[146]
TiO_2_NPs	15	PVA/cellulose nanocrystals	Active packaging	Increased mechanical strength, water vapor barrier, and UV barrier properties	Garlic Prevented weight loss and spoilage	[132]
TiO_2_NPs	17.5	PVA/PLA	Active packaging	Enhanced thermal stability, contact angle, WVTR (%) Antibacterial activity against E. coli and S. aureus	–	[143]
TiO_2_NPs	10–40	PBAT/ thermoplastic cassava starch	Ethylene scavenging active packaging	Enhanced mechanical strength and reduced oxygen, carbon dioxide, and water vapor permeability	Banana slower darkening color change	[134]
TiO_2_NPs	–	Poly(L-lactide)/ poly(butylene succinate) (PLLA/PBSU)	Active packaging	Enhanced mechanical performance, antimicrobial activity, and photodegradability	–	[145]

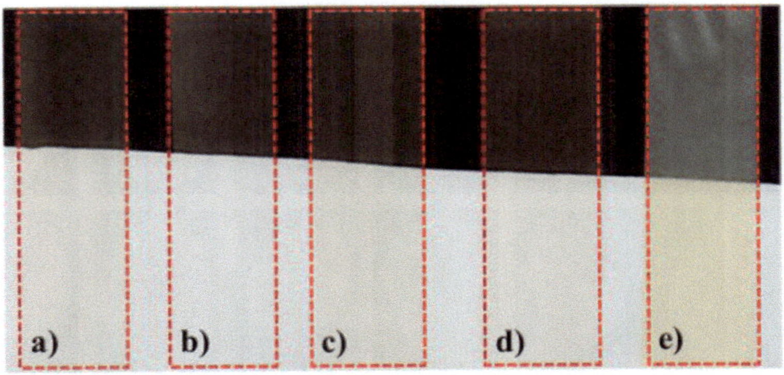

Fig. 4 Photo taken of films **a** Chitosan, **b** Starch T, **c** Chitosan/Starch, **d** Chitosan/Starch/TiO_2-NPs 0.25%, and **e** Chitosan/Starch/TiO_2-NPs 1%) on a standard black and white background. Reprinted from Menezes et al. [131], Copyright (2022), with permission from Elsevier

addition of TiO_2-NPs (20–25 nm) into chitosan/starch matrix to advance the fundamental understanding of the influence of nanoparticle ratio on functional features of bionanocomposite films [131]. Results demonstrated that more homogeneous surface morphology was obtained when The TiO_2-NPs concentration increased from 0.25% to 1%. The addition of 1% of TiO_2-NPs provided a reduction in solubility of the biopolymeric blend. Additionally, tensile strength significantly reduced with 0.25% TiO_2-NPs addition, while 1% TiO_2-NPs addition caused an important increase. Another remarkable was observed in the color of the films (Fig. 4). When TiO_2-NPs were incorporated into the blends, the bionanocomposites became whiter owing to the natural whiteness of the TiO_2-NPs.

In the study of Nguyen and Lee [132], garlic samples without any defect that also have similar appearance, size, color, and maturity degree were packaged with PVA/ Cellulose nanocrystals/TiO_2-NPs films to determine the effect of TiO_2-NPs concentration (0–10 wt%) on the shelf-life of garlic. Fresh garlic packaged in bags that developed with 5% and 10% TiO_2-NPs (Fig. 5e and f, respectively) nearly protected its original color and shape for up to 14 days without recognizable spoilage

Light is a primary external reason for food spoilage [132]. Furthermore, UV light causes negative effects on food including free radical generation, lipid oxidation, and color change [133]. The UV-shielding potential of TiO_2-NPs makes them suitable additives for light-sensitive goods packages [134]. With TiO_2-NPs incorporation into biopolymer films, a large difference between the absorption across a wide range of UV and visible regions occurred compared to neat films [131]. It was reported that chitosan/starch films containing 1% TiO_2-NPs lead to a reduction in light transmittance (%T) in UV-A (360 nm), UV-B (300 nm), UV-C (240 nm), and visible regions (600 nm) [131]. In the work by Mohr et al. [135], Ecovio® (45% of PLA, 55% of Ecoflex®) based biodegradable polymer film was developed by extrusion with different TiO_2-NPs contents (1 wt% and 5 wt%) to modify the light transmission characteristics. As expected, TiO_2-NPs acted as ultraviolet radiation

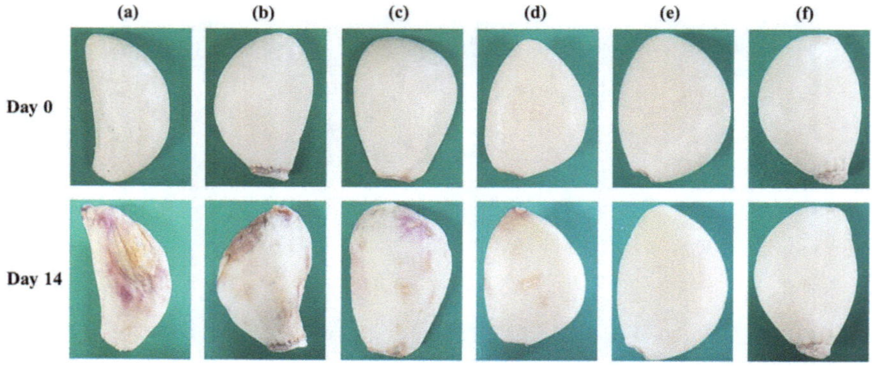

Fig. 5 Food preservation test of fresh garlic using **a** PVA, **b** PVA/Cellulose nanocrystals, **c** PVA/Cellulose nanocrystals/TiO_2-NPs 1%, **d** PVA/Cellulose nanocrystals/TiO_2-NPs 3%, **e** PVA/Cellulose nanocrystals TiO_2-NPs 5%, and **f** PVA/Cellulose nanocrystals/TiO_2-NPs 10% biocomposite films. Reprinted from Nguyen and Lee [132], Copyright (2022), with permission from Elsevier

blockers. The polymeric films were revealed to have efficient blocking efficiency against UV radiation. Ecovio® based films with 5 wt% TiO_2-NPs exhibited better blocking performance. These enhancements could be related to the light absorption, reflection, or scattering of TiO_2 [132].

Active packaging enhances food organoleptic features, prolongs food shelf life, and increases food safety [136]. It has also been shown in several studies that TiO_2-NPs are suitable additives to prepare antimicrobial active food packages [130, 132]. The use of TiO_2-NPs was found to be successful in the antimicrobial activities of the final biocomposite. The TiO_2-NPs loaded biocomposites were revealed to have antimicrobial activity in contrast to numerous Gram-positive (*S. aureus*) and Gram-negative (*E. coli, Salmonella typhimurium, and P. aeruginosa*) bacteria and fungi (*Aspergillus and Penicillium*) [130]. TiO_2-NPs can generate ROS with UV light activation, which can lead to oxidative damage to biomolecules like lipids, polysaccharides, proteins, and nucleic acids, causing cell death [132].

Active coatings are used to sustain apple quality by arranging water and gas exchange during storage, improving the appearance of the surface, and reducing mechanical damage and microbial deterioration. A study to evaluate the TiO_2-NPs and litchi peel extract to develop chitosan-based innovative active packaging films and watercolored apple coatings was carried out by Liu et al. [137]. The authors identified that the incorporation of TiO_2-NPs and/or litchi peel extract improved the mechanical strength, thermal stability, and water vapor barrier properties of chitosan films. Coating treatment respectably provided a lower respiration rate as well as inhibited weight loss and softening of Fuji apples. In another study, the performance of TiO_2-NPs on chitosan matrix with the addition of thymol and tween-80 agents for mushroom preservation was investigated. This study concluded that the developed coating has positive effects on button mushroom preservation by inhibiting respiration rate, weight loss, microbial contaminations, and lowering browning degree [138]. Similar results were obtained for green bell pepper coated with carboxymethyl cellulose and guanidinylated chitosan enriched with 5 wt% TiO_2-NPs (20 nm) for 15 days [133].

During the postharvest handling and storage process, fresh food products undergo flavor and color changes due to ripening which is related to the phytohormone ethylene (C_2H_4) [139]. This situation causes considerable quality loss and delimitates the storage life of fresh food. Hence, controlling excessive ethylene levels is a key parameter. TiO_2 is an effective photocatalyst that can be a potential additive as an ethylene scavenger for the active package applications of fruits and vegetables. It generates strong oxidizing power under UV light (< 385 nm) [140]. In some recent studies, like cellulose nanofiber /TiO_2-NPs nanocomposite films [141], chitosan/TiO_2-NPs [130, 139], gelatin/TiO_2-NPs/thymol@β-cyclodextrins [142] and PBAT/thermoplastic cassava starch/TiO_2-NPs [134] showed feasibility of TiO_2-NPs in ethylene scavenging biodegradable films. In these previous studies, the prepared packages were found efficient for prolonging the storage life of microbial-susceptible and ethylene-sensitive products such as tomato [139, 141, 142], Cavendish bananas [134], and strawberries [142]. Phothisarattana and colleagues [134] reported that TiO_2-NPs containing PBAT/thermoplastic cassava starch films also demonstrated efficient oxygen-scavenging activity by removing residual oxygen from the package headspace.

The degradability of polymer-based bionanocomposites is a crucial functionality. The degradation behavior of the composites was related to their basic composition and chemical structure [77]. Soil burial tests are generally applied to evaluate the biodegradation behavior of biocomposites. The mechanism of degradation in the soil burial test was explained with a two-stage process. Initially, the water diffusion into the film samples leads to the swelling of the samples and letting the microorganisms growth on the samples. It was followed by enzymatic and other secreted degradation resulting in weight loss and fragmentation of the samples. Several applications have been assumed for the biodegradation of TiO_2-NPs containing biocomposites [135, 143–145]. Biodegradation tests of PVA/PLA/TiO_2-NPs nanocomposite films were examined by soil burial degradation test for 20 weeks at room temperature. It was reported that in the first 12 weeks' nanocomposites exhibited rapid biodegradation rates and then a slower degradation rate was observed. The biodegradation was finalized with 41% weight loss (%) at the end of 20 weeks. Moreover, the color of nanocomposites was changed from white to yellow and then brown through degradation. Another important change was observed in the morphology of PVA/PLA/TiO_2-NPs films. The surface of films was smooth and homogeneous initially, while the surface took a rough, brittle, and easily crushed form during biodegradation. The disruption of film structure occurred via the formation of big cavities, small holes, and cracks followed by the formation of channels which led to the entry of water and microorganisms [143]. In another study, the biodegradation of PLA/TiO_2-NPs (Ecovio®) films was investigated by measuring the mass of produced and released CO_2 for 90 days [135]. The final product of biodegradation is CO_2.

The content of CO_2 is directly connected to the microorganism activity during the polymer biodegradation processes. The higher amount of CO_2 can be accepted as a higher microorganism's metabolic activity. It was noted that the PLA/TiO_2-NPs degradation in soil was not finished within 90 days. However, a greater amount of CO_2 generation was determined with the films containing TiO_2-NPs which is possibly related to the TiO_2-NPs photocatalytic activity that promotes the degradation of the films.

4 Migration

One of the most crucial steps for the fabrication of novel food contact materials is the evaluation of the overall and specific migration to determine if any unwanted or even harmful particles migrate to the packaged product [147]. From this view, an important parameter in the design of food packages containing metal oxide nanoparticles is the particle stabilization in the polymer matrix. This influences the bactericidal activity and migration of nanoparticles from packaging and then interaction with the packed food [148]. Migration of nanoparticles from packaging to food can happen via diffusion, dissolution, erosion, and abrasion of the packaging material [6]. Some metal oxides have proven to be of concern due to their harmful effects on human health in excessive amounts. The properties of nanomaterials such as dimension, charge, composition, morphology, physicochemical characteristics, and stability may influence their toxicology [149]. Hence, it is important to determine the toxicology and risk assessment of metal oxide nanoparticles containing biodegradable packaging. Moreover, it is also crucial to investigate the migration level of nanoparticles from packaging when in contact with foodstuffs, and the effects of their ingestion on the digestive system and other organs. On the other hand, the migration of packaging substances to food leads to chemical contamination, which can influence the organoleptic features. Additionally, the migration of nanoparticles from packaging is of concern because they could pollute the environment [6].

Published research addressing the migration evaluation dealing with metal oxide nanoparticles is given in Table 5. The migration of NPs is generally based on their size, shape, amount, distribution, and aggregation, as well as the film structure and type of food. The release of NPs is mostly determined experimentally by the use of food stimulants instead of particular food matrices [10]. Gvozdenko and co-workers observed that the migrated amount of CuO-NPs from methylcellulose film to cheese was under the toxic concentration after 7 days of storage at 35 °C [148]. It was also suggested that the migration of CuO-NPs could be decreased when cheese was stored in the refrigerator at lower temperatures (0–4 °C). In the research of Enescu et al. [150], it was determined that the migration of nano-sized TiO_2 was higher than micro-sized TiO_2 from chitosan matrix to olive oil. It was stated that the low migration can be explained by a physical mechanism that influences mass transfer due to dissolution from the cut edges of the chitosan film, however, the diffusion from the film was not seen. Because TiO_2 was not physically embedded in polymer

Table 5 Migration studies of metal oxide nanoparticles incorporated biodegradable packaging materials

Nanomaterial	Polymer matrix	Test procedure	Main findings	References
Gelatine stabilized CuO-NPs	Methylcellulose	Cheese storage in a thermostat at 35 ± 1 °C for 7 days	0.12 μg/mg CuO-NPs migrated (Not a toxic concentration)	[148]
TiO_2-NPs (10–40 nm)	PBAT/TPS	In distilled water, 10% ethanol and 3% acetic acid at 5 °C	0.2–1.3 mg/dm^2 overall migration of TiO_2-NPs The highest specific migration was distilled water > ethanol > acetic acid	[126]
TiO_2-NPs	α-chitosan	European normative EN 13,130–1:2004	The amount of migrated titanium was in negligible percent	[150]
Ag-NPs	PLA	Isooctane food simulant at 40 °C for 49 days	The migration of Ag-NPs increased with the migration time Highest migration was 28 mg/kg	[152]
ZnO-NPs	PLA	Meat fillets at 4 °C for 16 days	Overall migration for Zn^{2+} was below the migration limit of the National Institute of Health (40 mg/day)	[151]
ZnO-NPs	PLA/acetylated cellulose nanocrystals	Distilled water and 3% (v/v) acetate solution at 40 °C for 10 days	Migration for Zn^{2+} was below the specific migration limit (5 mg/kg)	[153]

matrix, it was coordinated by the chitosan functional groups which might decelerated the migration process. ZnO-NPs loaded PLA films were suggested as safe composite films to be used as packaging for seafood products due to showing low migration values (max 1.55 mg/kg) [151]. As a result, the primary challenge in designing the metal oxide NPs loaded biodegradable polymer-based system is decelerating the migration rate to stay within the migration limits. However, studies on the migration and risk assessment of the bionanocomposites are still not enough.

5 Conclusions and Future Perspective

The food industry challenges focused on the development of novel packages to provide tasty and convenient food with extended shelf life and long-time high-quality preservation. One of the other main challenges of this industry is to develop environment-friendly packages. Recently, biodegradable polymer-based food packaging materials have been adopted as potential alternatives for non-biodegradable and non-renewable petroleum-based ones. The growing scientific and industrial interest in biodegradable food packaging has brought about the progress of various methods for preparing bionanocomposites using nanoparticle fillers. Metal and metal oxide nanoparticles have established a significant place in various industries. Metal and metal oxide nanoparticles also emerged as suitable inorganic fillers for biodegradable polymer-based packaging materials due to their availability, low cost, functionality, and excellent antimicrobial properties. The incorporation of metal and metal oxide nanoparticles can significantly enhance the performance of biodegradable polymers in terms of bioactivity, gas, and water vapor barrier properties, physical properties, mechanical properties, thermal properties, UV light-blocking ability, and water sensitivity related to the interaction between the nanoparticles and polymer matrix. There are numerous studies on the development of active and smart biocomposite food packaging and coatings including metal and metal oxide nanoparticles. Both from technical and social points of view, metal and metal oxide nanoparticles incorporated biodegradable polymer-based composite food packaging will be in high demand for sustainable preparation of materials in order to show better performance than traditional food packaging systems. It is expected that the technological innovations related to bionanocomposites will be increased in the next years. However, despite the various benefits of metal and metal oxide nanoparticles in biodegradable food packaging systems, environmental, health, and safety aspects would be evaluated extensively before their further industrial-scale usage. The migration of nanoparticles from food packaging and its impact on foodstuffs and human health are among the primary concerns that limit the commercialization of bionanocomposites.

References

1. Jayakumar A et al (2022) Recent innovations in bionanocomposites-based food packaging films—a comprehensive review. Food Packag Shelf Life 33:100877. https://doi.org/10.1016/j.fpsl.2022.100877
2. Terzioğlu P et al (2021) Biowaste orange peel incorporated chitosan/polyvinyl alcohol composite films for food packaging applications. Food Packag Shelf Life 30:100742. https://doi.org/10.1016/j.fpsl.2021.100742
3. Terzioğlu P, Sıcak Y (2021) Citrus limon L. peel powder incorporated polyvinyl alcohol/corn starch antioxidant active films. J Instit Sci Technol:1311–1320. https://doi.org/10.21597/jist.825560
4. Yadav S, Mehrotra GK, Dutta PK (2021) Chitosan based ZnO nanoparticles loaded gallic-acid films for active food packaging. Food Chem 334:127605. https://doi.org/10.1016/j.foodchem.2020.127605
5. Jafarzadeh S, Jafari SM (2020) Impact of metal nanoparticles on the mechanical, barrier, optical and thermal properties of biodegradable food packaging materials. Crit Rev Food Sci Nutr 61(16):2640–2658. https://doi.org/10.1080/10408398.2020.1783200
6. Garcia CV, Shin GH, Kim JT (2018) Metal oxide-based nanocomposites in food packaging: applications, migration, and regulations. Trends Food Sci Technol 82:21–31. https://doi.org/10.1016/j.tifs.2018.09.021

7. Kumar S et al (2021) Plant extract mediated silver nanoparticles and their applications as antimicrobials and in sustainable food packaging: a state-of-the-art review. Trends Food Sci Technol 112:651–666. https://doi.org/10.1016/j.tifs.2021.04.031
8. Mittag A et al (2019) Toxicological assessment of magnesium oxide nanoparticles in HT29 intestinal cells. Arch Toxicol 93(6):1491–1500. https://doi.org/10.1007/s00204-019-02451-4
9. Spirescu VA et al (2021) Inorganic nanoparticles and composite films for antimicrobial therapies. Int J Mol Sci 22(9):4595. https://doi.org/10.3390/ijms22094595
10. Nikolic MV et al (2021) Metal oxide nanoparticles for safe active and intelligent food packaging. Trends Food Sci Technol 116:655–668. https://doi.org/10.1016/j.tifs.2021.08.019
11. Carbone M et al (2016) Silver nanoparticles in polymeric matrices for fresh food packaging. J King Saud Univ Sci 28(4):273–279. https://doi.org/10.1016/j.jksus.2016.05.004
12. ltd R (no date) Metal nanoparticles: global strategic business report, research and markets—market research reports—welcome. https://www.researchandmarkets.com/reports/5030157/metal-nanoparticles-global-strategic-business#tag-pos-4. Accessed 18 Jan 2023
13. ltd R (no date) Metal oxide nanoparticles market by type and end use industry: global opportunity analysis and industry forecast, 2021–2030, research and markets—market research reports—welcome. https://www.researchandmarkets.com/reports/5561192/metal-oxide-nanoparticles-market-by-type-and-end. Accessed 18 Jan 2023
14. Cameron S, Hosseinian F, Willmore W (2018) A current overview of the biological and cellular effects of nanosilver. Int J Mol Sci 19(7):2030. https://doi.org/10.3390/ijms19072030
15. Toker RD, Kayaman-Apohan N, Kahraman MV (2013) UV-curable nano-silver containing polyurethane based organic–inorganic hybrid coatings. Prog Org Coat 76(9):1243–1250. https://doi.org/10.1016/j.porgcoat.2013.03.023
16. Bang YJ, Shankar S, Rhim JW (2019) In situ synthesis of multi-functional gelatin/resorcinol/silver nanoparticles composite films. Food Packag Shelf Life 22:100399. https://doi.org/10.1016/j.fpsl.2019.100399
17. Kim YH et al (2022) Pine needle (pinus densiflora) extract-mediated synthesis of silver nanoparticles and the preparation of carrageenan-based antimicrobial packaging films. J Nanomater:1–15. https://doi.org/10.1155/2022/8395302
18. Liu J et al (2022) Soluble soybean polysaccharide films containing in-situ generated silver nanoparticles for antibacterial food packaging applications. Food Packag Shelf Life 31:100800. https://doi.org/10.1016/j.fpsl.2021.100800
19. Okcu Z, Yavuz Y, Kerse S (2018) Edible film and coating applications in fruits and vegetables. Alınteri Zirai Bilimler Dergisi:221–226. https://doi.org/10.28955/alinterizbd.368362
20. Giray Tufan E, Akpinar Borazan A, Koçkar ÖM (2021) A review on edible film and coating applications for fresh and dried fruits and vegetables. Bilecik Şeyh Edebali Üniversitesi Fen Bilimleri Dergisi 8(2):1073–1085. https://doi.org/10.35193/bseufbd.996827
21. Martínez-Molina EC et al (2021) Development and characterization of alginate-based edible film from Sargassum fluitans incorporated with silver nanoparticles obtained by green synthesis. J Food Meas Charact 16(1):126–136. https://doi.org/10.1007/s11694-021-01156-6
22. Çağrı Mehmetoğlu A, Sezer E, Erol S (2020) Development of antimicrobial whey protein-based film containing silver nanoparticles biosynthesised by aspergillus niger. Int J Food Sci Technol 56(2):965–973. https://doi.org/10.1111/ijfs.14749
23. Bahrami A et al (2019) Physico-mechanical and antimicrobial properties of tragacanth/hydroxypropyl methylcellulose/beeswax edible films reinforced with silver nanoparticles. Int J Biol Macromol 129:1103–1112. https://doi.org/10.1016/j.ijbiomac.2018.09.045
24. Jahantigh F (2019) Edible film based on vicia faba shell gum, nano silver and helichrysum arenarium essence. Nutr Food Sci 50(2):280–287. https://doi.org/10.1108/nfs-04-2019-0136
25. Shankar S, Khodaei D, Lacroix M (2021) Effect of chitosan/essential oils/silver nanoparticles composite films packaging and gamma irradiation on shelf life of strawberries. Food Hydrocolloids 117:106750. https://doi.org/10.1016/j.foodhyd.2021.106750
26. Motelica L et al (2021) Antibacterial biodegradable films based on alginate with silver nanoparticles and lemongrass essential oil–innovative packaging for cheese. Nanomaterials 11(9):2377. https://doi.org/10.3390/nano11092377

27. Phan DN et al (2021) Investigation of mechanical, chemical, and antibacterial properties of electrospun cellulose-based scaffolds containing orange essential oil and silver nanoparticles. Polymers 14(1):85. https://doi.org/10.3390/polym14010085
28. Wu Z et al (2019) Multifunctional chitosan-based coating with liposomes containing laurel essential oils and nanosilver for pork preservation. Food Chem 295:16–25. https://doi.org/10.1016/j.foodchem.2019.05.114
29. Gao H, Yang H (2020) Preparation and characterization of cinnamaldehyde/polyvinyl alcohol/silver nanoparticles ternary composite films. Int J Polym Anal Charact 26(1):24–36. https://doi.org/10.1080/1023666x.2020.1837587
30. Kwon S, Ko S (2022) Colorimetric freshness indicator based on cellulose nanocrystal–silver nanoparticle composite for ıntelligent food packaging. Polymers 14(17):3695. https://doi.org/10.3390/polym14173695
31. Zhai X et al (2019) A colorimetric hydrogen sulfide sensor based on gellan gum-silver nanoparticles bionanocomposite for monitoring of meat spoilage in intelligent packaging. Food Chem 290:135–143. https://doi.org/10.1016/j.foodchem.2019.03.138
32. You S et al (2022) Development of highly stable color indicator films based on κ-carrageenan, silver nanoparticle and red grape skin anthocyanin for marine fish freshness assessment. Int J Biol Macromol 216:655–669. https://doi.org/10.1016/j.ijbiomac.2022.06.206
33. Qin Y et al (2019) Preparation and characterization of antioxidant, antimicrobial and ph-sensitive films based on chitosan, silver nanoparticles and purple corn extract. Food Hydrocolloids 96:102–111. https://doi.org/10.1016/j.foodhyd.2019.05.017
34. Saenjaiban A et al (2020) Novel color change film as a time–temperature indicator using polydiacetylene/silver nanoparticles embedded in carboxymethyl cellulose. Polymers 12(10):2306. https://doi.org/10.3390/polym12102306
35. Cheng H et al (2022) Recent advances in intelligent food packaging materials: principles, preparation and applications. Food Chem 375:131738. https://doi.org/10.1016/j.foodchem.2021.131738
36. Mathew S et al (2019) Biodegradable and active nanocomposite pouches reinforced with silver nanoparticles for improved packaging of chicken sausages. Food Packag Shelf Life 19:155–166. https://doi.org/10.1016/j.fpsl.2018.12.009
37. Roy S, Rhim JW (2022) Starch/agar-based functional films integrated with enoki mushroom-mediated silver nanoparticles for active packaging applications. Food Biosci 49:101867. https://doi.org/10.1016/j.fbio.2022.101867
38. Zhang DY et al (2022) Soft and elastic silver nanoparticle-cellulose sponge as fresh-keeping packaging to protect strawberries from physical damage and microbial invasion. Int J Biol Macromol 211:470–480. https://doi.org/10.1016/j.ijbiomac.2022.05.092
39. Yahia R et al (2021) Synthesis and characterization of thermoplastic starch/pva/cardanol oil composites loaded with in-situ silver nanoparticles. J Appl Polym Sci 139(3):51511. https://doi.org/10.1002/app.51511
40. Pérez-Marroquín XA et al (2022) Green synthesis of silver nanoparticles for preparation of gelatin films with antimicrobial activity. Polymers 14(17):3453. https://doi.org/10.3390/polym14173453
41. Alqahtani T (2022) Silver nanoparticles impregnated in tapioca starch biofilm made using the electrospinning technique: a cutting-edge material for food packaging. Biomass Convers Biorefinery. https://doi.org/10.1007/s13399-022-03344-w
42. Ganesh Babu A, Saravanakumar SS, Sai Balaji MA (2022) Influence of silver nanoparticles on mechanical, thermal and antibacterial properties of poly (vinyl alcohol)/rice hull powder hybrid biocomposite films. Polym Compos 43(5):2824–2837. https://doi.org/10.1002/pc.26579
43. Taha IM et al (2022) Impact of starch coating embedded with silver nanoparticles on strawberry storage time. Polymers 14(7):1439. https://doi.org/10.3390/polym14071439
44. Zhao X et al (2022) Multifunctional chitosan/grape seed extract/silver nanoparticle composite for food packaging application. Int J Biol Macromol 207:152–160. https://doi.org/10.1016/j.ijbiomac.2022.02.180

45. Bumbudsanpharoke N, Ko S (2018) The green fabrication, characterization and evaluation of catalytic antioxidation of gold nanoparticle-lignocellulose composite papers for active packaging. Int J Biol Macromol 107:1782–1791. https://doi.org/10.1016/j.ijbiomac.2017.10.046
46. Karthick Raja Namasivayam S et al (2022) Green chemistry principles for the synthesis of anti fungal active gum scacia-gold nanocomposite—natamycin (Ga-AuNC–NT) against food spoilage fungal strain aspergillus ochraceopealiformis and its marked congo red dye adsorption efficacy. Environ Res 212:113386. https://doi.org/10.1016/j.envres.2022.113386
47. Chowdhury S et al (2020) Poly(vinyl) alcohol crosslinked composite packaging film containing gold nanoparticles on shelf life extension of banana. Food Packag Shelf Life 24:100463. https://doi.org/10.1016/j.fpsl.2020.100463
48. Virgili AH et al (2021) Nanocomposite film with antimicrobial activity based on gold nanoparticles, chitosan and aminopropylsilane. Surf Coat Technol 415:127086. https://doi.org/10.1016/j.surfcoat.2021.127086
49. Kumar TS et al (2020) Characterization, thermal and antimicrobial properties of hybrid cellulose nanocomposite films with in-situ generated copper nanoparticles in tamarindus indica nut powder. J Polym Environ 29(4):1134–1142. https://doi.org/10.1007/s10924-020-01939-w
50. Muthulakshmi L et al (2019) Preparation of cellulose/copper nanoparticles bionanocomposite films using a bioflocculant polymer as reducing agent for antibacterial and anticorrosion applications. Compos Part B Eng 175:107177. https://doi.org/10.1016/j.compositesb.2019.107177
51. Godoy-Gallardo M et al (2021) Antibacterial approaches in tissue engineering using metal ions and nanoparticles: from mechanisms to applications. Bioactive Mater 6(12):4470–4490. https://doi.org/10.1016/j.bioactmat.2021.04.033
52. Sougandhi PR, Ramanaiah S, Shobha Rani T (2021) Preparation and characterization of bio-nanocomposite films incorporating copper nanoparticles. Inorganic Nano-Metal Chem 52(5):661–667. https://doi.org/10.1080/24701556.2021.1946697
53. Gvozdenko AA et al (2022) Synthesis of cuo nanoparticles stabilized with gelatin for potential use in food packaging applications. Sci Rep 12(1). https://doi.org/10.1038/s41598-022-16878-w
54. Asdagh A et al (2020) Production and characterization of nanocomposite film based on whey protein isolated/copper oxide nanoparticles containing coconut essential oil and paprika extract. J Polym Environ 29(1):335–349. https://doi.org/10.1007/s10924-020-01882-w
55. Hasheminya SM et al (2018) Physicochemical, mechanical, optical, microstructural and antimicrobial properties of novel kefiran-carboxymethyl cellulose biocomposite films as influenced by copper oxide nanoparticles (cuonps). Food Packag Shelf Life 17:196–204. https://doi.org/10.1016/j.fpsl.2018.07.003
56. Roy S, Rhim JW (2019) Melanin-mediated synthesis of copper oxide nanoparticles and preparation of functional agar/cuo NP nanocomposite films. J Nanomater:1–10. https://doi.org/10.1155/2019/2840517
57. Fathi M et al (2022) Fabrication and characterization of multifunctional bio-safe films based on carboxymethyl chitosan and saffron petal anthocyanin reinforced with copper oxide nanoparticles for sensing the meat freshness. J Polym Environ 30(11):4538–4549. https://doi.org/10.1007/s10924-022-02490-6
58. Hasheminya SM et al (2019) Influence of simultaneous application of copper oxide nanoparticles and Satureja khuzestanica essential oil on properties of kefiran–carboxymethyl cellulose films. Polym Test 73:377–388. https://doi.org/10.1016/j.polymertesting.2018.12.002
59. Ebrahimi Y et al (2019) Development of antibacterial carboxymethyl cellulose-based nanobiocomposite films containing various metallic nanoparticles for food packaging applications. Journal of Food Science, 84(9), pp. 2537–2548. Available at: https://doi.org/10.1111/1750-3841.14744.
60. Saravanakumar K et al (2020) Physical and bioactivities of biopolymeric films incorporated with cellulose, sodium alginate and copper oxide nanoparticles for food packaging application. Int J Biol Macromol 153:207–214. https://doi.org/10.1016/j.ijbiomac.2020.02.250

61. Marandi MS et al (2022) Production of biodegradable films of Zein containing mentha asiatica essential oil and copper oxide nanoparticles: Investigation of physicochemical, antimicrobial, and antioxidant properties. J Polym Environ 30(10):4114–4129. https://doi.org/10.1007/s10924-022-02486-2
62. Al-Hayali JH, Cagrı-Mehmetoglu A (2022) Development of nanocomposite biodegradable film containing iron nanoparticles biosynthesised by saccharomyces cerevisiae. Int J Food Sci Technol 57(6):3616–3623. https://doi.org/10.1111/ijfs.15685
63. Lizundia E et al (2020) Synergic effect of nanolignin and metal oxide nanoparticles into poly(L-Lactide) bionanocomposites: material properties, antioxidant activity, and antibacterial performance. ACS Appl Bio Mater 3(8):5263–5274. https://doi.org/10.1021/acsabm.0c00637
64. Ligaj M et al (2020) Efficiency of novel antimicrobial coating based on iron nanoparticles for dairy products' packaging. Coatings 10(2):156. https://doi.org/10.3390/coatings10020156
65. Mehmood Z, Sadiq MB, Khan MR (2020) Gelatin nanocomposite films incorporated with magnetic iron oxide nanoparticles for shelf life extension of grapes. J Food Saf 40(4). https://doi.org/10.1111/jfs.12814
66. Guleria G et al (2022) Synthesis of α-Fe_2O_3/ethyl cellulose-based nanocomposites to extend the shelf-life of capsicum annuum L. var. Grossum. Mater Today Proc. https://doi.org/10.1016/j.matpr.2022.09.222
67. Mary TR, Jayavel R (2022) Fabrication of chitosan/cashew nut shell liquid/plant extracts-based bio-formulated nanosheets with embedded iron oxide nanoparticles as multi-functional barrier resist eco-packaging material. Appl Nanosci 12(5):1719–1730. https://doi.org/10.1007/s13204-022-02377-x
68. Alagu T (2020) Synthesis and impregnation of Fe_2O_3 nanoparticles on cellulose paper and sodium alginate films for the preservation of fruit and vegetables. J Microbiol Biotechnol Food Sci 9(6):1166–1169. https://doi.org/10.15414/jmbfs.2020.9.6.1166-1169
69. Chandrakumara GT et al (2019) Eco-friendly, green packaging materials from akaganeite and hematite nanoparticle-reinforced chitosan nanocomposite films. J Nanomater:1–11. https://doi.org/10.1155/2019/1049142
70. Kalyani P, Muthupandeeswari T (2022) Investigation on the altered properties of PVA filled magnesium oxide composite (pva@xmgo) thin films. Polym Bull 79(11):10115–10134. https://doi.org/10.1007/s00289-021-04004-1
71. Koshy RR et al (2022) MgONP/chitin nanowhisker-based hybrid filler bound soy protein thin films for bioactive packaging applications. Int J Food Sci Technol 57(9):5650–5662. https://doi.org/10.1111/ijfs.15805
72. Swaroop C, Shukla M (2018) Nano-magnesium oxide reinforced polylactic acid biofilms for food packaging applications. Int J Biol Macromol 113:729–736. https://doi.org/10.1016/j.ijbiomac.2018.02.156
73. Wang Y et al (2020) MgO/carboxymethyl chitosan nanocomposite improves thermal stability, waterproof and antibacterial performance for food packaging. Carbohydr Polym 236:116078. https://doi.org/10.1016/j.carbpol.2020.116078
74. El-Shaer A et al (2020) Correlation between photoluminescence and positron annihilation lifetime spectroscopy to characterize defects in calcined MgO nanoparticles as a first step to explain antibacterial activity. J Alloys Compd 817:152799. https://doi.org/10.1016/j.jallcom.2019.152799
75. Wang Y et al (2021) Gelatin-based nanocomposite film with bacterial cellulose–MgO nanoparticles and its application in packaging of preserved eggs. Coatings 11(1):39. https://doi.org/10.3390/coatings11010039
76. Naeeji N, Shahbazi Y, Shavisi N (2020) Effect of gamma irradiation on physico-mechanical and structural properties of Basil Seed mucilage-chitosan films containing *ziziphora clinopodioides* essential oil and MgO nanoparticles for rainbow trout packaging. J Food Process Preserv 44(10). https://doi.org/10.1111/jfpp.14781
77. Luo D et al (2022) Potato starch films by incorporating tea polyphenol and MgO nanoparticles with enhanced physical, functional and preserved properties. Int J Biol Macromol 221:108–120. https://doi.org/10.1016/j.ijbiomac.2022.09.010

78. Swaroop C, Shukla M (2019) Development of blown polylactic acid-MgO nanocomposite films for food packaging. Compos Part A Appl Sci Manuf 124:105482. https://doi.org/10.1016/j.compositesa.2019.105482
79. Amaregouda Y, Kamanna K, Gasti T (2022) Biodegradable polyvinyl alcohol/carboxymethyl cellulose composite incorporated with L-alanine functionalized MgO nanoplates: physico-chemical and food packaging features. J Inorg Organomet Polym Mater 32(6):2040–2055. https://doi.org/10.1007/s10904-022-02261-9
80. Hosseini SN, Pirsa S, Farzi J (2021) Biodegradable nano composite film based on modified starch-albumin/MgO; antibacterial, antioxidant and structural properties. Polym Test 97:107182. https://doi.org/10.1016/j.polymertesting.2021.107182
81. Wang X et al (2021) Biodegradable poly(butylene adipate-co-terephthalate) antibacterial nanocomposites reinforced with MgO nanoparticles. Polymers 13(4):507. https://doi.org/10.3390/polym13040507
82. Kalemtas A et al (2021) Mechanical and antibacterial properties of ZnO/chitosan bio-composite films. J Polym Eng 42(1):35–47. https://doi.org/10.1515/polyeng-2021-0143
83. Kim I et al (2020) ZnO nanostructures in active antibacterial food packaging: preparation methods, antimicrobial mechanisms, safety issues, future prospects, and challenges. Food Rev Int 38(4):537–565. https://doi.org/10.1080/87559129.2020.1737709
84. Anugrahwidya R, Armynah B, Tahir D (2022) Composites bioplastic film for various concentration of zinc oxide (Zno) nanocrystals towards physical properties for high biodegradability in soil and seawater. J Polym Environ 30(6):2589–2601. https://doi.org/10.1007/s10924-021-02363-4
85. Hasheminya SM et al (2022) Development and structural characterization of novel biomaterial polymeric films based on the mucilage extracted from salvia mirzayanii seed gum incorporated with zinc oxide nanoparticles. J Food Meas Charact 16(4):3042–3053. https://doi.org/10.1007/s11694-022-01405-2
86. Amjadi S et al (2019) Application of reinforced ZnO nanoparticle-incorporated gelatin bionanocomposite film with chitosan nanofiber for packaging of chicken fillet and cheese as food models. Food Bioprocess Technol 12(7):1205–1219. https://doi.org/10.1007/s11947-019-02286-y
87. Baek SK, Song KB (2018) Development of gracilaria vermiculophylla extract films containing zinc oxide nanoparticles and their application in smoked salmon packaging. LWT 89:269–275. https://doi.org/10.1016/j.lwt.2017.10.064
88. Channa IA et al (2022) UV blocking and oxygen barrier coatings based on polyvinyl alcohol and zinc oxide nanoparticles for packaging applications. Coatings 12(7):897. https://doi.org/10.3390/coatings12070897
89. Batool M et al (2021) Quality control of nano-food packing material for grapes (vitis vinifera) based on ZnO and polylactic acid (PLA) biofilm. Arab J Sci Eng 47(1):319–331. https://doi.org/10.1007/s13369-021-05361-9
90. Kumar S et al (2019) Bionanocomposite films of agar incorporated with ZnO nanoparticles as an active packaging material for shelf life extension of green grape. Heliyon 5(6). https://doi.org/10.1016/j.heliyon.2019.e01867
91. Zheng K et al (2022) Properties and biological activity of chitosan-coix seed starch films incorporated with nano zinc oxide and artemisia annua essential oil for pork preservation. LWT 164:113665. https://doi.org/10.1016/j.lwt.2022.113665
92. Phothisarattana D et al (2022) Blown film extrusion of PBAT/TPS/ZnO nanocomposites for shelf-life extension of meat packaging. Colloids Surf B Biointerfaces 214:112472. https://doi.org/10.1016/j.colsurfb.2022.112472.
93. Amjadi S et al (2020) Multifunctional betanin nanoliposomes-incorporated gelatin/chitosan nanofiber/ZnO nanoparticles nanocomposite film for fresh beef preservation. Meat Sci 167:108161. https://doi.org/10.1016/j.meatsci.2020.108161
94. Mohammadi H et al (2019) Nanocomposite films with CMC, okra mucilage, and ZnO nanoparticles: extending the shelf-life of chicken breast meat. Food Packag Shelf Life 21:100330. https://doi.org/10.1016/j.fpsl.2019.100330

95. Hu W et al (2022) Fabrication of highly transparent and multifunctional polyvinyl alcohol/starch based nanocomposite films using zinc oxide nanoparticles as compatibilizers. Int J Biol Macromol 204:284–292. https://doi.org/10.1016/j.ijbiomac.2022.02.020
96. Kumar S et al (2020) Biodegradable hybrid nanocomposite of chitosan/gelatin and green synthesized zinc oxide nanoparticles for food packaging. Foods 9(9):1143. https://doi.org/10.3390/foods9091143
97. Zhang R et al (2021) Development of polylactic acid/zno composite membranes prepared by ultrasonication and electrospinning for food packaging. LWT 135:110072. https://doi.org/10.1016/j.lwt.2020.110072
98. Díez-Pascual AM (2021) Poly(3-hydroxybutyrate-co-3-hydroxyhexanoate) with zinc oxide nanoparticles for food packaging. J Food Process Eng 45(7). https://doi.org/10.1111/jfpe.13814
99. Liu J et al (2021) Chitosan-based films with antioxidant of bamboo leaves and ZnO nanoparticles for application in active food packaging. Int J Biol Macromol 189:363–369. https://doi.org/10.1016/j.ijbiomac.2021.08.136
100. Priyadarshi R, Kim SM, Rhim JW (2021) Carboxymethyl cellulose-based multifunctional film combined with zinc oxide nanoparticles and grape seed extract for the preservation of high-fat meat products. Sustain Mater Technol 29. https://doi.org/10.1016/j.susmat.2021.e00325
101. Alamdari S et al (2022) Green synthesis of multifunctional zno/chitosan nanocomposite film using wild mentha pulegium extract for packaging applications. Surf Interfaces 34:102349. https://doi.org/10.1016/j.surfin.2022.102349
102. Motelica L et al (2020) Innovative antimicrobial chitosan/ZnO/Ag NPs/citronella essential oil nanocomposite—potential coating for grapes. Foods 9(12):1801. https://doi.org/10.3390/foods9121801
103. Babapour H, Jalali H, Mohammadi Nafchi A (2021) The synergistic effects of zinc oxide nanoparticles and fennel essential oil on physicochemical, mechanical, and antibacterial properties of potato starch films. Food Sci Nutr 9(7):3893–3905. https://doi.org/10.1002/fsn3.2371
104. Alizadeh-Sani M et al. (2020) Preparation of active nanocomposite film consisting of sodium caseinate, ZnO nanoparticles and rosemary essential oil for food packaging applications. J Polym Environ 29(2):588–598. https://doi.org/10.1007/s10924-020-01906-5
105. Shahbazi Y, Shavisi N (2018) Chitosan coatings containing mentha spicata essential oil and zinc oxide nanoparticle for shelf life extension of rainbow trout fillets. J Aquat Food Prod Technol 27(9):986–997. https://doi.org/10.1080/10498850.2018.1518945
106. Liu J et al (2021) pH-sensitive and antibacterial films developed by incorporating anthocyanins extracted from purple potato or roselle into chitosan/polyvinyl alcohol/nano-ZnO matrix: comparative study. Int J Biol Macromol 178:104–112. https://doi.org/10.1016/j.ijbiomac.2021.02.115
107. Lu M et al (2022) Colorimetric indicator based on chitosan/gelatin with nano-zno and black peanut seed coat anthocyanins for application in intelligent packaging. Food Res Int 160:111664. https://doi.org/10.1016/j.foodres.2022.111664
108. Ran R et al (2022) Preparation of ph-colorimetric films based on soy protein isolate/zno nanoparticles and grape-skin red for monitoring pork freshness. Food Control 137:108958. https://doi.org/10.1016/j.foodcont.2022.108958
109. Kim HJ, Roy S, Rhim JW (2022) Gelatin/agar-based color-indicator film integrated with clitoria ternatea flower anthocyanin and zinc oxide nanoparticles for monitoring freshness of shrimp. Food Hydrocolloids 124:107294. https://doi.org/10.1016/j.foodhyd.2021.107294
110. Yang Z et al (2021) Bilayer PH-sensitive colorimetric films with light-blocking ability and electrochemical writing property: application in monitoring crucian spoilage in smart packaging. Food Chem 336:127634. https://doi.org/10.1016/j.foodchem.2020.127634
111. Zheng T, Tang P, Li G (2023) Development of a pH-sensitive film based on collagen/chitosan/zno nanoparticles and mulberry extract for pork freshness monitoring. Food Chem 402:134428. https://doi.org/10.1016/j.foodchem.2022.134428

112. Hoque M, Sarkar P, Ahmed J (2022) Preparation and characterization of tamarind kernel powder/ZnO nanoparticle-based food packaging films. Ind Crops Prod 178:114670. https://doi.org/10.1016/j.indcrop.2022.114670.
113. Chen J et al (2022) The nano antibacterial composite film carboxymethyl chitosan/gelatin/nano ZnO improves the mechanical strength of food packaging. Int J Biol Macromol 220:462–471. https://doi.org/10.1016/j.ijbiomac.2022.08.005
114. El Fawal G et al (2020) Fabrication of antimicrobial films based on hydroxyethylcellulose and ZnO for food packaging application. Food Packag Shelf Life 23:00462. https://doi.org/10.1016/j.fpsl.2020.100462
115. Jayakumar A et al (2019) Starch-PVA composite films with zinc-oxide nanoparticles and phytochemicals as intelligent pH sensing wraps for food packaging application. Int J Biol Macromol 136:395–403. https://doi.org/10.1016/j.ijbiomac.2019.06.018
116. Hari KD et al (2021) Improvement of the UV barrier and antibacterial properties of crosslinked pectin/zinc oxide bionanocomposite films. Polymers 13(15):2403. https://doi.org/10.3390/polym13152403
117. Shankar S, Wang LF, Rhim JW (2018) Incorporation of zinc oxide nanoparticles improved the mechanical, water vapor barrier, UV-light barrier, and antibacterial properties of PLA-based nanocomposite films. Mater Sci Eng C 93:289–298. https://doi.org/10.1016/j.msec.2018.08.002
118. Xiao L et al (2022) Antioxidant and antibacterial PBAT/lignin-zno nanocomposite films for active food packaging. Ind Crops Prod 187:115515. https://doi.org/10.1016/j.indcrop.2022.115515
119. Din MI et al (2022) Facile synthesis of biodegradable corn starch-based plastic composite film reinforced with zinc oxide nanoparticles for packaging applications. Inorg Nano-Metal Chem:1–10. https://doi.org/10.1080/24701556.2022.2081190
120. Armynah B, Anugrahwidya R, Tahir D (2022) Composite cassava starch/chitosan/pineapple leaf fiber (palf)/zinc oxide (ZnO): bioplastics with high mechanical properties and faster degradation in soil and seawater. Int J Biol Macromol 213:814–823. https://doi.org/10.1016/j.ijbiomac.2022.06.038
121. Bujok S et al (2021) Sustainable microwave synthesis of biodegradable active packaging films based on polycaprolactone and layered zno nanoparticles. Polym Degrad Stab 190:109625. https://doi.org/10.1016/j.polymdegradstab.2021.109625
122. Patil MB et al (2022) Microwave-assisted synthesis of Poly (acrylamide-co-2-hydroxyethyl methacrylate)/chitosan semi-IPN ZnO nanocomposite membranes for food packaging applications. J Mater Res Technol 20:3537–3548. https://doi.org/10.1016/j.jmrt.2022.08.079
123. Akshaykranth A et al (2022) Antibacterial activity study of ZnO incorporated biodegradable poly (lactic acid) films for food packaging applications. Polym Bull. https://doi.org/10.1007/s00289-022-04126-0
124. Wang F et al (2022) Study on the barrier properties and antibacterial properties of cellulose-based multilayer coated paperboard used for fast food packaging. Food Biosci 46:101398. https://doi.org/10.1016/j.fbio.2021.101398
125. Tarabiah A et al (2022) Enhanced structural, optical, electrical properties and antibacterial activity of PEO/CMC doped ZnO nanorods for energy storage and food packaging applications. J Polym Res 29:167. https://doi.org/10.21203/rs.3.rs-1459938/v1
126. Phothisarattana D, Harnkarnsujarit N (2022) Migration, aggregations and thermal degradation behaviors of TiO_2 and ZnO incorporated PBAT/TPS nanocomposite blown films. Food Packag Shelf Life 33:100901. https://doi.org/10.1016/j.fpsl.2022.100901
127. Mesgari M, Aalami AH, Sahebkar A (2021) Antimicrobial activities of chitosan/titanium dioxide composites as a biological nanolayer for food preservation: a review. Int J Biol Macromol 176:530–539. https://doi.org/10.1016/j.ijbiomac.2021.02.099
128. Alizadeh Sani M et al (2022) Titanium dioxide nanoparticles as multifunctional surface-active materials for smart/active nanocomposite packaging films. Adv Colloid Interface Sci 300:102593. https://doi.org/10.1016/j.cis.2021.102593

129. Cao C et al (2020) Poly (butylene adipate-co-terephthalate)/titanium dioxide/silver composite biofilms for food packaging application. LWT 132:109874. https://doi.org/10.1016/j.lwt.2020.109874
130. Siripatrawan U, Kaewklin P (2018) Fabrication and characterization of chitosan-titanium dioxide nanocomposite film as ethylene scavenging and antimicrobial active food packaging. Food Hydrocolloids 84:125–134. https://doi.org/10.1016/j.foodhyd.2018.04.049
131. Gomes de Menezes FL et al (2021) TiO_2-enhanced chitosan/cassava starch biofilms for sustainable food packaging. Colloids Surf A Physicochem Eng Aspects 630:127661. https://doi.org/10.1016/j.colsurfa.2021.127661
132. Nguyen SV, Lee BK (2022) PVA/CNC/TiO_2 nanocomposite for food-packaging: Improved mechanical, UV/water vapor barrier, and antimicrobial properties. Carbohydr Polym 298:120064. https://doi.org/10.1016/j.carbpol.2022.120064
133. Salama HE, Abdel Aziz MS (2020) Optimized carboxymethyl cellulose and guanidinylated chitosan enriched with titanium oxide nanoparticles of improved UV-barrier properties for the active packaging of green bell pepper. Int J Biol Macromol 165:1187–1197. https://doi.org/10.1016/j.ijbiomac.2020.09.254
134. Phothisarattana D et al (2021) Biodegradable poly(butylene adipate-co-terephthalate) and thermoplastic starch-blended TiO_2 nanocomposite blown films as functional active packaging of fresh fruit. Polymers 13(23):4192. https://doi.org/10.3390/polym13234192
135. Mohr LC et al (2019) Titanium dioxide nanoparticles applied as ultraviolet radiation blocker in the polylactic acid bidegradable polymer. Polym Test 77:105867. https://doi.org/10.1016/j.polymertesting.2019.04.014
136. Hosseinzadeh S et al (2020) Chitosan/TiO_2 nanoparticle/*cymbopogon citratus* essential oil film as food packaging material: physico-mechanical properties and its effects on microbial, chemical, and organoleptic quality of minced meat during refrigeration. J Food Process Preserv 44(7):e14536. https://doi.org/10.1111/jfpp.14536
137. Liu Z et al (2021) Chitosan films incorporating litchi peel extract and titanium dioxide nanoparticles and their application as coatings on watercored apples. Prog Org Coat 151:106103. https://doi.org/10.1016/j.porgcoat.2020.106103
138. Sami R et al (2020) Effect of titanium dioxide nanocomposite material and antimicrobial agents on mushrooms shelf-life preservation. Processes 8(12):1632. https://doi.org/10.3390/pr8121632.
139. Kaewklin P et al (2018) Active packaging from chitosan-titanium dioxide nanocomposite film for prolonging storage life of tomato fruit. Int J Biol Macromol 112:523–529. https://doi.org/10.1016/j.ijbiomac.2018.01.124
140. Ramesh T et al (2016) Application of ultraviolet light assisted titanium dioxide photocatalysis for food safety: a review. Innov Food Sci Emerg Technol 38:105–115. https://doi.org/10.1016/j.ifset.2016.09.015
141. Riahi Z et al (2022) Cellulose nanofiber-based ethylene scavenging antimicrobial films incorporated with various types of titanium dioxide nanoparticles to extend the shelf life of fruits. ACS Appl Polym Mater 4(7):4765–4773. https://doi.org/10.1021/acsapm.2c00338
142. Du H et al (2022) Multifunctional film based on gelatin with titanium dioxide and thymol@β-cyclodextrins for fresh-keeping packaging. Food Biosci 50:102168. https://doi.org/10.1016/j.fbio.2022.102168
143. Ali HE et al (2022) Preparation and characterization of polyvinyl alcohol/polylactic acid/titanium dioxide nanocomposite films enhanced by γ-irradiation and its antibacterial activity. J Appl Polym Sci 139(24):52344. https://doi.org/10.1002/app.52344
144. Alizadeh Sani M et al (2022) Development of green halochromic smart and active packaging materials: TiO_2 nanoparticle- and anthocyanin-loaded gelatin/κ-carrageenan films. Food Hydrocolloids 124:107324. https://doi.org/10.1016/j.foodhyd.2021.107324
145. Gu XY et al (2021) Reactive TiO_2 nanoparticles compatibilized PLLA/PBSU blends: fully biodegradable polymer composites with improved physical, antibacterial and degradable properties. Chin J Polym Sci 39(12):1645–1656. https://doi.org/10.1007/s10118-021-2632-x

146. Vlad-Bubulac T et al (2022) Fabrication of poly(vinyl alcohol)/chitosan composite films strengthened with titanium dioxide and polyphosphonate additives for packaging applications. Gels 8(8):474. https://doi.org/10.3390/gels8080474
147. Souza VG, Fernando AL (2016) Nanoparticles in food packaging: biodegradability and potential migration to food-a review. Food Packag Shelf Life 8:63–70. https://doi.org/10.1016/j.fpsl.2016.04.001
148. Gvozdenko AA et al (2022) Synthesis of cuo nanoparticles stabilized with gelatin for potential use in food packaging applications. Sci Rep 12:12843. https://doi.org/10.1038/s41598-022-16878-w
149. Wyser Y et al (2016) Outlook and challenges of nanotechnologies for food packaging. Packag Technol Sci 29(12):615–648. https://doi.org/10.1002/pts.2221
150. Enescu D et al (2020) Evaluation of the specific migration according to EU standards of titanium from chitosan/metal complexes films containing TiO_2 particles into different food simulants: a comparative study of the nano-sized vs micro-sized particles. Food Packag Shelf Life 26:100579. https://doi.org/10.1016/j.fpsl.2020.100579
151. Heydari-Majd M et al (2019) A new active nanocomposite film based on PLA/ZnO nanoparticle/essential oils for the preservation of refrigerated Otolithes ruber fillets. Food Packag Shelf Life 19:94–103. https://doi.org/10.1016/j.fpsl.2018.12.002
152. Fan C et al (2019) Effect of high pressure treatment on properties and nano–Ag migration of PLA-based food packaging film. Polym Test 76:73–81. https://doi.org/10.1016/j.polymertesting.2019.03.005
153. Yu F et al (2021) Poly(lactic acid)-based composite film reinforced with acetylated cellulose nanocrystals and zno nanoparticles for active food packaging. Int J Biol Macromol 186:770–779. https://doi.org/10.1016/j.ijbiomac.2021.07.097

Carbon-Based Nanomaterials in Food Packaging and Preservation

Ajahar Khan, Parya Ezati, Sai Kumar Tammina, Zohreh Riahi, and Ruchir Priyadarshi

Abstract Food packaging has made great progress due to global trends and consumer demands. These advances require the development of packaging materials that can provide a longer shelf life of the packed food, along with monitoring its quality and safety based on international protocols. Nanotechnology can address all these requirements while implementing the principal packaging functions, i.e., containment, preservation, protection, communications, and marketing. In contrast, the unique characteristics of carbon-based nanomaterials to obtain improved food quality and safety have aroused interest in their potential application in the food industry. Carbon-based nanomaterials offer the prospects for finding solutions to the long-standing challenges in the food industry to maintain food quality, assess food safety, improve mechanical, antimicrobial, and barrier properties, extend the shelf life, and monitor food condition during transportation and storage. This chapter focuses on recent developments/investigations in carbon-based nanomaterials for food safety and packaging applications. The characteristic features of nano-materials impacting food products and their packaging are discussed. Risk assessment of the consumer safety issues and possible environmental and public health concerns of carbon-based nanomaterial applications in food disciplines are also summarized. It concludes with an outlook of challenges and future perspectives of carbon-based nanomaterials in the food safety sector.

Keywords Carbon-based nanomaterials · Antimicrobial · Food packaging · Food safety

---

A. Khan (✉) · Z. Riahi · R. Priyadarshi
Department of Food and Nutrition, BioNanocomposite Research Center, Kyung Hee University, 26 Kyungheedae-Ro, Dongdaemun-Gu, Seoul 02447, Republic of Korea
e-mail: arkhan.029@gmail.com

P. Ezati
Department of Food Science, University of Guelph, ON, N1G2W1, Canada

S. K. Tammina
Department of Chemical and Petroleum Engineering, SAN Campus, Khalifa University of Science and Technology (KUST), Abu Dhabi, United Arab Emirates

M. Alle et al. (eds.), *Functional Nanomaterials and Nanocomposites for Biodegradable Food Packaging*, Smart Nanomaterials Technology,
https://doi.org/10.1007/978-981-97-9084-5_5

1 Introduction

With the world's population growing in both rural and urban areas, there is an urgent demand to ensure a steady supply of processed food products with preserved nutritional values. However, food loss during processing, transportation, and consumer supply has a significant negative impact on the food market [1]. Controlling the contamination of food to assure food safety and quality is critical for both suppliers and consumers [2]. Modern societies are becoming increasingly concerned about food safety, as billions of people risk consuming contaminated food globally [3].

Degradation of food resulting from oxidation and microbial contaminations are key issues arising during its processing, storage, and distribution. The spoilage of processed products causes massive economic debts to the food supplier [4]. As a result, the first step is to improve our awareness about the reactions that cause food deterioration to create and select appropriate processing methods to reduce quality loss [5]. In this regard, packaging is an important food processing method for food suppliers since it ensures safety from unwanted factors and maintains the quality of packaged food items. Therefore, packaged products must provide a favorable atmosphere based on food type to ensure excellent quality throughout the distribution process. Packaging is also significant for improving shelf life by limiting biochemical and microbiological deterioration of food [4].

Packaging has been categorized into an active and intelligent form that can be employed for several causes in a variety of food products [6]. Intelligent packaging can track the health of the goods being packaged or the environment inside the package, such as moisture level, total volatile basic nitrogen (TVBN), gaseous composition, temperature change, pH, and toxic contaminants to provide consumers with information about the microbiological, physical, and chemical quality of the packaged food [6]. On the other hand, an active form of packaging comprise antioxidant and antibacterial agents that can protect packed items against lipid oxidation and microbial contamination, and extend their shelf life [6].

Aside from that, an array of new detection technologies and methods have been established that will enhance food safety and quality monitoring, such as time–temperature indicators, pathogen and contamination biosensors, and bacterial spoilage sensing probes for packaged foods, all of which have produced promising results [7]. Despite extensive development and research, advancement in this sector is still in the beginning stages. In this regard, nanotechnology and nanoscience are cutting-edge scientific advances that, compared to pharmaceuticals and medical distribution, are nearly modern in their use in farm and food processing. Nanotechnology has emerged as a technical breakthrough that has the potential to transform the whole agri-food sector, including global food production, quality, nutritional value, and food safety [8].

A variety of possibilities have been proposed in the literature, all of which revolve around the employment of a combination of materials. For example, the addition of (nano-)particles into the packaging material generates specific qualities such as mechanical strength, UV barrier properties, anti-microbial effects, and so on [9]. When compared to conventional detection technologies, nanomaterials-based approaches offer target recognition as well as significantly better detection, selectivity, and sensitivity information [10]. To monitor quality changes in packaged food, a variety of efficient nanomaterials, including metallic nanoparticles [11], semiconductors [12], and carbon-based materials [13], have been developed. Meanwhile, carbon nanostructures have several advantages over inorganic ones, such as the fact that they can be typically derived from food processing wastes, are compatible with biodegradable polymers, and are edible non-dose-limited and nontoxic [9].

This chapter focuses on recent investigations and the basic concepts regarding the use of carbon-based nanomaterials and their derivatives for the determination of food risk, extending their shelf-life, and monitoring quality changes to ensure the safety of processed food products. The focus is on summarizing food analysis, active and intelligent food packaging, assessment of the possible public concerns and associated risks, quality control, health issues, and environmental implications of carbon-based nanomaterial applications in food-related disciplines. Finally, the future directions and challenges for carbon-based nanomaterials in food technology are discussed.

2 Nanotechnology in Food Sciences

Nanotechnology combines engineering, chemistry, physics, biology, and other related fields and uses nanomaterials with a size range of 1–100 nm [14]. Due to its incredibly small size, the material acquires special qualities that were not present in the bulk state. By recognizing and making use of these characteristics, nanotechnology produces novel products with improved functionality. From product formulation, processing, nature of ingredients, and packaging, to product functionality and food safety, nanotechnology has completely changed the subject of food science. Commercial applications of nanotechnology in the food industry can be applied to every stage of food manufacture, supervision, processing, storage, and supply (Fig. 1). Harvested food adopting nanotechnology in the packaging, storage, processing, and production of food is stated to a nanofood. Nanomaterials are used in many foods to improve the flow properties, color, stability, and flavor during processing and extend the food's shelf life. For instance, anticaking agents like aluminosilicate compounds are frequently used in processed foods that are granulated or powdered. On the other hand, Anatase titanium dioxide is a common food whitener and brightener component found in cheeses, sauces, and desserts [15].

In non-food and food items, nanomaterials are typically employed as an anticaking agent, to thicken pastes in powdered products to maintain flow properties, and as a carrier of flavors or perfumes. Nanomaterials at threshold concentrations are often employed in food items and are permitted as food additives by the European Union (E551) [16]. Flavor and fragrance are among the important elements of the food system, which provide sensory awareness of taste and aroma to enhance the

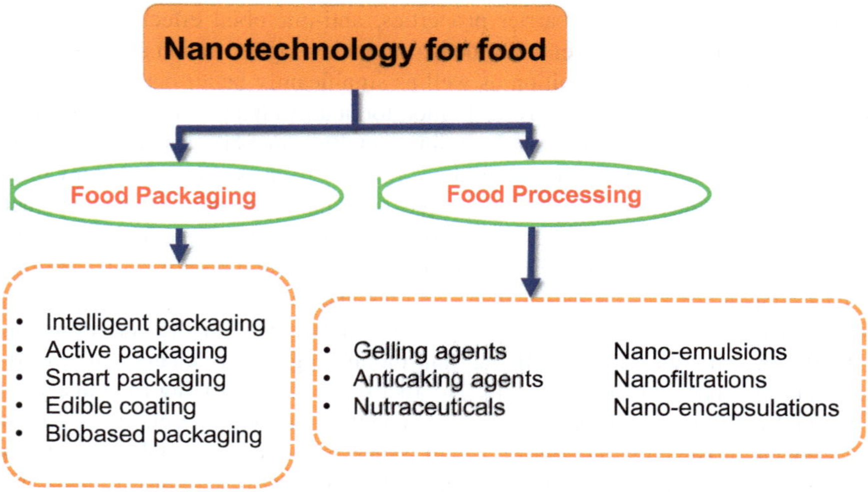

Fig. 1 Possible uses of nanotechnology in food applications

entire eating experience. Nanoencapsulation strategies have been widely used to enhance flavor retention and release as well as to provide a balanced flavor profile [17]. Additionally, it has been widely discussed that nanomaterials serve as fragrance and flavor carriers in both non-food and food goods [18].

Nanotechnology can be utilized to increase the products' shelf life by preserving the foodstuffs in packaging materials that restrict the exchange of air and moisture and control the release of antimicrobials or antioxidant agents within the packaging surroundings. Food packaging needs precise biological, chemical, and physical properties in addition to tamper resistance and protection. The product's label and any nutritional data for the packaged meal are also visible on the packaging [19]. The use of nanotechnology in packaging is categorized based on how it performs. The majority of nanoparticles employed for packaging purposes have antimicrobial properties, serve as carriers for antimicrobial polypeptides, and prevent degradation due to microbial contamination. Packaging materials are composed of an antimicrobial agent-loaded layer of starch colloids which operate as a barrier to bacteria by the controlled release of antimicrobials from the packaged product [14].

Innovative nanocarbon-based antimicrobial substances have also shown the potential to prolong shelf life by reducing food spoilage from bacterial contaminations. There have been several different carbon-based nanocomposites discussed as promising antimicrobial packaging systems. Their intrinsic physicochemical characteristics cause oxidative stress by encouraging the production of reactive oxygen species and subsequently damaging bacterial cells [20]. Additionally, releasing active agents at the cell surface, from the outside, or from within the cell may alter cellular structure or function. The most often used nanoparticles in the food sector as antimicrobials are carbon dots incorporated in biopolymer films [21]. An overview of the

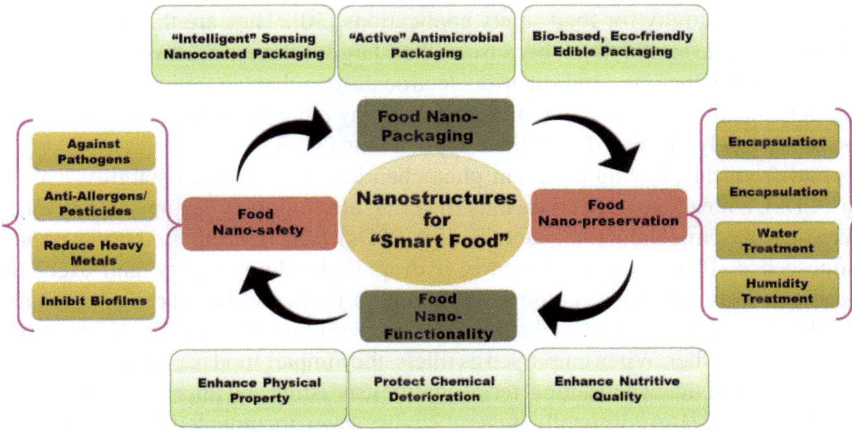

Fig. 2 Systematic representation of the application of nanoparticles in various areas of the food industry. Adopted from [22]

significant applications of nanomaterials, nanoparticles, and nanostructures in the food industry is shown in Fig. 2.

3 Carbon Nanomaterials

The unique characteristics that carbon nanomaterials and their derivatives possess allow them to be successfully employed in a wide range of research domains [23]. Carbon is one of the most abundant non-metallic substances that is widely distributed in the biosphere and is extensively used for various purposes owing to its unique characteristics and ease of modification. The main characteristic features of carbon nanomaterials include their stability in high heat, thermal conductivity, quantum tunneling effects, tunable size, and surface area [24]. Another important factor supporting carbon's wide range of applications is the fact that it may exist in a variety of allotropic forms, ranging from 0-D to 3-D structures [25, 26]. For instance, carbon-based nanomaterials can take a variety of forms based on their spatial dimensions, including fullerenes (zero-dimensional), carbon nanotubes (one-dimensional), graphene (two-dimensional), graphene coil (multidimensional), and so on [24]. Carbon has a special hybridization characteristic and is sensitive toward its structure, which can easily vary with synthesis methods and lead to tuned manipulation in carbon-based nanocomposites [27]. This characteristic of carbon and its derivatives has not been observed in other inorganic nanofillers [28].

Since the process for constructing carbon nanomaterials has a lengthy history, its uses are likewise well recognized and accurately defined [29]. Carbon nanostructures have been widely investigated for their outstanding chemical, physical, and electrical conductivity for food safety applications [30]. They are the perfect materials for electrode modification because of their large surface area, diverse functional groups, high conductivity, and superior electrocatalytic characteristics. Whether it be electrochemical, biomedical, environmental, or any other discipline, these materials have been widely discussed with their applications [31, 32].

Apart from this, having excellent photochemical and chemical stability, carbon dots (CDs), a novel type of carbon nanomaterial, also uniquely feature the key characteristics of semiconductor quantum dots (QDs) [33]. Due to its strong photoluminescence (PL), good fluorescence, hydrophilicity, low toxicity, and excellent biocompatibility, CD has become a viable option in the field of carbon nanomaterials [34]. Polymers for food packaging have been reinforced with CDs as a nontoxic and sustainable filler. When employed as fillers, they impart food packaging materials with strong antioxidant and antibacterial qualities, assisting and improving the packed food shelf life [27]. Additionally, several studies have documented the synthesis of CDs from sustainable sources and their use in applications related to food packaging. According to previous investigations, coating with CD-based formulations enhanced the fresh-cut fruits and vegetables' physicochemical qualities and reduced the microbial contaminations during storage, affecting malondialdehyde content, respiration rate, firmness, weight loss, fungal infection, total bacterial count, total soluble solids, peroxidases activity, water status, flavor, taste, and polyphenol oxidase activity [35]. Therefore, CDs present rich prospects for use in applications involving food packaging and preservation.

4 Carbon Nanomaterials-Based Sensing Tools for Food Safety Analysis

Nanomaterials' extremely small size enables rapid identification of soil, air, and water contaminants, which is essential for ensuring food safety [8]. Additionally, numerous reports support the use of sustainable nanomaterials in food processing. For the purpose of food safety, chemistry, and food analysis, carbon-based nanomaterials and their derivatives have been extensively studied as reliable biosensors and promising analytical tools [36, 37]. The supply of safe food with actual nutritional values is a global concern of the food industry [38]. Using food sensors to identify chemical contaminants like pathogens, pollutants, pesticides, and toxins can be determined in real-time using nanotechnology, which primarily benefits the food sector [39].

According to a published report, graphene and reduced graphene oxide (rGO) composite's field-effect capacitive can be utilized as an immunosensor (biosensor) to determine subfemtomolar levels of aflatoxin B in food [40]. Further, using the dispersive solid-phase extraction (DSPE) procedure, multiwalled carbon nanotubes (MWCNT) were employed as sorbent material in a new approach investigated by da Silva et al. to extract aflatoxins from baby food samples [41]. Aflatoxin adsorption onto MWCNT was accomplished using 6.25% acetonitrile, and toxin removal from MWCNT was accomplished using dichloromethane. Recovery rates were within the permissible limit for the extraction using 20 mg MWCNT, an extractive solution desorbed with 10 ml of dichloromethane without acidification. Aflatoxin B1 was determined using this new strategy in baby food samples at an 80 ng/kg level.

By employing CD-coated molecularly imprinted polymer (MIP), Liang et al. suggested a wonderful approach for accurately detecting aflatoxin B1 in peanut samples [42]. This technique demonstrated extremely good sensitivity ranging from 0.5 to 2000 ng/mL (a broad range of linear response) with a 0.118 ng/mL detection limit [43]. Nagarajan et al. employed CDs-coated amine functionalized cellulose sponges to detect and separate atrazine (toxic herbicide). With satisfactory sorption and a maximum adsorption capacity of 32.06 mg/g, the produced CD complex followed the Freundlich isotherm model and pseudo-first-order kinetics [44].

Shekarbeygia et al. investigated the effects of rose pigment extraction techniques on CDs' optical characteristics. The detection of diazinon in water was then made using a highly accurate nanosensor based on the fluorescence quenching of the as-synthesized CDs [45]. Lin et al. developed a universal approach for determining pesticides based on the iron ions' redox capacity, in which the fluorescence intensity of CDs was proportional to the concentration of the pesticide [46]. For the quick and effective detection of thiabendazole, various optical sensors based on eco-friendly MIPs coated on CDs have been developed [47].

Further, nitrogen-doped CDs demonstrated high aqueous solution stability and were used for the inner filter effect mechanism-based fluorometric detection of tetracyclines. With an acceptable recovery and a good relative standard deviation (RSD), these CDs were successfully used to detect residual tetracyclines in pork and fish samples [48]. The presence of carbendazim in samples of strawberries and apples was detected using a fluorescence "turn on" ratiometric fluorescent sensor. In this approach, the sensing approach was exhibited by strong fluorescence resonance energy transfer between gold nanoclusters and N-doped CDs. It was possible to ratiometrically detect carbendazim by capturing the second-order Rayleigh scattering and signal photoluminescence of the CDs/gold nanocluster system [49].

Chatzimitakos et al. [50] employed CDs derived from citrus lemon peel to detect tartrazine concentrations. They practically validated the purpose of the fabricated sensor by applying the same technique with energy drinks, juice, and ice cream as real foods. The outcomes demonstrated that adding 80.0 μM of tartrazine decreased the fluorescence intensity of CDs through a quenching mechanism. A linear association between the PL spectra of CDs and rising tartrazine concentrations in the range of 0.6–23.5 μM was observed [50]. Tartrazine in food samples was also successfully detected by the fluorescence quenching of CDs in the concentration range of 0.25–32.50 mM [51]. According to previous investigations, prolonged tartrazine consumption is injurious to human health [52]. A CD-MnO_2 sensing probe was designed by Liu et al. and utilized to find ascorbic acid in juices, fruits, and vegetables [53]. Likewise, Ahmed et al. prepared CDs to quantify tannic acid based on their ability to quench the fluorescence of CDs, and they showed that this approach successfully detected 0.018 mg L^{-1} of tannic acid in both white and red wine [54].

Tang et al. [55] engineered enzyme-nano silica-doped carbon nanotubes for the quantitative staphylococcal enterotoxin B detection in food, whereas Wang et al. [56] constructed graphene-based CDs for determining high-sensitivity fluorescent ochratoxin A. Additionally, Wang et al. [43] developed novel fluorescent CDs with highly sensitive aflatoxin B1 detection. On aptamer/gold nanoparticles, positively charged N-CDs (nitrogen-doped carbon dots) were assembled for the determination of aflatoxin B1. With a detection limit of 5 pg mL^{-1} (16 pM), the assay solution displayed a linear response to aflatoxin B1 in the range of 5 pg mL^{-1} to 2.00 ng mL^{-1}.

Glyphosate residue is a hazardous herbicide that causes significant adverse ecological effects. Wang et al. prepared CDs derived from glyphosate antibodies for glyphosate determination with an 8 ng mL^{-1} detection limit [57]. The produced CDs' PL spectra displayed a linear relationship as glyphosate concentration rose from 0.01 to 80 g mL^{-1}. Satisfactory recovery rates (87.4–103.7%) were obtained using CD to identify glyphosate in tea [57]. The CDs-based fluorescent probe was also employed for the sensitive detection of Hg^{2+}; the detection limit was 0.014 mol L^{-1}, and a decrease in CD emission intensity was seen with increasing Hg^{2+} ion concentration [58]. For the actual detection of Cr^{6+} ions, Athika et al. [59] utilized a CD sensor. With a limit of detection for Cr^{6+} ions of 14 M, the CD sensor demonstrated linear detection performance in relation to Cr^{6+} ion concentration. According to Huang et al. [60], adding Fe^{3+} to the CD solution caused the CDs to lose their fluorescence. Additionally, they discovered that CDs play a significant part in CD-mediated fluorescence ion sensing applications and demonstrate a synergistic effect of binding capacity due to surface functional groups. Based on the above discussion, it can be concluded that carbon nanomaterial-based sensing probes are very promising for the detection of a variety of contaminants in food (Table 1), including toxins, veterinary drugs, metal ions, hazardous residues, and pesticides.

Table 1 Optical sensors based on CD for the application of food safety analysis

Materials	Food type	Analyte	Limit of detection	Refs.
Hapten-labeled CDs and WS_2 nanosheets	Chicken breast	Amantadine	0.10 ng/g	[61]
CDs@Zeolite	Water	Picric acid	0.36 μM	[62]
Rose pigment-CDs	Water	Diazinon	0.01 μM	[45]
N-acetyl-l-cysteine-CDs	Fish samples	Histamine	13.0 ppb	[63]
CDs@mMIP	Milk	Penicillin-G	0.34 nM	[64]
N-CQDs/AuNCs	Strawberry/apple	Carbendazim	37.25/0.83 μM	[49]
B-CDs/P-CDs@ZIF-8	Fruit juice	Triticonazole	4.0 nM	[65]
r-QDs/CDs	Milk	Cephalexin	0.7 μM	[66]
White pepper-CDs	Pig liver	Coenzyme A	8.75 nM	[67]
NCDs/AuNPs	Peanut/corn	Aflatoxin B1	16 pM	[43]

5 Carbon Nanomaterials-Based Active Food Packaging

Providing consumers with increasingly nutritious foods will be one of the biggest issues being faced by the food sector in the upcoming years. In order to achieve this, packaging technology must be improved with respect to food quality. The primary and most evident purpose of packaging is to contain and safeguard food by external factors, and preserve its quality from the time it leaves the production facility until it reaches the consumer [68]. In this regard, carbon nanomaterial and their derivatives seem to be promising candidates owing to their unique physicochemical characteristics that provide desired features to a wide range of possible packaging applications. These materials, as active agents, can help to develop new antioxidant, antibacterial, anti-ultraviolet, biocompatible, and biodegradable films for food packaging to retain the freshness and quality of the foodstuff, and also to enhance its shelf-life [69]. The use of carbon nanomaterials as an active ingredient improves the chemical, mechanical, and physical characteristics of packaging without negatively affecting the other characteristics of the film or the food [70]. The use of carbon nanomaterials in producing active and intelligent food packaging platforms is shown in Table 2.

The carbon nanomaterials, such as carbon dots used in food packaging, can improve the quality of packaged foodstuffs [71]. Besides, the applications for packaging that prevent counterfeiting can be assumed using carbon dots. Anti-counterfeit packaging is the use of secure packaging to guarantee the safety and authenticity of processed food products. Carbon dots' photoluminescent characteristics can be used to offer anti-counterfeiting characteristics. For instance, the packing material could contain carbon dots that emit a particular fluorescence signal that could be detected by flashing fluorescent lights. Carbon dots can either be deliberately coated onto the polymer matrices' surfaces or implanted into the interior of the polymer matrix. A more difficult task for counterfeiters is to replicate costly items. These labels are useful for high-end products that might be locally produced [69].

Various studies have also explored the functional properties of CDs for food preservation. For instance, using a chitosan coating solution made of kelp-derived CDs significantly inhibited peroxidase activity, decreased weight loss, and maintained the freshness of cucumber stored at 4 °C [72]. Yeast and mold development was prevented by the 4.5% CD-loaded chitosan coating during 15 days of storage. Similarly to this, chitosan and CD coating retained the quality and extended the shelf life of fresh cucumbers [35]. Zhao et al. reported the prolonged shelf life of soymilk on adding chitosan-based thickeners having different concentrations of banana-derived CDs (4–8%) [73]. The sample with 8% CQD and 0.16% chitosan successfully extended the shelf life of milk by 4 days, whereas it inhibited the development of *B. subtilis*, *S. aureus*, and *E. coli*.

Patil et al. studied the impregnation of CD (derived from spent tea residue) in poly(vinyl alcohol) (PVA) solution to fabricate a UV-barrier packaging film [74]. The film with 3 mg/mL CDs was able to completely block the light transmittance in the 230–315 nm range. The film, due to the presence of CD, completely blocked UV-B and UV-C radiation, whereas the film's UV barrier ability improved the quality

Table 2 Active and intelligent food packaging systems based on carbon dots and biopolymer composite matrix

Carbon source	Biopolymer	Properties of films	Refs.
Waste acorn cups	Chitosan/polyvinyl alcohol	The UV absorption capacity is enhanced by the composite film	[75]
Waste tea residue	Polyvinyl alcohol	Good UV blocking, high mechanical strength, and flexible capabilities	[74]
Cotton cellulose	Collagen	Good UV barrier and high ionic strength	[76]
Cow milk CDs	Polymethylmethacrylate	Flexible nature and good antimicrobial action	[77]
The residue of radiata pine	Polyvinyl alcohol-cellulose nanofiber	Anti-counterfeiting, mechanical properties, good UV barrier, and water resistance properties	[78]
Lactose	Regenerated cellulose	Highly transparent with enhanced mechanical properties	[79]
Kelp (*Laminaria japonica*)	Chitosan	Effectively inhibits microorganisms, decreases water mobility and increases the shelf life, and improves the quality of food	[72]
Banana	Chitosan	Excellent microbial action and improve the soy milk's shelf life	[73]
Banana paste	Polyvinyl alcohol	Increases the shelf life of fried meatballs and protects against UV radiation	[80]
1,4 Dihydroxyanthraquinone	Polyvinyl alcohol	Good sensing abilities	[81]

of the grapes when it was used to package them. Therefore, owing to some unique characteristics, including tunable photoluminescence, strong biocompatibility, and better stability, carbon dots can be applied as additives, coating agents, antimicrobial, and antioxidant agents in food packaging. As a result, carbon dots are successfully used in a variety of industries, including nano-level additives, coating agents, and active packaging. Due to their strong functional qualities, CDs are highly sought-after in the food packaging industry [69].

6 Carbon Nanomaterials for Intelligent Packaging

Developing the ability to convey information on food freshness, intelligent packaging is an additional extension to active packaging [82]. Intelligent labels or sensing technology that monitor changes in internal environmental conditions can be used to convey real-time information about food change. Because of its benefits, including ease of handling, precise outcomes, and high sensitivity, intelligent label detection technology stands out among other detection methods [83]. An intelligent label connected to the container changes color to indicate freshness [84]. Food waste can be decreased by accurately tracking the optimum consumption date and discarding items that are unfit for consumption when the indicator film's color changes. The indicator in the intelligent label is sensitive to ambient pH changes caused primarily by amines and volatile nitrogen compounds released by the food over time in the packing environment [85]. This is the fundamental principle behind intelligent label detection of meat products. To achieve real-time meat freshness detection, such sensitive changes can be translated into response values, which are often color changes that are visible to the unaided eye [86]. Koshy et al. prepared a pH-sensitive biodegradable starch film (starch/CD/CTE film) by combining starch with CDs and anthocyanins extracted from Clitoria ternatea flower (CTE) [87]. They used this film as an intelligent monitoring system to determine the freshness of packaged pork samples. They also found that CTE-containing starch films exhibit poor mechanical strength and high hydrophilicity. These drawbacks can be fixed by including CDs which strengthen the starch backbone. The synergistic effect of CTE and CDs resulted in the film's exceptional oxidation resistance. Additionally, the indicator films showed color variations at various pH levels due to the presence of CTE. Pork samples held at 25 °C produced large quantities of total volatile basic nitrogen as a result of the action of microbes, which in turn altered the pH of the gaseous composition within the package. As storage time increased, the pork went from being fresh to spoiled, and the indicator film's color shifted visibly from purple to green.

Due to their perishability, seasonality, and regionality, fresh agricultural products confront significant difficulties in storage, market supply, and product development. An essential component of the safe production of agricultural goods is preservation. The change in oxygen level in packaging is a key indicator of food quality change during storage since oxygen is a byproduct of food respiration. Using the electrospinning method, Xu et al. developed an oxygen-sensitive nanofiber film (RuDPP@SiCDs/SiO_2 film) using tris [4, 7-diphenyl-1, 10-phenanthroline] ruthenium (II) dichloride (RuDPP), which is a oxygen-sensitive probe, and silane-functionalized CDs (SiCDs), which acted as oxygen-sensitive luminophore [88]. By monitoring the emission intensity ratio of ruthenium complexes to CDs in the film to reflect the oxygen concentration in the environment, the freshness of fruits in packaging was measured. The fruit storage test used a straightforward grape preservation system for oxygen monitoring, where the indicator film was attached to the container used to package grapes above the headspace. They found that, with increasing the storage time, the fluorescent hue of the nanofiber film altered. In the beginning, the nanofiber sheet fluoresced blue. Due to the grapes' respiration on the third day of storage, the fluorescent color turned purple. The fluorescence turned bright red on the sixth and seventh days of storage as the respiration progressed, and there was no discernible variation in color, suggesting that the oxygen had been exhausted in the conical flask. According to the aforementioned results, RuDPP@SiCDs/SiO_2

nanofiber film containing CDs can be utilized to monitor the respiration of packaged agricultural goods remotely. Further, chlorogenic acid was employed by Zhang et al. to develop affordable and eco-friendly photoluminescent CDs [89]. A specific functional film with improved mechanical and physical characteristics was engineered using PVA doped with CDs. The produced functional film demonstrated strong antioxidant activity and photobleaching resistance due to the presence of CDs. Besides, the films were sensitive to both pH and Al^{3+} ions. Using the film-forming solution in the form of coatings in bananas demonstrated their ability to increase the shelf life of the fruit by retarding oxidative browning. The functional film can also be utilized to detect the quality of stored milk and eggs over the course of storage, particularly by responding to the increase in amine concentration during the purification process and the resulting increase in the pH inside the package. The films were also used to determine the Al^{3+} residue in samples of fried dough sticks. This film has the potential to be a very effective material in intelligent food packaging that not only determine additives and contaminants but also extend the shelf life of packaged products. The above discussion suggested that carbon nanomaterials seem to be a reliable intelligent/active food packaging material that may be used to track changes in food quality over time and extend their shelf life.

7 Carbon Nanomaterials as Coating Agent in Food Preservation

Due to its biodegradable and nontoxic nature, sustainable materials-based edible coatings are considerably used in preserving a wide variety of food products [90]. By constructing a semi-permeable barrier, edible coatings on food surfaces could regulate gas exchange, enzymatic browning, and water loss and suppress microbiological deterioration [91]. Due to its high antioxidant capacity, CDs have been successfully employed as a coating agent for vegetable and fruit storage. Eskalen et al. investigated the ability of water-soluble fluorescent rosemary leaves-derived CDs as active carbon fillers (antioxidants) in banana preservation [92]. The coating solution was made by combining CDs and PVA solutions. PVA-CDs and pure-PVA solutions were applied for coating on bananas of identical size, whereas bananas without coating with the polymer solutions were designated as the control for comparison. The presence of black spots and the number of black regions on the banana peels determined the degree of banana oxidation. They observed that PVA solutions coated bananas were noticed to be full of black spots on the 19th day of storage, whereas PVA-CDs coating was effective in preserving bananas in their natural appearance compared to the others (Fig. 3). In Zhang et al. follow-up investigation, PVA solution and chlorogenic acid-CDs/PVA (CA-CDs/PVA) solution were also coated on bananas, whereas bananas without coating served as the control [89]. The bananas coated with CA-CDs/PVA showed the best preservation effect with the fewest black spots after 5 days of storage at room temperature (25 °C). In a similar

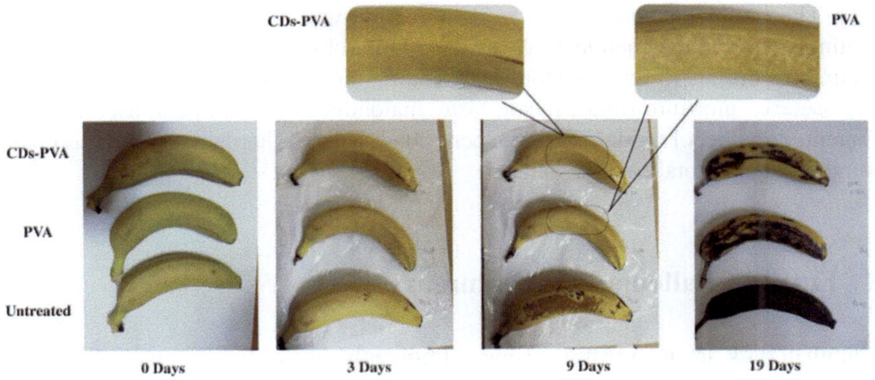

Fig. 3 Actual view of PVA+CDs (top), PVA (middle) coated bananas, and untreated (bottom) bananas as a function of days. Adapted with permission from [92]

manner, the coating solution of PVA added with coffee bean shell-CDs (CS-CDs/PVA) was applied to the bananas by Zhang et al. [93]. CS-CDs/PVA-coated bananas displayed fewer black spots on the peels after three days of storage at room temperature (23 °C) than bananas without coating and those coated with pure-PVA solution. In another storage experiment, Xiao et al. coated the mixture solution of CQDs/chitosan on the fresh mango surface [94]. The findings showed that fungi and bacteria on the mango surface were effectively inhibited by the CQDs/chitosan coating. It successfully delayed the conversion of acid or sugar, and reduced the vitamin C oxidation in mango, while considerably decreasing the water loss and decay rate. It also retarded the respiration of mango during storage, thereby extending the shelf life of the fruit.

These findings demonstrated that due to their antioxidant capacity, CDs could be used to preserve fruit. Previous investigations discussed the CD-based coating as a potential preserving agent for fresh-cut vegetables. Fresh-cut fruits and vegetables are quite popular in daily diets due to their convenience, freshness, nutrition, and other qualities. Cucumber, in particular, is one of the most common vegetables in human diets. Fresh-cut cucumbers' shelf life is significantly shortened by cutting and processing because it causes nutritional loss, browning, and tissue damage. A CDs/chitosan (CDs/CH) coating was employed on fresh-cut cucumbers in the study by Fan et al. to increase their fresh-keeping duration in packaging with a modified atmosphere [95]. The cucumbers coated with CDs/CH were stored for 15 days at 4 °C. Coated cucumber's quality was substantially enhanced, and its microbial population was effectively reduced during storage. The 4.5% CDs/CH coating maintains the levels of total soluble solids and ascorbic acid with a positive impact on preserving flavor, obstructing peroxidase activity, and preventing weight loss. The 4.5% CDs/CH coating could also lessen the fluidity of water in freshly cut cucumbers, according to LF-NMR results. Additionally, they investigated how CD coating in conjunction with ultrasound treatment could lengthen the fresh-keeping time of freshly cut cucumbers

[35]. The findings demonstrated, in combination with ultrasound treatment, CDs coating enhanced the chemical, physical, and microbiological quality of freshly cut cucumbers during 15 days of storage, including the total number of yeasts, molds, and bacteria, moisture status, flavor, taste, malondialdehyde content, ascorbic acid content, polyphenol oxidase activity, peroxidase activity, hardness, respiration rate, weight loss, and total soluble solids.

8 Future Challenges and Summary

Nanomaterials are being utilized more frequently in the food industry and food science, but because they can potentially cause harmful consequences, care must be taken while using them. Future humankind may be on the path of a nanotoxicity catastrophe, claims research from the British Royal Society [96]. Although the application of nanoparticles in food nanotechnology has advanced significantly, the awareness about the toxicity of these materials is very little. The two main safety issues when employing nanoparticles are heavy metals and allergen leakage. Without the required information and restrictions, nanoparticles are already being added to food products at a rate that is faster than desired, endangering human health and the environment [97].

Only if we have an in-depth knowledge of the properties of nanomaterials, such as surface chemistry, composition, size, and solubility, then it is easy to find their suitable applications in the preservation of food products. While many applications are attracted to nanoparticles because of their distinctive properties, the safety of human health in the case of food applications raises questions about this [98]. Therefore, more study is required to comprehend the effects of ingesting nanoparticles by consumers as well as the causes and effects of bacterial responses to nanoparticles. Regular assessments of the safety of nanoparticles in food must take into account both their nature and the properties of the food composition in which they are disseminated.

In this regard, analytical approaches are needed to determine and describe nanoparticles and their properties in matrices like consumer food products, water, air, and soil to which humans and ecosystems interact. These analytical techniques are required to be utilized to characterize the nanomaterials in ecotoxicological and toxicological evaluation; only then can an adequate risk valuation be performed, and the characteristics of toxic nanoparticles utilized, regulated, or identified in standardized testing.

For innovative or advanced food packaging, carbon-based nanomaterials such as carbon dots have a number of benefits, including barrier properties (gas and radiation blocking), thermal activity, water barrier, improved mechanical qualities, and antioxidant and antimicrobial potentials. In combination with biological–chemical indicators, carbon dots make accurate and quick packages for evaluating the quality and originality of food. Depending on the processing circumstances and source materials, CDs can be synthesized with a variety of functionalities and compositions. They have a significant potential for use as intelligent agents, nanoscale food additives, active ingredients, and coating agents to extend the shelf life of food products. Additionally, the assessment of CDs as functional food additives to prolong the shelf life of foodstuffs is still lacking. Therefore, realistic food products needed to be employed to demonstrate the usefulness of CDs as functional fillers in packaging materials. Additionally, metal or heteroatom-doped carbon dots are widely discussed to fabricate multifunctional packaging films. Therefore, it would also be essential to look into the toxicity or potential risks of different kinds of CDs in order to ensure that they may be utilized safely in food applications. In this context, additional investigation is required to ascertain the impact of particular carbon nanomaterial characteristics, such as form, functional groups, charge, and composition, on the characteristics of packaging materials engineered from various polymers.

References

1. Wrona M, Silva F, Salafranca J, Nerín C, Alfonso MJ, Caballero MÁ (2021) Design of new natural antioxidant active packaging: screening flowsheet from pure essential oils and vegetable oils to ex vivo testing in meat samples. Food Control 1(120):107536
2. Jeddi MZ, Boon PE, Cubadda F, Hoogenboom R, Mol H, Verhagen H et al (2022) A vision on the "foodture" role of dietary exposure sciences in the interplay between food safety and nutrition. Trends Food Sci Technol 1(120):288–300
3. Atta OM, Manan S, Shahzad A, Ul-Islam M, Ullah MW, Yang G (2022) Biobased materials for active food packaging: a review. Food Hydrocoll 1(125):107419
4. Lorenzo JM, Batlle R, Gómez M (2014) Extension of the shelf-life of foal meat with two antioxidant active packaging systems. LWT Food Sci Technol 59(1):181–188
5. Fang Z, Zhao Y, Warner RD, Johnson SK (2017) Active and intelligent packaging in meat industry. Trends Food Sci Technol 1(61):60–71
6. Soltani Firouz M, Mohi-Alden K, Omid M (2021) A critical review on intelligent and active packaging in the food industry: research and development. Food Res Int 1(141):110113
7. Liu D, Yang L, Shang M, Zhong Y (2019) Research progress of packaging indicating materials for real-time monitoring of food quality. Mater Express 9(5):377–396
8. Thiruvengadam M, Rajakumar G, Chung IM (2018) Nanotechnology: current uses and future applications in the food industry. 3 Biotech [Internet] 8(1):74. https://doi.org/10.1007/s13205-018-1104-7
9. Alfei S, Marengo B, Zuccari G (2020) Nanotechnology application in food packaging: a plethora of opportunities versus pending risks assessment and public concerns. Food Res Int 1(137):109664
10. Ashrafizadeh M, Mohammadinejad R, Kailasa SK, Ahmadi Z, Afshar EG, Pardakhty A (2020) Carbon dots as versatile nanoarchitectures for the treatment of neurological disorders and their theranostic applications: a review. Adv Colloid Interface Sci 1(278):102123
11. dos Santos CA, Ingle AP, Rai M (2020) The emerging role of metallic nanoparticles in food. Appl Microbiol Biotechnol [Internet] 104(6):2373–2383 [cited 2022 Mar 28]. https://doi.org/10.1007/s00253-020-10372-x
12. Reshma VG, Mohanan PV (2019) Quantum dots: applications and safety consequences. J Lumin 1(205):287–298
13. Teleanu DM, Chircov C, Grumezescu AM, Volceanov A, Teleanu RI (2019) Contrast agents delivery: an up-to-date review of nanodiagnostics in neuroimaging. Nanomaterials [Internet] 9(4):542 [cited 2022 Mar 28]. https://www.mdpi.com/2079-4991/9/4/542/htm

14. Nile SH, Baskar V, Selvaraj D, Nile A, Xiao J, Kai G (2020) Nanotechnologies in food science: applications, recent trends, and future perspectives. Nano-Micro Lett [Internet] 12(1):45. https://doi.org/10.1007/s40820-020-0383-9
15. Samal D (2017) Use of nanotechnology in food industry: a review. Int J Environ Agric Biotechnol [Internet] 2(4):2270–2278. http://ijeab.com/detail/use-of-nanotechnology-in-food-industry-a-review/
16. He X, Deng H, Hwang H (2019) The current application of nanotechnology in food and agriculture. J Food Drug Anal [Internet] 27(1):1–21. https://linkinghub.elsevier.com/retrieve/pii/S102194981830173X
17. Nakagawa K (2014) Nano- and microencapsulation of flavor in food systems. In: Nano- and microencapsulation for foods [Internet]. John Wiley & Sons, Ltd, Chichester, UK, pp 249–271. https://doi.org/10.1002/9781118292327.ch10
18. Dekkers S, Krystek P, Peters RJB, Lankveld DPK, Bokkers BGH, van Hoeven-Arentzen PH et al (2011) Presence and risks of nanosilica in food products. Nanotoxicology [Internet] 5(3):393–405. https://doi.org/10.3109/17435390.2010.519836
19. Chellaram C, Murugaboopathi G, John AA, Sivakumar R, Ganesan S, Krithika S et al (2014) Significance of nanotechnology in food industry. APCBEE Procedia [Internet] 8:109–113. https://linkinghub.elsevier.com/retrieve/pii/S2212670814000906
20. Wu H, Yin JJ, Wamer WG, Zeng M, Lo YM (2014) Reactive oxygen species-related activities of nano-iron metal and nano-iron oxides. J Food Drug Anal [Internet] 22(1):86–94. https://linkinghub.elsevier.com/retrieve/pii/S1021949814000088
21. He X, Hwang HM (2016) Nanotechnology in food science: functionality, applicability, and safety assessment. J Food Drug Anal [Internet] 24(4):671–681. https://linkinghub.elsevier.com/retrieve/pii/S1021949816300758
22. Bajpai VK, Kamle M, Shukla S, Mahato DK, Chandra P, Hwang SK et al (2018) Prospects of using nanotechnology for food preservation, safety, and security. J Food Drug Anal [Internet] 26(4):1201–1214. https://linkinghub.elsevier.com/retrieve/pii/S1021949818301169
23. Cui L, Ren X, Sun M, Liu H, Xia L (2021) Carbon dots: synthesis, properties and applications, nanomaterials. Nanomaterials 11:11
24. Nasir S, Hussein M, Zainal Z, Yusof N (2018) Carbon-based nanomaterials/allotropes: a glimpse of their synthesis, properties and some applications. Materials (Basel) [Internet] 11(2):295. http://www.mdpi.com/1996-1944/11/2/295
25. Pandey R, Siddiqui S, Saurabh S, Pankaj A, Singh PK, Parmar D et al (2021) Waste candle soot derived carbon nanoparticles: a competent alternative for the management of Helicoverpa armigera. Chemosphere [Internet] 264:128537. https://linkinghub.elsevier.com/retrieve/pii/S0045653520327326
26. Singh M, Singh S, Singh SP, Patel SS (2020) Recent advancement of carbon nanomaterials engrained molecular imprinted polymer for environmental matrix. Trends Environ Anal Chem [Internet] 27:e00092. https://linkinghub.elsevier.com/retrieve/pii/S2214158820300209
27. Liu H, Ding J, Zhang K, Ding L (2019) Construction of biomass carbon dots based fluorescence sensors and their applications in chemical and biological analysis. TrAC Trends Anal Chem 1(118):315–337
28. Mauter MS, Elimelech M (2008) Environmental applications of carbon-based nanomaterials. Environ Sci Technol [Internet] 42(16):5843–5958. https://doi.org/10.1021/es8006904
29. Meng D, Fan J, Ma J, Du SW, Geng J (2019) The preparation and functional applications of carbon nanomaterial/conjugated polymer composites. Compos Commun [Internet] 12:64–73. https://linkinghub.elsevier.com/retrieve/pii/S2452213918302298
30. Rajeshkumar L, Ramesh M, Bhuvaneswari V, Balaji D (2023) Carbon nano-materials (CNMs) derived from biomass for energy storage applications: a review. Carbon Lett [Internet] 33(3):661–690. https://doi.org/10.1007/s42823-023-00478-3
31. Sridharan R, Monisha B, Kumar PS, Gayathri KV (2022) Carbon nanomaterials and its applications in pharmaceuticals: a brief review. Chemosphere [Internet] 294:133731. https://linkinghub.elsevier.com/retrieve/pii/S0045653522002247

32. Zhang Y, Luo J, Zhang H, Li T, Xu H, Sun Y et al (2022) Synthesis and adsorption performance of three-dimensional gels assembled by carbon nanomaterials for heavy metal removal from water: a review. Sci Total Environ [Internet] 852:158201. https://linkinghub.elsevier.com/retrieve/pii/S0048969722053001
33. Travlou NA, Giannakoudakis DA, Algarra M, Labella AM, Rodríguez-Castellón E, Bandosz TJ (2018) S- and N-doped carbon quantum dots: surface chemistry dependent antibacterial activity. Carbon NY 1(135):104–111
34. Salimi Shahraki H, Ahmad A, Bushra R (2022) Green carbon dots with multifaceted applications–waste to wealth strategy. FlatChem 1(31):100310
35. Fan K, Zhang M, Chen H (2020) Effect of ultrasound treatment combined with carbon dots coating on the microbial and physicochemical quality of fresh-cut cucumber. Food Bioprocess Technol [Internet] 13(4):648–660 [cited 2022 Nov 17]. https://doi.org/10.1007/s11947-020-02424-x
36. Karimi-Maleh H, Beitollahi H, Senthil Kumar P, Tajik S, Mohammadzadeh Jahani P, Karimi F et al (2022) Recent advances in carbon nanomaterials-based electrochemical sensors for food azo dyes detection. Food Chem Toxicol [Internet] 164:112961. https://linkinghub.elsevier.com/retrieve/pii/S0278691522001594
37. Sabui P, Mallick S, Singh KR, Natarajan A, Verma R, Singh J et al (2023) Potentialities of fluorescent carbon nanomaterials as sensor for food analysis. Luminescence [Internet] 38(7):1047–1063. https://doi.org/10.1002/bio.4406
38. Kuswandi B, Wicaksono Y, Jayus, Abdullah A, Heng LY, Ahmad M (2011) Smart packaging: sensors for monitoring of food quality and safety. Sens Instrum Food Qual Saf [Internet] 5(3–4):137–146 [cited 2022 Mar 24]. https://doi.org/10.1007/s11694-011-9120-x
39. Cheng W, Tang X, Zhang Y, Wu D, Yang W (2021) Applications of metal-organic framework (MOF)-based sensors for food safety: enhancing mechanisms and recent advances. Trends Food Sci Technol [Internet] 112:268–282. https://linkinghub.elsevier.com/retrieve/pii/S0924224421002569
40. Basu J, Datta S, RoyChaudhuri C (2015) A graphene field effect capacitive Immunosensor for sub-femtomolar food toxin detection. Biosens Bioelectron [Internet] 68:544–549. https://linkinghub.elsevier.com/retrieve/pii/S0956566315000639
41. da Silva LP, Azevedo Vargas E, Madureira FD, Faria AF, Augusti R (2020) Development and validation of a novel analytical method to quantify aflatoxins in baby food samples by employing dispersive solid phase extraction with multiwalled carbon nanotubes. Food Anal Methods [Internet] 13(7):1530–1537. https://doi.org/10.1007/s12161-020-01778-4
42. Liang G, Zhai H, Huang L, Tan X, Zhou Q, Yu X et al (2018) Synthesis of carbon quantum dots-doped dummy molecularly imprinted polymer monolithic column for selective enrichment and analysis of aflatoxin B1 in peanut. J Pharm Biomed Anal [Internet] 149:258–264. https://linkinghub.elsevier.com/retrieve/pii/S0731708517319313
43. Wang B, Chen Y, Wu Y, Weng B, Liu Y, Lu Z et al (2016) Aptamer induced assembly of fluorescent nitrogen-doped carbon dots on gold nanoparticles for sensitive detection of AFB1. Biosens Bioelectron 15(78):23–30
44. Nagarajan D, Varada OM, Venkatanarasimhan S (2021) Carbon dots coated on amine functionalized cellulose sponge for the adsorption of the toxic herbicide atrazine. Mater Today Proc [Internet] 47:790–799. https://linkinghub.elsevier.com/retrieve/pii/S2214785320359332
45. Shekarbeygi Z, Farhadian N, Khani S, Moradi S, Shahlaei M (2020) The effects of rose pigments extracted by different methods on the optical properties of carbon quantum dots and its efficacy in the determination of Diazinon. Microchem J [Internet] 158:105232. https://linkinghub.elsevier.com/retrieve/pii/S0026265X20305129
46. Lin B, Yan Y, Guo M, Cao Y, Yu Y, Zhang T et al (2018) Modification-free carbon dots as turn-on fluorescence probe for detection of organophosphorus pesticides. Food Chem 15(245):1176–1182
47. Kazemifard N, Ensafi AA, Rezaei B (2020) Green synthesized carbon dots embedded in silica molecularly imprinted polymers, characterization and application as a rapid and selective fluorimetric sensor for determination of thiabendazole in juices. Food Chem 25(310):125812

48. Xing X, Huang L, Zhao S, Xiao J, Lan M (2020) S,N-Doped carbon dots for tetracyclines sensing with a fluorometric spectral response. Microchem J [Internet] 157:105065. https://linkinghub.elsevier.com/retrieve/pii/S0026265X20311097
49. Yang Y, Xing X, Zou T, Wang Z, Zhao R, Hong P et al (2020) A novel and sensitive ratiometric fluorescence assay for carbendazim based on N-doped carbon quantum dots and gold nanocluster nanohybrid. J Hazard Mater [Internet] 386:121958. https://linkinghub.elsevier.com/retrieve/pii/S0304389419319120
50. Chatzimitakos T, Kasouni A, Sygellou L, Avgeropoulos A, Troganis A, Stalikas C (2017) Two of a kind but different: luminescent carbon quantum dots from Citrus peels for iron and tartrazine sensing and cell imaging. Talanta 175:305–312
51. Xu H, Yang X, Li G, Zhao C, Liao X (2015) Green synthesis of fluorescent carbon dots for selective detection of tartrazine in food samples. J Agric Food Chem [Internet] 63(30):6707–6714 [cited 2022 Mar 24]. https://doi.org/10.1021/acs.jafc.5b02319
52. Tanaka T (2006) Reproductive and neurobehavioural toxicity study of tartrazine administered to mice in the diet. Food Chem Toxicol 44(2):179–187
53. Liu J, Chen Y, Wang W, Feng J, Liang M, Ma S et al (2016) "Switch-on" fluorescent sensing of ascorbic acid in food samples based on carbon quantum dots-MnO2 probe. J Agric Food Chem [Internet] 64(1):371–380 [cited 2022 Mar 24]. https://doi.org/10.1021/acs.jafc.5b05726
54. Ahmed GHG, Laíño RB, Calzón JAG, García MED (2015) Fluorescent carbon nanodots for sensitive and selective detection of tannic acid in wines. Talanta 15(132):252–257
55. Tang D, Tang J, Su B, Chen G (2010) Ultrasensitive electrochemical immunoassay of staphylococcal enterotoxin B in food using enzyme-nanosilica-doped carbon nanotubes for signal amplification. J Agric Food Chem 58(20):10824–10830
56. Wang S, Zhang Y, Pang G, Zhang Y, Guo S (2017) Tuning the aggregation/disaggregation behavior of graphene quantum dots by structure-switching aptamer for high-sensitivity fluorescent ochratoxin A sensor. Anal Chem 89(3):1704–1709
57. Wang D, Lin B, Cao Y, Guo M, Yu Y (2016) A highly selective and sensitive fluorescence detection method of glyphosate based on an immune reaction strategy of carbon dot labeled antibody and antigen magnetic beads. J Agric Food Chem 64(30):6042–6050
58. Zhang Y, Gao Z, Yang X, Chang J, Liu Z, Jiang K (2019) Fish-scale-derived carbon dots as efficient fluorescent nanoprobes for detection of ferric ions. RSC Adv 9(2):940–949
59. Athika M, Prasath A, Duraisamy E, Sankar Devi V, Selva Sharma A, Elumalai P (2019) Carbon-quantum dots derived from denatured milk for efficient chromium-ion sensing and supercapacitor applications. Mater Lett 241:156–159
60. Huang G, Chen X, Wang C, Zheng H, Huang Z, Chen D et al (2017) Photoluminescent carbon dots derived from sugarcane molasses: synthesis, properties, and applications. RSC Adv 7(75):47840–47847
61. Dong B, Li H, Sun J, Mujtaba Mari G, Ai J, Han D et al (2020) Homogeneous fluorescent immunoassay for the simultaneous detection of chloramphenicol and amantadine via the duplex FRET between carbon dots and WS2 nanosheets. Food Chem [Internet] 327:127107. https://linkinghub.elsevier.com/retrieve/pii/S0308814620309699
62. Wang B, Mu Y, Zhang C, Li J (2017) Blue photoluminescent carbon nanodots prepared from zeolite as efficient sensors for picric acid detection. Sensors Actuators B Chem [Internet] 253:911–917. https://linkinghub.elsevier.com/retrieve/pii/S0925400517312546
63. Shi R, Feng S, Park CY, Park KY, Song J, Park JP et al (2020) Fluorescence detection of histamine based on specific binding bioreceptors and carbon quantum dots. Biosens Bioelectron [Internet] 167:112519. https://linkinghub.elsevier.com/retrieve/pii/S095656632030511X
64. Jalili R, Khataee A, Rashidi MR, Razmjou A (2020) Detection of penicillin G residues in milk based on dual-emission carbon dots and molecularly imprinted polymers. Food Chem [Internet] 314:126172. https://linkinghub.elsevier.com/retrieve/pii/S0308814620300182
65. Shokri R, Amjadi M (2021) A ratiometric fluorescence sensor for triticonazole based on the encapsulated boron-doped and phosphorous-doped carbon dots in the metal organic framework. Spectrochim Acta Part A Mol Biomol Spectrosc [Internet] 246:118951. https://linkinghub.elsevier.com/retrieve/pii/S1386142520309306

66. Hao AY, Wang XQ, Mei YZ, Nie JF, Yang YQ, Dai CC (2021) A smartphone-combined ratiometric fluorescence probe for specifically and visibly detecting cephalexin. Spectrochim Acta Part A Mol Biomol Spectrosc [Internet] 249:119310. https://linkinghub.elsevier.com/retrieve/pii/S1386142520312890
67. Long R, Guo Y, Xie L, Shi S, Xu J, Tong C et al (2020) White pepper-derived ratiometric carbon dots for highly selective detection and imaging of coenzyme A. Food Chem [Internet] 315:126171. https://linkinghub.elsevier.com/retrieve/pii/S0308814620300170
68. Barlow CY, Morgan DC (2013) Polymer film packaging for food: an environmental assessment. Resour Conserv Recycl [Internet] 78:74–80. https://linkinghub.elsevier.com/retrieve/pii/S0921344913001432
69. Qu J-H, Wei Q, Sun D-W (2018) Carbon dots: principles and their applications in food quality and safety detection. Crit Rev Food Sci Nutr 58(14):2466–2475
70. Zhao L, Zhang M, Mujumdar AS, Wang H (2022) Application of carbon dots in food preservation: a critical review for packaging enhancers and food preservatives. Crit Rev Food Sci Nutr [Internet] [cited 2022 Jul 19]. https://doi.org/10.1080/10408398.2022.2039896
71. Radnia F, Mohajeri N, Zarghami N (2020) New insight into the engineering of green carbon dots: possible applications in emerging cancer theranostics. Talanta 1(209):120547
72. Fan K, Zhang M, Fan D, Jiang F (2019) Effect of carbon dots with chitosan coating on microorganisms and storage quality of modified-atmosphere-packaged fresh-cut cucumber. J Sci Food Agric [Internet] 99(13):6032–6041. https://doi.org/10.1002/jsfa.9879
73. Zhao L, Zhang M, Wang H, Devahastin S (2020) Effect of carbon dots in combination with aqueous chitosan solution on shelf life and stability of soy milk. Int J Food Microbiol [Internet] 326:108650. https://linkinghub.elsevier.com/retrieve/pii/S0168160520301446
74. Patil AS, Waghmare RD, Pawar SP, Salunkhe ST, Kolekar GB, Sohn D et al (2020) Photophysical insights of highly transparent, flexible and re-emissive PVA @ WTR-CDs composite thin films: a next generation food packaging material for UV blocking applications. J Photochem Photobiol A Chem [Internet] 400:112647. https://linkinghub.elsevier.com/retrieve/pii/S1010603020304469
75. Xu N, Gao S, Xu C, Fang Y, Xu L, Zhang W (2021) Carbon quantum dots derived from waste acorn cups and its application as an ultraviolet absorbent for polyvinyl alcohol film. Appl Surf Sci [Internet] 556:149774. https://linkinghub.elsevier.com/retrieve/pii/S0169433221008503
76. Pinto T da S, Rodrigues PNS, Marinho LES, Verly RM, Bretas Roa JP, de Oliveira LCA et al (2019) Self-assembled hybrid nanocomposite films of carbon dots and hydrolyzed collagen. Mater Chem Phys [Internet] 230:44–53. https://linkinghub.elsevier.com/retrieve/pii/S0254058419302603
77. Han S, Zhang H, Xie Y, Liu L, Shan C, Li X et al (2015) Application of cow milk-derived carbon dots/Ag NPs composite as the antibacterial agent. Appl Surf Sci [Internet] 328:368–373. https://linkinghub.elsevier.com/retrieve/pii/S0169433214027627
78. Xu L, Zhang Y, Pan H, Xu N, Mei C, Mao H et al (2019) Preparation and performance of radiata-pine-derived polyvinyl alcohol/carbon quantum dots fluorescent films. Materials (Basel) [Internet] 13(1):67 [cited 2022 May 30]. https://www.mdpi.com/1996-1944/13/1/67/htm
79. Cuevas A, Campos BB, Romero R, Algarra M, Vázquez MI, Benavente J (2019) Eco-friendly modification of a regenerated cellulose based film by silicon, carbon and N-doped carbon quantum dots. Carbohydr Polym [Internet] 206:238–244. https://linkinghub.elsevier.com/retrieve/pii/S0144861718312712
80. Zhao L, Zhang M, Mujumdar AS, Adhikari B, Wang H (2022) Preparation of a novel carbon dot/polyvinyl alcohol composite film and its application in food preservation. ACS Appl Mater Interfaces [Internet] 14(33):37528–37539. https://doi.org/10.1021/acsami.2c10869
81. Wu J, Chen ·X, Zhang Z, Zhang J (2022) "Off-on" fluorescence probe based on green emissive carbon dots for the determination of Cu2+ ions and glyphosate and development of a smart sensing film for vegetable packaging. Microchim Acta [Internet] 189(3):1–9 [cited 2022 Mar 24]. https://doi.org/10.1007/s00604-022-05241-5

82. Chitrakar B, Zhang M, Bhandari B (2021) Improvement strategies of food supply chain through novel food processing technologies during COVID-19 pandemic. Food Control [Internet] 125:108010. https://linkinghub.elsevier.com/retrieve/pii/S0956713521001481
83. Sun Q, Zhang M, Mujumdar AS (2019) Recent developments of artificial intelligence in drying of fresh food: a review. Crit Rev Food Sci Nutr [Internet] 59(14):2258–2275. https://doi.org/10.1080/10408398.2018.1446900
84. Wu LT, Tsai IL, Ho YC, Hang YH, Lin C, Tsai ML et al (2021) Active and intelligent gellan gum-based packaging films for controlling anthocyanins release and monitoring food freshness. Carbohydr Polym [Internet] 254:117410. https://linkinghub.elsevier.com/retrieve/pii/S0144861720315836
85. Chen H, Zhang M, Bhandari B, Yang C (2019) Development of a novel colorimetric food package label for monitoring lean pork freshness. LWT [Internet] 99:43–49. https://linkinghub.elsevier.com/retrieve/pii/S0023643818307801
86. Chen H, Zhang M, Bhandari B, Yang C (2020) Novel pH-sensitive films containing curcumin and anthocyanins to monitor fish freshness. Food Hydrocoll 100:105438
87. Koshy RR, Koshy JT, Mary SK, Sadanandan S, Jisha S, Pothan LA (2021) Preparation of pH sensitive film based on starch/carbon nano dots incorporating anthocyanin for monitoring spoilage of pork. Food Control 126:108039
88. Xu Y, Yang D, Huo S, Ren J, Gao N, Chen Z et al (2021) Carbon dots and ruthenium doped oxygen sensitive nanofibrous membranes for monitoring the respiration of agricultural products. Polym Test [Internet] 93:106957. https://linkinghub.elsevier.com/retrieve/pii/S0142941820321863
89. Zhang X, Wang H, Niu N, Chen Z, Li S, Liu SX et al (2020) Fluorescent poly(vinyl alcohol) films containing chlorogenic acid carbon nanodots for food monitoring. ACS Appl Nano Mater [Internet] 3(8):7611–7620. https://doi.org/10.1021/acsanm.0c01229
90. Duan X, Zhang M, Li X, Mujumdar AS (2008) Microwave freeze drying of sea cucumber coated with nanoscale silver. Dry Technol [Internet] 26(4):413–419. https://doi.org/10.1080/07373930801929136
91. Luo X, Han Y, Chen X, Tang W, Yue T, Li Z (2020) Carbon dots derived fluorescent nanosensors as versatile tools for food quality and safety assessment: a review. Trends Food Sci Technol [Internet] 95:149–161. https://linkinghub.elsevier.com/retrieve/pii/S0924224419302110
92. Eskalen H, Çeşme M, Kerli S, Özğan Ş (2021) Green synthesis of water-soluble fluorescent carbon dots from rosemary leaves: applications in food storage capacity, fingerprint detection, and antibacterial activity. J Chem Res [Internet] 45(5–6):428–435. https://doi.org/10.1177/1747519820953823
93. Zhang X, Wang H, Ma C, Niu N, Chen Z, Liu S et al (2018) Seeking value from biomass materials: preparation of coffee bean shell-derived fluorescent carbon dots via molecular aggregation for antioxidation and bioimaging applications. Mater Chem Front [Internet] 2(7):1269–1275. http://xlink.rsc.org/?DOI=C8QM00030A
94. Xiao D, Pu H, Tian H, Yang D, Yang Y, Li H (2019) Application of carbon dots-chitosan coating in preservation of mango. Food Ferment Ind 45(22):130–135
95. Fan K, Zhang M, Jiang F (2019) Ultrasound treatment to modified atmospheric packaged fresh-cut cucumber: influence on microbial inhibition and storage quality. Ultrason Sonochem [Internet] 54:162–170. https://linkinghub.elsevier.com/retrieve/pii/S1350417719300781
96. Amini SM, Gilaki M, Karchani M (2014) Safety of nanotechnology in food industries. J Electron 6(4):962–968
97. Ranjan S, Dasgupta N, Chakraborty AR, Melvin Samuel S, Ramalingam C, Shanker R et al (2014) Nanoscience and nanotechnologies in food industries: opportunities and research trends. J Nanopart Res [Internet] 16(6):2464. https://doi.org/10.1007/s11051-014-2464-5
98. Ameta SK, Rai AK, Hiran D, Ameta R, Ameta SC (2020) Use of nanomaterials in food science. In: Biogenic nano-particles and their use in agro-ecosystems [Internet]. Springer, Singapore, pp 457–488. https://doi.org/10.1007/978-981-15-2985-6_24

Cellulose-Based Nanomaterials in Food Packaging

Ahmed A. Oun, Manal A. Sorour, and Amira S. El-Mahrouky

Abstract Nanocellulose has emerged as a promising material for various applications in food packaging, particularly as a carrier for active materials and color indicators. This chapter has explored the versatility and potential of nanocellulose in improving the functionality and performance of food packaging materials. The unique properties of nanocellulose, such as its high surface area, biocompatibility, and tunable surface chemistry, facilitated the incorporation of active ingredients, such as antimicrobial agents, antioxidants, and flavoring compounds into nanocellulose-based matrices. Nanocellulose materials can play a direct role in enhancing the mechanical strength, water vapor resistance, gas barrier properties, or thermal stability of packaging films. They can also play an indirect role as carriers for other functional materials, such as active ingredients and colorants. Cellulose-loaded materials with active and intelligent components offer enhanced shelf life, quality maintenance, and can provide visual color change to monitor the freshness and spoilage of packaged food products.

Keywords Nanocellulose · Food packaging · Active packaging · Smart packaging

1 Introduction

Food packaging plays a crucial role in preserving food quality and safety by protecting it from physical, chemical, and biological hazards. Additionally, it is used to ensure food quality and safety, extend shelf life, minimize food waste, provide essential information to consumers, and enhance convenience [1]. The increasing

A. A. Oun (✉)
Nanotechnology and Advanced Materials Central Lab, Regional Center for Food and Feed, Agricultural Research Center, Giza, Egypt
e-mail: aahmedoun2013@gmail.com

M. A. Sorour · A. S. El-Mahrouky
Food Engineering and Packaging Department, Food Technology Research Institute, Agricultural Research Center, Giza, Egypt

M. Alle et al. (eds.), *Functional Nanomaterials and Nanocomposites for Biodegradable Food Packaging*, Smart Nanomaterials Technology,
https://doi.org/10.1007/978-981-97-9084-5_6

emphasis on environmental sustainability demands packaging materials that are renewable, biodegradable, and have a reduced carbon footprint. Cellulosic materials offer a renewable, biodegradable, and customizable alternative to traditional packaging materials, contributing to a more sustainable and environmentally conscious packaging industry [2, 3].

Cellulose is the primary structural component of plant cell walls and can be found in various sources. These sources include wood, which is the most abundant source of cellulose and encompasses both hardwood and softwood. Additionally, cellulose can be found in materials such as cotton, hemp, fruit and vegetable waste, agricultural residues (such as straw, corn husks, sugarcane bagasse, rice husks, and wheat bran), and even algae and microorganisms, which offer sustainable cellulose production through biotechnological processes [4–6]. Cellulosic materials have been used in commercial packaging materials in the form of paperboard, paper, corrugated fiberboard, cellophane, and cellulose acetate [7, 8].

For more advanced properties, cellulose nanofibrils (CNFs) and cellulose nanocrystals (CNCs) are isolated from cellulosic sources using mechanical methods, chemical methods, biological methods, or a combination of these techniques [9–12]. Isolated nanocellulose materials exhibit unique properties such as high strength, large surface area, and excellent barrier properties, making them promising candidates for enhancing food packaging. Cellulose-based materials have found various applications in food packaging, serving multiple roles. Directly, they enhance specific properties such as mechanical strength, water vapor resistance, gas barrier properties, and thermal stability. Indirectly, they can act as carriers for other functional materials, including antimicrobial agents, antioxidants, UV barriers, colorants for colorimetric indicators, and even Pickering emulsion. For instance, nanocellulose materials have been used as reinforcing agents to improve the mechanical, barrier, and biodegradability properties of film polymers such as polylactic acid (PLA) or polyhydroxyalkanoates (PHA) [13–15]. Additionally, it is used to improve the mechanical, thermal, gas, and water barrier properties of natural biodegradable films such as carboxymethyl cellulose (CMC) [16], polyvinyl alcohol/chitosan [17, 18], gelatin/polyvinyl alcohol [19], starch [20], agar [21]. On the other hand, cellulosic materials are used as carriers for colorants in colorimetric pH-sensitive indicators [17, 22–26], active materials [27, 28], metal and metal oxide NPs [29], Pickering agents in nanoemulsion [30]. This is evident from the growing number of published articles focusing on the applications of cellulosic materials in the food packaging sector [31].

2 Cellulose-Based Materials

2.1 Sources and Properties of Cellulose

Cellulose ($C_6H_{10}O_5$) stands as one of the most abundant natural raw materials on Earth, constituting the primary component of plant cell walls along with hemicellulose, lignin, and a relatively small amount of extractives. Cellulose is predominantly present in plant-based materials, and there are several rich sources of cellulose, including wood, cotton fibers, hemp, bamboo, straw, jute, flax, sugarcane bagasse, bacterial cellulose, and algae (Fig. 1) [32]. The cellulose content and properties may vary among different sources, and each source may have specific applications based on its unique characteristics [33]. While wood remains the primary source of cellulose fiber, constituting approximately 40–50% of the dry mass of wood. However, the environmental concerns and reduction of forests resulting from the increased demand for wood resources have generated greater interest in the study of non-wooden cellulosic materials. As a result, non-wood cellulose sources, such as plants and agricultural wastes, have gained significant attention [9, 34]. Generally, cellulose isolated from non-wood cellulose sources is first washed to remove dirt and aqueous soluble substances. Then, the source is dried and mechanically cut into small parts. Next, extractives such as wax, pectin, and pigments are removed using solvents. Lastly, lignin and hemicellulose are eliminated through chemical processes such as alkalization and bleaching to obtain pure cellulose for subsequent applications [10, 34, 35]. Cellulose in different forms and properties can be obtained from cellulosic sources through various isolation and pretreatment methods [32, 36, 37]. Depending on the treatment method, a wide range of cellulose products can be obtained, including cellulose-based hydrogels for absorbents, cellulose membranes for filters, cellulose aerogels with low density and excellent thermal insulation properties, cellulose fibers for textiles and reinforcements, and cellulose-based films for packaging [38].

Chemically, cellulose is a polymer composed of repeating units of glucose molecules, which are connected by β-1,4 glycosidic bonds [7]. Each glucose unit in cellulose is inverted with its neighboring units, creating a linear and elongated chain structure. Assembling of cellulose chains through a combination of extensive intra- and intermolecular hydrogen bonds and van der Waals forces, resulting in the formation of elementary fibrils. These fibrils further aggregate to form microfibrils with lengths that can extend up to several micrometers and diameters ranging from 5 to 50 nm as shown in Fig. 2 [5]. Cellulose is insoluble in water and most organic solvents due to its highly crystalline structure and strong hydrogen bonding between cellulose chains. Although insoluble, cellulose has a high affinity for water. It can absorb and retain water, contributing to its ability to form gels and provide moisture regulation in various applications [39]. Hydrogen bonds play an important role in the properties of cellulose. Cellulose monomers possess three hydroxyl groups located at positions 2C-OH, 3C-OH, and 6C-OH. These hydroxyl groups have the ability to create hydrogen bonds, fostering interchain bonding between adjacent chains and

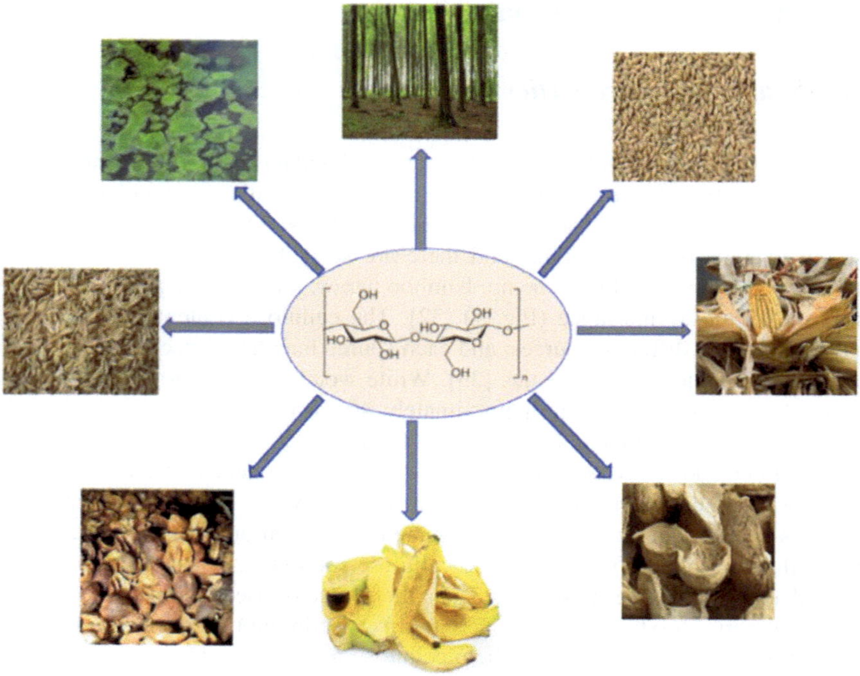

Fig. 1 Different sources of cellulose [34]

intrachain bonding between glucan units within the same chain. This molecular arrangement facilitates the formation of fibrillar packing [40]. Cellulose exists in six different polymorphs, namely cellulose I, II, III_I, III_{II}, IV_I, and IV_{II}. These polymorphs can undergo transformation from one form to another through chemical or thermal treatments [41, 42]. Cellulose I is the native form of cellulose that exists in plants, providing rigidity and strength to plant cell walls. Native cellulose contains a mixture of crystal structures Cellulose Iα and Cellulose Iβ, exhibit differences in their crystal structures and hydrogen bonding arrangements depending on the origin of cellulose. Cellulose Iα has a triclinic crystal structure and is typically present in the primary cell walls of plants. In contrast, cellulose Iβ displays a monoclinic crystal structure and is commonly found in the secondary cell walls of plants [40, 43]. The hydrogen bonds in cellulose Iβ form a more extended and parallel arrangement. Additionally, cellulose Iβ exhibits a higher degree of crystallinity compared to cellulose Iα due to its more organized and closely packed structure [32, 44]. Cellulose II, also known as regenerated cellulose, is a modified form of cellulose obtained by treating cellulose I with a mixture of chemicals (e.g. NaOH/urea aqueous treatment). In Cellulose II, the cellulose chains are organized in a more disordered, less tightly packed manner, more soluble in certain solvents, more susceptible to chemical modifications, possess good tensile strength, and have hygroscopic properties.

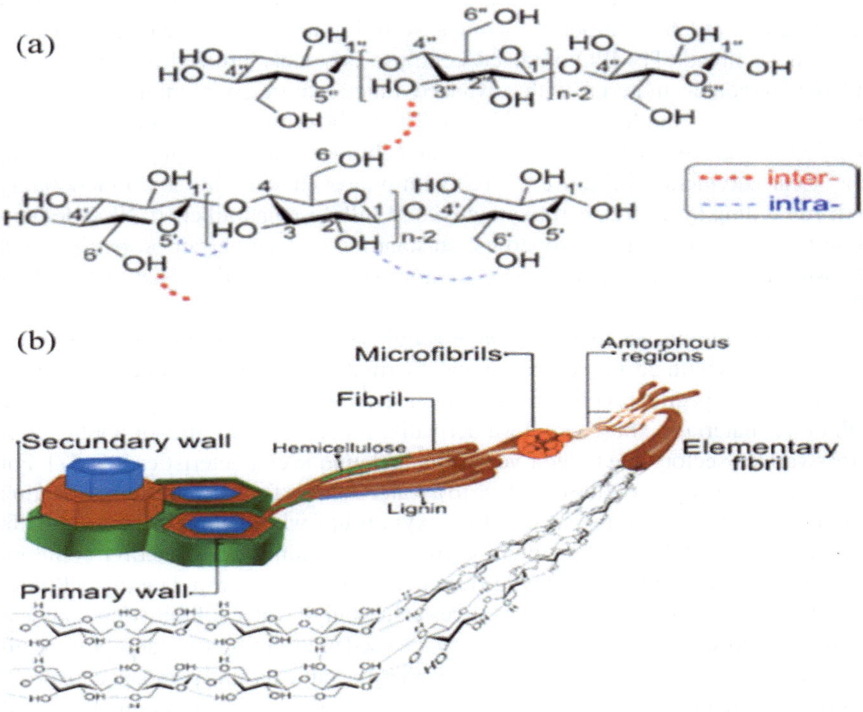

Fig. 2 **a** Chemical structure and **b** fibers and microfibers arrangement of cellulose [47]

Cellulose II finds applications in various industries, including textiles, films, hydrogels/aerogels, and other cellulose-based products [29, 45, 46]. Cellulose III_I and III_{II} can be obtained by treating cellulose I and cellulose II with ammonia, respectively. While, cellulose IV_I, and IV_{II} are derived from III_I and III_{II}, respectively, through heat treatment at 260 °C [47].

2.2 Cellulose Derivatives

Cellulose derivatives are obtained through the surface modification of the original cellulose to enhance specific properties or enable new functionalities [43, 48]. The modification of cellulosic materials can be achieved through two main approaches: by targeting hydroxyl groups or by utilizing ring-opening techniques. Hydroxyl group-based chemical modifications of cellulose involve methods such as TEMPO-oxidation, carboxymethylation, phosphorylation, sulfoethylation, or cationization. On the other hand, ring-opening chemical modifications rely on the periodate oxidation method, which enables the introduction of dialdehyde groups using various techniques [49]. For instance, the TEMPO-oxidation method is employed to convert the OH group at C6 to carboxylic groups, resulting in negatively charged surfaces of cellulose fibers. TEMPO-oxidized cellulose is widely utilized as a renewable and biodegradable material with advanced functionalities, including films, coatings, membranes, and hydrogels [50–54]. The carboxymethylation method is used to introduce carboxymethyl groups onto the cellulose chains, leading to the formation of carboxymethyl cellulose (CMC) with improved solubility and functionality [55]. According to reports, carboxymethylated cellulose nanofibers (CNFs) demonstrate favorable mechanical properties, outstanding oxygen barrier properties, high viscosity, and a greater degree of fibrillation when compared to enzymatic CNFs [49, 56].

In general, the surface modification of cellulosic materials aims to generate cellulose derivatives with desirable mechanical, thermal, and chemical properties, as well as to introduce specific functionalities, enhance compatibility, and improve adhesion with other materials. These cellulose derivatives offer a wide range of applications across various sectors due to their versatility and unique characteristics [49, 57]. For instance, converting insoluble cellulose to a water-soluble form through etherification reactions by substituting the surface hydroxyl groups with methyl groups to obtain methylcellulose (MC), carboxymethyl groups to obtain carboxymethyl cellulose (CMC), and hydroxypropyl methyl groups to obtain hydroxypropyl methylcellulose (HPMC) [40, 58, 59]. Cellulose derivatives exhibit high water solubility and possess thickening and stabilizing properties. They are used as thickeners, binders, and emulsifiers in various industries, including food, pharmaceuticals, and personal care products [60, 61]. In the food industry, cellulose derivatives such as carboxymethyl cellulose are used as thickeners, stabilizers, and emulsifiers. They improve the texture and stability of food products, enhance mouthfeel, and prevent syneresis [62].

Cellulose acetate is formed by modifying the structure of cellulose through the acetylation process, which involves replacing hydroxyl groups on the cellulose chain with acetyl groups. Cellulose acetate is widely used in food packaging, textiles, and film production. The exceptional barrier properties, transparency, printability, and environmental benefits of cellulose acetate render it highly suitable for a diverse range of food products, including snacks, confectionery, dried fruits, bakery items, and more [63, 64].

3 Fabrication Methods of Nanocellulose Materials

Nanocellulose material is an extraordinary substance obtained by reducing cellulose down to the nanometer scale. Obtained nanocellulose exhibited several attractive characteristics such as a large aspect ratio (~70), excellent mechanical properties, high surface area, flexibility, and biodegradability properties [47]. Cellulose nanocrystals (CNCs), also known as cellulose whiskers, and cellulose nanofibrils

(CNFs), alternatively referred to as nanofibrillated cellulose (NFC) or microfibrillated cellulose (MFC), can be obtained from cellulosic materials by different isolation method. In general, cellulose nanocrystals (CNCs) are isolated using chemical methods such as acid hydrolysis, while cellulose nanofibrils (CNFs) are obtained through mechanical methods. However, researchers have also employed a combination of mechanical and chemical techniques to obtain nanocellulose materials [6, 10, 65].

Mechanical methods involve physically disintegrating cellulose fibers to achieve nanoscale dimensions using equipment such as homogenizers, microfluidizers, grinders, extruders, blenders, ultrasonic devices, cryocrushers, and ball milling machines. Mechanical methods involve passing cellulose suspensions through narrow channels at high velocities to generate nanoscale fibrils. On the other hand, chemical methods involve treatments to selectively remove non-cellulosic components and isolate nanocellulose. Acid hydrolysis is a well-established chemical method in which cellulose is treated with acid, typically sulfuric acid, to hydrolyze the amorphous regions and isolate cellulose nanocrystals (CNCs). This process results in highly crystalline nanocrystals with a rod-like shape [35, 66, 67].

Both mechanical and chemical methods have their respective advantages and disadvantages. Mechanical methods are simpler, scalable, and generally do not require the use of harsh chemicals. However, they may necessitate significant energy input and can lead to a wide size distribution of the resulting nanocellulose. On the other hand, chemical methods, although more complex, allow for precise control over the properties of nanocellulose and can produce homogeneous nanocellulose products. However, they often involve the use of chemicals that require careful handling and disposal and may result in lower yields [35, 68, 69].

4 Applications of Cellulose-Based Materials in Food Packaging

Plastic-based materials have been widely used in the food packaging industry in recent years due to their low cost, ease of molding, and good mechanical, physical, and barrier properties. The serious environmental and health risks of low plastic decomposition forced researchers to seek out suitable biodegradable polymers to replace or reduce plastic [70]. Cellulose is one of the most sustainable, renewable, and abundant biodegradable polymers on the planet. The global cellulose market was valued at USD 219.53 billion in 2018 and is expected to grow to USD 305.08 billion by 2026 [71]. Recently, Garrido-Romero et al. [31] published a bibliometric study about the application of cellulose in food packaging. They found a rapidly expanding interest in this field, particularly during the past ten years, as evidenced by the 1029 documents published in high-impact journals [31]. Nanocellulose is a nanoscale material derived from cellulose that can be obtained in different forms such as cellulose nanofibrils (CNFs) and cellulose nanocrystals (CNCs). Nanocellulose possesses unique properties such as high strength, large surface area, and excellent barrier properties, making it a promising candidate for improving food packaging [72]. Compared to other nanomaterials and advanced materials, nanocellulose is non-toxic, renewable, abundant, inexpensive, extremely strong, and lightweight. The global market for nanocellulose was estimated to be worth USD 0.4 billion in 2022 and is expected to reach USD 2.0 billion by 2030, rising at a CAGR of 21.9% from 2022 to 2030 [73].

Commercial packaging materials are made of cellulosic materials including cellophane (regenerated cellulose), cellulose acetate (acetylated cellulose), paperboard, paper, corrugated fiberboard, and pulp containers. These materials can be utilized in various forms such as rigid containers, wrapping materials, as well as flexible primary and secondary packages [7]. Cellulose-based materials have been used in food packaging materials serving various roles. The direct role is to enhance specific properties, such as mechanical strength, water vapor resistance, gas barrier properties, and thermal stability. They can also serve an indirect role as carriers for other functional materials, such as antimicrobial, antioxidant, UV-barrier, colorants for colorimetric indicators, or Pickering emulsion.

Nanocellulose can be incorporated into food packaging materials using various techniques, depending on the desired properties and applications. It is important to note that the incorporation of nanocellulose into food packaging materials requires careful consideration of factors such as compatibility, processing techniques, and the potential migration of nanoparticles into the food. The unique chemical structure and the existence of abundant hydroxyl groups especially at the carbon (C) positions of C2, C3, and C6 within the cellulose chain enables its surface modification, improving the interaction, interfacial bonding, and compatibility between cellulosic materials and polymer matrices [74]. Nanocellulose can be modified via physical and chemical (e.g., alkali treatment, silane treatment, acetylation, benzoylation, isocyanate treatment, etc.) methods [8, 74, 75].

Reportedly, techniques like solution casting method, a coating method, layer-by-layer assembly (LBL), extrusions, polymeric hydrogel, spraying (spray drying), electrospinning, nanoemulsion, and adsorption methods are used for adding cellulose-based materials as reinforcing agents for food packaging to improve the mechanical, physical, and barrier properties of biodegradable films [58]. However, among them, the solution casting method (coatings and nanocellulose film), nanocellulose composite, nanocellulose-based active packaging, and nanocellulose-based sensors are the most commonly used methods. The solution casting method is one of the most common and simplest techniques. It is based on blending cellulosic materials with one or more polymers, with or without functional materials [47, 76]. It is usually used for preparing coating or nanocellulose composite film solutions. Nanocellulose can be applied directly over the surface of food as a coating or as a thin film on existing packaging materials, such as paper or plastic films. The nanocellulose coating can improve the barrier properties, mechanical strength, and durability of the packaging, enhancing its performance in terms of food preservation and shelf life extension [77, 78]. Nanocellulose-based composite films depend on combined nanocellulose with other polymers or materials to improve specific properties. For example, nanocellulose can be incorporated into biodegradable polymers, such as polylactic acid (PLA) or polyhydroxyalkanoates (PHA), to enhance their mechanical properties, biodegradability, and barrier performance. The resulting nanocellulose composites can be used for various packaging formats, including trays, bottles, and pouches.

Nanocellulose-based active packaging is based on the incorporation of active materials such as antimicrobial agents, antioxidants, or oxygen scavengers into nanocellulose-based films or coatings to actively inhibit the growth of microorganisms, delay oxidation, or remove oxygen from the packaging environment. Also, cellulosic materials have been used as a carrier for varying pigments used in preparing colorimetric indicators for smart packaging systems. It has been used in the fabrication of pH color indicator sensors. It showed a high possibility of monitoring and indicating the freshness or spoilage of the food in real time. These sensors can provide valuable information to consumers and aid in reducing food waste.

It can be concluded that nanocellulose materials are used in food packaging materials for the following properties.

4.1 Improving Mechanical Properties

Nanocellulose materials played important roles in improving the mechanical properties of different packaging materials, ensuring the protection of food during transportation and storage. For example, the PBAT/PLA blended film reinforced with CNCs had a 52% higher tensile strength, 29% higher elongation at break, and 118% higher Young's modulus than control films. Also, the obtained mechanical properties are comparable to those of commercially used plastic materials in food packaging [79]. Cellulose nanocrystals isolated from grain straws (rice straw, wheat straw, and barley straw) and blended with carboxymethyl cellulose (CMC) biopolymer showed an improvement in the mechanical and barrier properties of CMC/CNCs-based films [16]. Also, the incorporation of cellulose nanocrystals (CNCs) led to an increase in the tensile strength and film stiffness of neat PVA-chitosan film by 74.4% and 156.5%, respectively, without sacrificing film flexibility [17]. This probably occurs because of the high aspect ratio and abundance of free hydroxyl groups on nanocellulose. As a consequence, nanocellulose acts as a cross-linking agent, forming additional hydrogen bonds within the polymer chain matrix. This phenomenon ultimately enhances the tensile strength (TS) of the material [70]. On the other hand, exceeding the optimal percentage of nanocellulose can lead to a decrease in mechanical properties, which may be attributed to the high interaction level of the particle–particle rather than the interaction between the nanofillers and the fiber or matrix. The high concentration of nanocellulose has the potential to create microspaces between the nanofiller and polymer matrix. Moreover, it may tend to aggregation, thereby weakening the interaction between polymer chains and causing a reduction in the mechanical properties [21]. For instance, the incorporation of nanocellulose into chitosan film resulted in an increase in tensile strength (TS) from 8.4 to 12.6 MPa when the nanocellulose content was raised from 2 to 7%. However, the TS decreased to 8.9 MPa when the nanocellulose content reached 12% [80]. Also, the TS of the gelatin/polyvinyl alcohol blend increased from 21.8 MPa to 26.5 MPa by increasing the concentration of bacterial nanocellulose up to 7.5% and decreased again to 21.1 when the concentration reached 10% [19]. Not only concentration, but also the source, nanocellulose properties (size, shape, zeta potential, etc.), and isolation methods play an important role in the mechanical properties of packaging materials. Oun and Rhim [76] examined the effect of oxidized nanocellulose derived from cotton linter (CL) and microcrystalline cellulose (MCC) on the TS of carboxymethyl cellulose (CMC) films. The results demonstrated that the TS of CMC films increased with an increase in the content of CNCs up to 10 wt%. Notably, CNCs obtained from CL exhibited greater effectiveness in enhancing the TS of CMC-based composite films compared to CNCs derived from MCC [76]. According to the findings of Wang et al. [81], the size of nanocellulose has a significant impact on the tensile strength (TS) properties of alginate. The TS of the film exhibited an increase from 45.6 MPa to 54.9 MPa when the cellulose nanowhiskers (with a length of 200–400 nm and a diameter of 15–30 nm) were incorporated at a 4% concentration. However, it decreased to 47 MPa when the concentration was increased to 6%. Interestingly, the TS decreased further after the addition of cellulose fibers, which were longer and had a diameter of 10–20 μm. The researchers suggested that the small size of cellulose nanowhiskers resulted in a higher number of -OH groups available for interaction with the alginate polymer matrix through hydrogen bonding [81]. Based on the available evidence, it can be inferred that various properties of nanocellulose, including its source, size, isolation method, surface modification, as well as the preparation method of nanocellulose-polymer composites, and the type of polymer employed, have a substantial impact on the mechanical properties of packaging films.

4.2 Improving Water Vapor and Oxygen Barrier Properties

The water vapor and oxygen barrier properties of biodegradable food packaging films play a critical role in preserving the freshness and quality of packaged food products. However, high water vapor and oxygen permeability are considered one of the drawbacks of these films. To address this issue, researchers have utilized nanocellulose and other nanomaterials to achieve low water vapor permeability.

One of the challenges in using poly(lactic acid) (PLA) film for food packaging is its low water and gas barrier properties. To address this issue, researchers have investigated the impact of different sources, shapes, concentrations, isolation methods, and surface modifications of nanocellulose on enhancing the barrier properties of PLA. The results have demonstrated improvements in the water and gas barrier properties of PLA film following the incorporation of nanocellulose materials, particularly in relation to oxygen gas [82]. For instance, nanofibrillated cellulose (NFC) or carboxymethyl cellulose (CMC) with cationic polyethyleneimine (PEI) was deposited onto PLA using the layer-by-layer deposition method. The oxygen permeability of the uncoated PLA film was 151.5 7 cm^3/m^2·day·atm, which decreased to 9.0 and 10.7 cm^3/m^2·day·atm after the deposition of 50 bilayers of NFC/PEI and CMC/PEI, respectively. Additionally, the water vapor permeability (WVP) of the uncoated PLA decreased from 4.8 to 2.3 and 2.5 g/m^2·day·atm after the deposition of 50 bilayers of NFC/PEI and CMC/PEI, respectively [83]. In another study, PLA/cellulose nanocrystal (CNC) films were prepared using the reactive extrusion method, and the obtained films exhibited a reduction in both water vapor barrier properties (27–50%) and oxygen permeability (20–65%), depending on the type of filler [84]. The water vapor permeability of the gelatin-polyvinyl alcohol blend film decreased by 14% after the incorporation of bacterial cellulose nanowhiskers (7.5%) [19]. Chen et al. [20] found that the water vapor of starch film decreased by 27.6%, 21.9%, and 18.9% and oxygen permeability decreased by 32.9%, 23.6%, and 15.3% after adding 4 wt.% of bamboo, cotton linter, and sisal nanocellulose, respectively [20]. Also, the water vapor permeability of pectin film supported by 5% nanocellulose significantly decreased from 1.28×10^{-10} g/m s Pa to 9.3×10^{-11} g/m s Pa [85]. Generally, the improvement in water and gas barrier properties of films after incorporation of nanocellulose materials is attributed to the well-distributed nanocellulose within the polymer matrix, which results in the formation of a tortuous diffusion path [16]. Interestingly, the coating of paperboard with TEMPO-oxidized cellulose nanofibers-polypyrrole led to the preservation of the texture and quality of stored cherry tomatoes for 10 days, in comparison to an uncoated paperboard, as shown in Fig. 3. They revealed that to improve the moisture and oxygen barrier properties of paperboard by network formed by the combination of cellulose nanofibers-polypyrrole [86].

4.3 Improving Thermal Properties

Investigating the thermal characteristics of reinforcing agents like cellulose-based materials is critical for evaluating their potential in reinforced-polymer composite manufacturing. This is particularly significant considering that the processing temperature of numerous polymeric materials exceeds 200 °C. The thermal stability of cellulosic materials isolated from agricultural wastes is increased after the removal of low thermal stability components like hemicellulose, lignin, and pectin [75, 87, 88]. It has been reported that chemical and mechanical treatments significantly affect the thermal stability of isolated cellulosic materials [10]. For example, the alkaline treatment showed an improvement in thermal stability after the alkalization process. It decreased with further acid and oxidation treatments due to the presence of sulfate ester groups and carboxyl groups, respectively [21, 89]. Nanocellulose materials have been used to improve the thermal stability of polymer-biocomposite films. The thermal stability of polyvinyl alcohol/chitosan blended film has been improved by the incorporation of cellulose nanocrystals (CNC) and the maximum degradation temperature (T_{max}) of control film was 300 °C and increased to 350, 356, and 367 °C with the addition of CNC by 1, 3, and 5 wt%, respectively [90]. Also, Cao et al. found that the T_{max} of pure waterborne polyurethane (WPU) film increased from 337

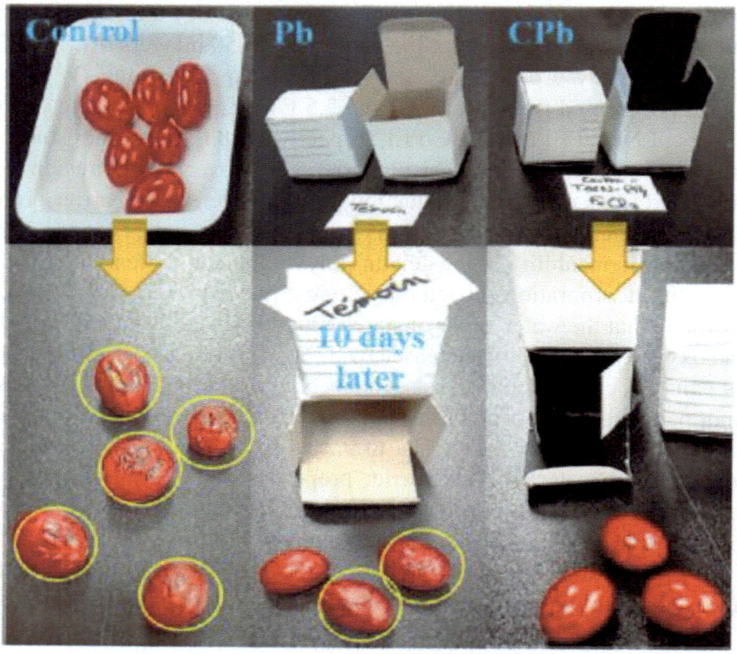

Fig. 3 Cherry tomatoes packaged for 10 days in control, with paperboard (Pb) and coated paperboard (CPb) boxes [86]

to 400 °C after incorporation of CNC [91]. However, nanocellulose prepared using the regeneration method with NaOH/Urea showed only a negligible enhancement in the thermal stability of agar film [92]. On the other hand, the thermal stability of starch-based composite films was reduced upon the incorporation of TEMPO-oxidized nanocellulose isolated from different sources such as cotton linter, bamboo, and sisal [20]. To address this issue, surface modification of cellulosic materials has been utilized as a method to enhance the thermal stability of cellulose-based materials. For example, malic acid and silane coupling agents have been employed to modify the hydroxyl groups of cellulose nanofibrils (CNFs). As a result, the thermal stability of cassava starch has been improved by 36.9% through the incorporation of modified-CNFs [93]. Based on the previous results, it is evident that the impact of cellulosic materials on the thermal stability of packaging films exhibits various behaviors, which depend on factors such as the source of cellulose, isolation methods employed, properties of nanocellulose, as well as the type of polymers involved.

4.4 Nanocellulose-Based Active Packaging

Active packaging refers to an advanced type of packaging system that is based on the incorporation of functional materials that actively interact with the packaged product or its surrounding environment to extend shelf life, enhance safety, and improve overall quality [94, 95]. In this regard, various cellulosic materials have been used to improve the dispersibility and slow release of functional materials like essential oils and extracts [96], metal and metal oxide nanoparticles [97] with high antimicrobial and antioxidant properties, UV-barrier, water vapor, and gas barrier, and good mechanical properties (Fig. 4). For this, cellulosic materials in various forms and properties have been used as a carrier for those functional materials. For instance, modified regenerated cellulose film prepared from cotton fiber and anchored with nanosilver exhibited antimicrobial activity and kept on the freshness of cherry tomatoes for nine days [98]. Cellulose-based film loaded with oregano essential oil showed strong antibacterial activity and reached 99.9% against *E. coli* and *Listeria monocytogenes* [99]. Cellulose film loaded with poacic acid has been shown to effectively suppress the growth of *Staphylococcus aureus* [100]. Recently, researchers have been intrigued by the distinctive properties of nanocellulose, such as non-toxicity, high surface area, and abundant OH groups. These properties have led researchers to utilize nanocellulose as a reducing and capping agent in the synthesis of inorganic nanoparticles [47]. In this regard, cellulosic materials have been loaded with metal and metal oxide NPs such as Ag NPs, CuO NPs, ZnO NPs, etc. [29, 47]. TEMPO-oxidized cellulose nanocrystals were used as a green reducing and capping agent for preparing AgNPs [101]. Also, various types of nanocellulose like bacterial nanocellulose, cellulose nanofibrils, and cellulose nanocrystals are used for the synthesis of ZnO NPs [102, 103]. Cellulose/metal and metal oxide nanoparticles (NPs) are employed in the preparation of multifunctional packaging materials. Biodegradable films, including PLA, PHBV, PLA-PBAT, PVOH, and PVOH-chitosan, supported with cellulose/metal and metal oxide NPs, exhibit enhanced UV-barrier capabilities, antimicrobial properties as well as mechanical strength of active packaging films [47]. An active film made from bacterial cellulose coated with a polypyrrole-zinc oxide nanocomposite was used for packaging chicken thigh meat. The results showed a decrease in the growth of microbial load, an increase in the shelf life, and the stabilization of the rheological properties of the chicken thigh meat [104].

4.5 Nanocellulose-Based Sensors

Nanocellulose-based food sensors have emerged as a promising technology for food quality monitoring and safety assurance. Nanocellulose can be functionalized with specific materials to detect and quantify various analytes, including gases, volatile compounds, pathogens, and contaminants. The nanocellulose matrix provides a stable and porous platform for immobilizing sensing elements, enabling rapid and

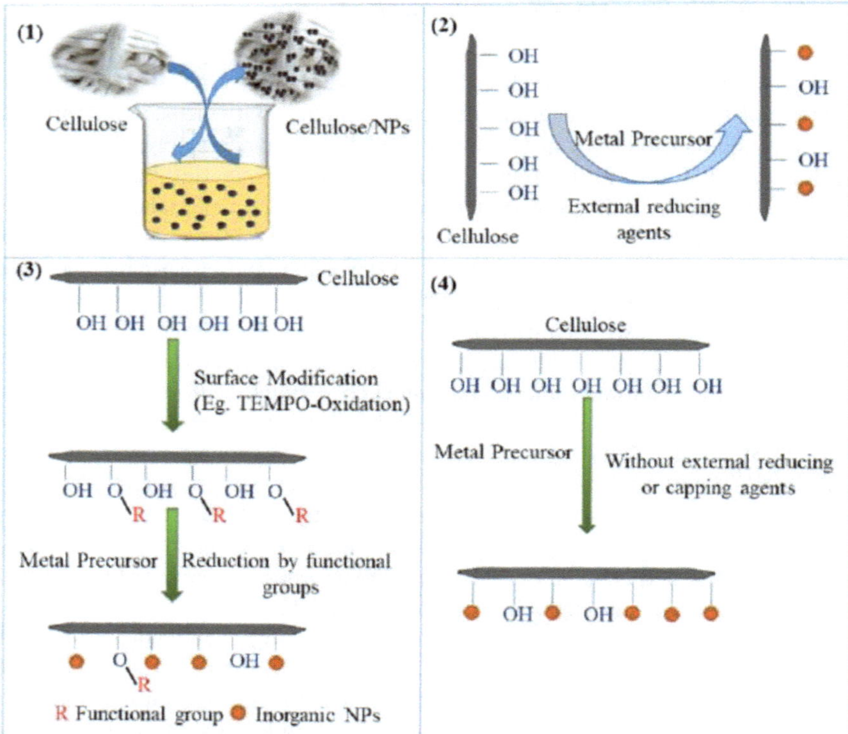

Fig. 4 Synthesis methods of nanocellulose/metal NPs hybrid composite. (1) Simple blinding of cellulose and prepared inorganic NPs, (2) synthesis of NPs using an external reducing agent, (3) synthesis of NPs via nanocellulose surface modification, and (4) synthesis of NPs using nanocelluloses itself without external reducing agents [47]

sensitive detection. They have the potential to revolutionize the food industry by providing real-time, on-site monitoring, ensuring food quality, and enhancing consumer safety [105, 106].

Cellulose materials have been utilized as matrices or carriers to prepare colorimetric pH-sensitive indicators for intelligent food packaging applications. For instance, the freshness of milk and dairy products was determined using colorimetric pH-sensitive indicators made from cellulose/chitosan/black carrot anthocyanins [26], bacterial cellulose/red cabbage extract [107], cellulose filter paper/starch/black carrot anthocyanins [108], cellulose nanocrystal/tara gum/grape skins extract [109], and polyvinylpyrrolidone-carboxymethyl cellulose-bacterial cellulose-guar gum/essential oil/red cabbage anthocyanin [110].

The freshness of pork based on observations of ammonia sensitivity and color change determined using TEMPO-oxidized cellulose nanofibrils loaded with various crosslinkers and anthocyanins, including aronia (*Aronia melanocarpa*), red cabbage

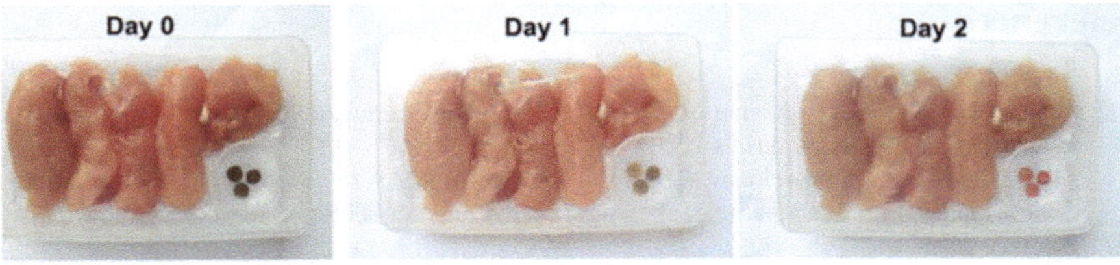

Fig. 5 Color change of hydrogel indicators for chicken breast freshness during storage [113]

(*Brassica oleracea*), and Liriope (*Liriope platyphylla*) [111]. The freshness of rainbow trout fillet was tracked using cellulose paper as a carrier to starch-alizarin solution [22], and bacterial nanocellulose loaded with black carrot anthocyanins [112].

The freshness of chicken breast samples was determined using nanocellulose-based hydrogel isolated from sugarcane bagasse and loaded with pH-responsive dyes (methyl red/ bromothymol blue) [113]. The color changes from green to red when the level of microorganisms exceeds the acceptable limit for human consumption, as demonstrated in Fig. 5. While a cellulose-chitosan film incorporated with alizarin was employed to track the freshness of minced beef. The color of the film converted from brown to purple when the total volatile basic nitrogen level reached the critical threshold of 20.53 mg N/100 g [114]. Moreover, cellulose has been loaded with various types of natural pigments, including water-soluble pigments such as aronia, red cabbage, and butterfly pea, as well as alcohol-soluble pigments like alizarin and shikonin. These cellulose-colorant indicators have been utilized to monitor the ripening process of Kimchi [115].

5 Conclusions

Cellulosic materials offer several advantages, including renewability, biodegradability, and excellent barrier properties. Their utilization in food packaging promotes environmental sustainability and contributes to the development of more eco-friendly and effective packaging solutions. Cellulosic materials find applications in barrier coatings, edible films, active packaging, intelligent packaging, sustainable packaging solutions, and nanocellulose-based packaging. However, the cost of production, scalability, and compatibility with existing packaging processes and equipment of nanocellulose-based materials present challenges for commercial viability.

References

1. Silva FAGS, Dourado F, Gama M, Poças F (2020) Nanocellulose bio-based composites for food packaging. Nanomaterials 10:1–29
2. Abdul Khalil HPS, Davoudpour Y, Saurabh CK et al (2016) A review on nanocellulosic fibres as new material for sustainable packaging: process and applications. Renew Sustain Energy Rev 64:823–836
3. Cheng H, Xu H, Julian McClements D, Chen L, Jiao A, Tian Y, Miao M, Jin Z (2022) Recent advances in intelligent food packaging materials: principles, preparation and applications. Food Chem 375:131738
4. Lo HY (2013) Cellulose nanocrystals and self-assembled nanostructures from cotton, rice straw and grape skin: a source perspective. J Mater Sci 48:7837–7846
5. Tanpichai S, Boonmahitthisud A, Soykeabkaew N, Ongthip L (2022) Review of the recent developments in all-cellulose nanocomposites: properties and applications. Carbohydr Polym 286:119192
6. Trache D, Hussin MH, Haafiz MKM, Thakur VK (2017) Recent progress in cellulose nanocrystals: sources and production. Nanoscale 24:17–25
7. Qasim U, Osman AI, Al-Muhtaseb AH, Farrell C, Al-Abri M, Ali M, Vo DVN, Jamil F, Rooney DW (2021) Renewable cellulosic nanocomposites for food packaging to avoid fossil fuel plastic pollution: a review. Environ Chem Lett 19:613–641
8. Saedi S, Garcia CV, Kim JT, Shin GH (2021) Physical and chemical modifications of cellulose fibers for food packaging applications. Cellulose 28:8877–8897
9. Ventura-Cruz S, Tecante A (2021) Nanocellulose and microcrystalline cellulose from agricultural waste: review on isolation and application as reinforcement in polymeric matrices. Food Hydrocoll 118:106771
10. Oun AA, Rhim J-WW (2018) Isolation of oxidized nanocellulose from rice straw using the ammonium persulfate method. Cellulose 25:2143–2149
11. Rambabu N, Panthapulakkal S, Sain M, Dalai AK (2016) Production of nanocellulose fibers from pinecone biomass: evaluation and optimization of chemical and mechanical treatment conditions on mechanical properties of nanocellulose films. Ind Crops Prod 83:746–754
12. Yu S, Sun J, Shi Y, Wang Q, Wu J, Liu J (2021) Nanocellulose from various biomass wastes: its preparation and potential usages towards the high value-added products. Environ Sci Ecotechnol 5:100077
13. Zaaba NF, Jaafar M, Ismail H (2021) Tensile and morphological properties of nanocrystalline cellulose and nanofibrillated cellulose reinforced PLA bionanocomposites: a review. Polym Eng Sci 61:22–38
14. Wu H, Liu X, Hua X, Zhang J (2023) Thermal-oxidation degradation of polylactic acid/cellulose nanocrystal composites: effects of surface chemistry. Ind Crops Prod 202:117008
15. Kian LK, Saba N, Jawaid M, Sultan MTH (2019) A review on processing techniques of bast fibers nanocellulose and its polylactic acid (PLA) nanocomposites. Int J Biol Macromol 121:1314–1328
16. Oun AA, Rhim JW (2016) Isolation of cellulose nanocrystals from grain straws and their use for the preparation of carboxymethyl cellulose-based nanocomposite films. Carbohydr Polym 150:187–200
17. Oun AA, Shin GH, Kim JT (2022) Antimicrobial, antioxidant, and pH-sensitive polyvinyl alcohol/chitosan-based composite films with aronia extract, cellulose nanocrystals, and grapefruit seed extract. Int J Biol Macromol 213:381–393
18. Oun AA, Hwa G, Rhim J, Tae J (2022) Recent advances in polyvinyl alcohol-based composite films and their applications in food packaging. Food Packag Shelf Life 34:100991
19. Haghighi H, Gullo M, La China S, Pfeifer F, Siesler HW, Licciardello F, Pulvirenti A (2021) Characterization of bio-nanocomposite films based on gelatin/polyvinyl alcohol blend reinforced with bacterial cellulose nanowhiskers for food packaging applications. Food Hydrocoll 113:106454

20. Chen Q, Liu Y, Chen G (2019) A comparative study on the starch-based biocomposite films reinforced by nanocellulose prepared from different non-wood fibers. Cellulose 26:2425–2435
21. Oun AA, Rhim J-W (2015) Effect of post-treatments and concentration of cotton linter cellulose nanocrystals on the properties of agar-based nanocomposite films. Carbohydr Polym 134:20–29
22. Ezati P, Tajik H, Moradi M, Molaei R (2019) Intelligent pH-sensitive indicator based on starch-cellulose and alizarin dye to track freshness of rainbow trout fillet. Int J Biol Macromol 132:157–165
23. Freitas PAV, Silva RRA, de Oliveira TV, Soares RRA, Junior NS, Moraes ARF, Pires AC, Soares NFF (2020) Development and characterization of intelligent cellulose acetate-based films using red cabbage extract for visual detection of volatile bases. Lwt 132:109780
24. Aghaei Z, Emadzadeh B, Ghorani B, Kadkhodaee R (2018) Cellulose acetate nanofibres containing alizarin as a halochromic sensor for the qualitative assessment of rainbow trout fish spoilage. Food Bioprocess Technol 11:1087–1095
25. Boonsiriwit A, Lee M, Kim M, Inthamat P, Siripatrawan U, Lee YS (2021) Hydroxypropyl methylcellulose/microcrystalline cellulose biocomposite film incorporated with butterfly pea anthocyanin as a sustainable pH-responsive indicator for intelligent food-packaging applications. Food Biosci 44:101392
26. Ebrahimi Tirtashi F, Moradi M, Tajik H, Forough M, Ezati P, Kuswandi B (2019) Cellulose/chitosan pH-responsive indicator incorporated with carrot anthocyanins for intelligent food packaging. Int J Biol Macromol 136:920–926
27. Yang Y, Liu H, Wu M, Ma J, Lu P (2020) Bio-based antimicrobial packaging from sugarcane bagasse nanocellulose/nisin hybrid films. Int J Biol Macromol 161:627–635
28. Söğüt E (2019) Properties of solvent cast polycaprolactone films containing pomegranate seed oil stabilized with nanocellulose. Turk J Agric Food Sci Technol 7:220–233
29. Shankar S, Oun AA, Rhim J-W (2018) Preparation of antimicrobial hybrid nano-materials using regenerated cellulose and metallic nanoparticles. Int J Biol Macromol 107:17–27
30. Oun AA, Shin GH, Kim JT (2022) Multifunctional poly(vinyl alcohol) films using cellulose nanocrystals/oregano and cellulose nanocrystals/cinnamon pickering emulsions: effect of oil type and concentration. Int J Biol Macromol 194:736–745
31. Garrido-Romero M, Aguado R, Moral A, Brindley C, Ballesteros M (2022) From traditional paper to nanocomposite films: analysis of global research into cellulose for food packaging. Food Packag Shelf Life 31:100788
32. Baghaei B, Skrifvars M (2020) All-cellulose composites: a review of recent studies on structure, properties and applications. Molecules. https://doi.org/10.3390/molecules25122836
33. Sacui IA, Nieuwendaal RC, Burnett DJ, Stranick SJ, Jorfi M, Weder C, Foster EJ, Olsson RT, Gilman JW (2014) Comparison of the properties of cellulose nanocrystals and cellulose nanofibrils isolated from bacteria, tunicate, and wood processed using acid, enzymatic, mechanical, and oxidative methods. ACS Appl Mater Interfaces 6:6127–6138
34. Chopra L, Manikanika (2021) Extraction of cellulosic fibers from the natural resources: a short review. Mater Today Proc 48:1265–1270
35. Shaghaleh H, Xu X, Wang S (2018) Current progress in production of biopolymeric materials based on cellulose, cellulose nanofibers, and cellulose derivatives. RSC Adv 8:825–842
36. Jonoobi M, Oladi R, Davoudpour Y, Oksman K, Dufresne A, Hamzeh Y, Davoodi R (2015) Different preparation methods and properties of nanostructured cellulose from various natural resources and residues: a review. Cellulose 22:935–969
37. Rezanezhad S, Nazanezhad N, Asadpur G (2013) Isolation of nanocellulose from rice waste via ultrasonication. Lignocellulose 2:282–291
38. Wang J, Wang L, Gardner DJ, Shaler SM, Cai Z (2021) Towards a cellulose-based society: opportunities and challenges. Cellulose. https://doi.org/10.1007/s10570-021-03771-4
39. Zuppolini S, Salama A, Cruz-Maya I, Guarino V, Borriello A (2022) Cellulose amphiphilic materials: chemistry, process and applications. Pharmaceutics 14:386

40. Fritz C, Jeuck B, Salas C, Gonzalez R, Jameel H, Rojas OJ (2016) Cellulose Chemistry and properties: fibers, nanocelluloses and advanced materials
41. Gupta PK, Uniyal V, Naithani S (2013) Polymorphic transformation of cellulose I to cellulose II by alkali pretreatment and urea as an additive. Carbohydr Polym 94:843–849
42. Wang S, Lu A, Zhang L (2016) Recent advances in regenerated cellulose materials. Prog Polym Sci 53:169–206
43. He X, Lu W, Sun C, Khalesi H, Mata A, Andaleeb R, Fang Y (2021) Cellulose and cellulose derivatives: different colloidal states and food-related applications. Carbohydr Polym 255:117334
44. Dufresne A (2008) Cellulose-based composites and nanocomposites. In: Gandini A, Belgacem MN (eds) Monomers, polymers and composites from renewable resources. Elsevier, Oxford, UK, pp 401–418
45. Tu H, Zhu M, Duan B, Zhang L (2021) Recent progress in high-strength and robust regenerated cellulose materials. Adv Mater 33:1–22
46. Wan C, Jiao Y, Wei S, Zhang L, Wu Y, Li J (2019) Functional nanocomposites from sustainable regenerated cellulose aerogels: a review. Chem Eng J 359:459–475
47. Oun AA, Shankar S, Rhim J-W (2020) Multifunctional nanocellulose/metal and metal oxide nanoparticle hybrid nanomaterials. Crit Rev Food Sci Nutr 60:435–460
48. Hubbe MA, Ayoub A, Daystar JS, Venditti RA, Pawlak JJ (2013) Enhanced absorbent products incorporating cellulose and its derivatives: a review. BioResources 8:6556–6629
49. Rol F, Belgacem MN, Gandini A, Bras J (2019) Recent advances in surface-modified cellulose nanofibrils. Prog Polym Sci 88:241–264
50. Ben HK, Kaddami H, Dufresne A, Boufi S, Magnin A, Erchiqui F (2018) Impact of TEMPO-oxidization strength on the properties of cellulose nanofibril reinforced polyvinyl acetate nanocomposites. Carbohydr Polym 181:1061–1070
51. Koga H, Saito T, Kitaoka T, Nogi M, Suganuma K, Isogai A (2013) Transparent, conductive, and printable composites consisting of TEMPO-oxidized nanocellulose and carbon nanotube. Biomacromol 14:1160–1165
52. Zhang N, Zang GL, Shi C, Yu HQ, Sheng GP (2016) A novel adsorbent TEMPO-mediated oxidized cellulose nanofibrils modified with PEI: preparation, characterization, and application for Cu(II) removal. J Hazard Mater 316:11–18
53. Songfeng E, Ning D, Wang Y, Huang J, Jin Z, Ma Q, Yang K, Lu Z (2020) Ternary synergistic strengthening and toughening of bio-inspired TEMPO-oxidized cellulose nanofibers/borax/polyvinyl alcohol composite film with high transparency. ACS Sustain Chem Eng 8:15661–15669
54. Cheng D, Wen Y, An X, Zhu X, Ni Y (2016) TEMPO-oxidized cellulose nanofibers (TOCNs) as a green reinforcement for waterborne polyurethane coating (WPU) on wood. Carbohydr Polym 151:326–334
55. Ji H, Xiang Z, Qi H, Han T, Pranovich A, Song T (2019) Strategy towards one-step preparation of carboxylic cellulose nanocrystals and nanofibrils with high yield, carboxylation and highly stable dispersibility using innocuous citric acid. Green Chem 21:1956–1964
56. Naderi A, Lindström T, Sundström J, Flodberg G, Erlandsson J (2016) A comparative study of the properties of three nanofibrillated cellulose systems that have been produced at about the same energy consumption levels in the mechanical delamination step. Nord Pulp Pap Res J 31:364–371
57. Liyanage S, Acharya S, Parajuli P, Shamshina JL, Abidi N (2021) Production and surface modification of cellulose bioproducts. Polymers (Basel) 13:3433
58. Liu Y, Ahmed S, Sameen DE, Wang Y, Lu R, Dai J, Li S, Qin W (2021) A review of cellulose and its derivatives in biopolymer-based for food packaging application. Trends Food Sci Technol 112:532–546
59. Liesiene J (2010) Synthesis of water-soluble cationic cellulose derivatives with tertiary amino groups. Cellulose 17:167–172
60. Arancibia C, Navarro-Lisboa R, Zúñiga RN, Matiacevich S (2016) Application of CMC as thickener on nanoemulsions based on olive oil: physical properties and stability. Int J Polym Sci 2016:6280581

61. Meng Z, Qi K, Guo Y, Wang Y, Liu Y (2018) Effects of thickening agents on the formation and properties of edible oleogels based on hydroxypropyl methyl cellulose. Food Chem 246:137–149
62. Xu Y, Sun H, Lv J, Wang Y, Zhang Y, Wang F (2023) Effects of polysaccharide thickening agent on the preparation of walnut oil oleogels based on methylcellulose: characterization and delivery of curcumin. Int J Biol Macromol 232:123291
63. Vatanpour V, Pasaoglu ME, Barzegar H, Teber OO, Kaya R, Bastug M, Khataee A, Koyuncu I (2022) Cellulose acetate in fabrication of polymeric membranes: a review. Chemosphere 295:133914
64. Sayyed AJ, Deshmukh NA, Pinjari DV (2019) A critical review of manufacturing processes used in regenerated cellulosic fibres: viscose, cellulose acetate, cuprammonium, LiCl/DMAc, ionic liquids, and NMMO based lyocell. Cellulose 26:2913–2940
65. Tang Y, Yang H, Vignolini S (2022) Recent progress in production methods for cellulose nanocrystals: leading to more sustainable processes. Adv Sustain Syst. https://doi.org/10.1002/adsu.202100100
66. Sehaqui H, Mautner A, Perez De Larraya U, Pfenninger N, Tingaut P, Zimmermann T (2016) Cationic cellulose nanofibers from waste pulp residues and their nitrate, fluoride, sulphate and phosphate adsorption properties. Carbohydr Polym 135:334–340
67. Adel A, El-shafei A, Ibrahim A, Al-shemy M (2018) Extraction of oxidized nanocellulose from date palm (Phoenix Dactylifera L.) sheath fibers: Influence of CI and CII polymorphs on the properties of chitosan/bionanocomposite film. Ind Crop Prod 124:155–165
68. Noguchi Y, Homma I, Matsubara Y (2017) Complete nanofibrillation of cellulose prepared by phosphorylation. Cellulose 24:1–10
69. Qiu X, Hu S (2013) "Smart" materials based on cellulose: a review of the preparations, properties, and applications. Materials (Basel) 6:738–781
70. Bangar SP, Harussani MM, Ilyas RA, Ashogbon AO, Singh A, Trif M, Jafari SM (2022) Surface modifications of cellulose nanocrystals: processes, properties, and applications. Food Hydrocoll 130:107689
71. Market Research Report (2020) Cellulose market size, share, growth. Industry Report 2026. https://www.fortunebusinessinsights.com/cellulose-market-102062
72. Vilarinho F, Sanches Silva A, Vaz MF, Farinha JP (2017) Nanocellulose in green food packaging. Crit Rev Food Sci Nutr 8398:1–12
73. Market Research Report (2023) Nanocellulose market by type (Microfibrillated cellulose [MFC], Cellulose nanocrystals [CNC], Bacterial nanocellulose [BNC]), Application (composites & packaging, paper & board, paints & coatings, oil & gas, personal care & cosmetics), and region—global. https://www.marketsandmarkets.com/Market-Reports/nanocellulose-market-56392090.html
74. Punia S, Harussani MMM, Ilyas RAA et al (2022) Surface modifications of cellulose nanocrystals: processes, properties, and applications. Food Hydrocoll 130:107689
75. Amjad A, Anjang Ab Rahman A, Awais H, Zainol Abidin MS, Khan J (2022) A review investigating the influence of nanofiller addition on the mechanical, thermal and water absorption properties of cellulosic fibre reinforced polymer composite. J Ind Text 51:65S-100S
76. Oun AA, Rhim J-W (2017) Characterization of carboxymethyl cellulose-based nanocomposite films reinforced with oxidized nanocellulose isolated using ammonium persulfate method. Carbohydr Polym 174:484–492
77. Vilarinho F, Sanches Silva A, Vaz MF, Farinha JP (2018) Nanocellulose in green food packaging. Crit Rev Food Sci Nutr 58:1526–1537
78. Bharimalla AK, Deshmukh SP, Vigneshwaran N, Patil PG, Prasad V (2017) Nanocellulose-polymer composites for applications in food packaging: current status, future prospects and challenges. Polym Plast Technol Eng 56:805–823
79. Andrade MS, Ishikawa OH, Costa RS, Seixas MVS, Rodrigues RCLB, Moura EAB (2022) Development of sustainable food packaging material based on biodegradable polymer reinforced with cellulose nanocrystals. Food Packag Shelf Life. https://doi.org/10.1016/j.fpsl.2021.100807

80. Pereda M, Dufresne A, Aranguren MI, Marcovich NE (2014) Polyelectrolyte films based on chitosan/olive oil and reinforced with cellulose nanocrystals. Carbohydr Polym 101:1018–1026
81. Wang L, Shankar S, Rhim J (2017) Food hydrocolloids properties of alginate-based films reinforced with cellulose fibers and cellulose nanowhiskers isolated from mulberry pulp. Food Hydrocoll 63:201–208
82. Khosravi A, Fereidoon A, Khorasani MM, Naderi G, Ganjali MR, Zarrintaj P, Saeb MR, Gutiérrez TJ (2020) Soft and hard sections from cellulose-reinforced poly(lactic acid)-based food packaging films: a critical review. Food Packag Shelf Life 23:100429
83. Aulin C, Karabulut E, Tran A, Waìšgberg L, Lindström T (2013) Transparent nanocellulosic multilayer thin films on polylactic acid with tunable gas barrier properties. ACS Appl Mater Interfaces 5:7352–7359
84. Dhar P, Gaur SS, Soundararajan N, Gupta A, Bhasney SM, Milli M, Kumar A, Katiyar V (2017) Reactive extrusion of polylactic acid/cellulose nanocrystal films for food packaging applications: influence of filler type on thermomechanical, rheological, and barrier properties. Ind Eng Chem Res 56:4718–4735
85. Chaichi M, Hashemi M, Badii F, Mohammadi A (2017) Preparation and characterization of a novel bionanocomposite edible film based on pectin and crystalline nanocellulose. Carbohydr Polym 157:167–175
86. Bideau B, Loranger E, Daneault C (2018) Nanocellulose-polypyrrole-coated paperboard for food packaging application. Prog Org Coatings 123:128–133
87. Alemdar A, Sain M (2008) Biocomposites from wheat straw nanofibers: Morphology, thermal and mechanical properties. Compos Sci Technol 68:557–565
88. Chen X, Yu J, Zhang Z, Lu C (2011) Study on structure and thermal stability properties of cellulose fibers from rice straw. Carbohydr Polym 85:245–250
89. Mondragon G, Fernandes S, Retegi A, Peña C, Algar I, Eceiza A, Arbelaiz A (2014) A common strategy to extracting cellulose nanoentities from different plants. Ind Crops Prod 55:140–148
90. Perumal AB, Sellamuthu PS, Nambiar RB, Sadiku ER (2018) Development of polyvinyl alcohol/chitosan bio-nanocomposite films reinforced with cellulose nanocrystals isolated from rice straw. Appl Surf Sci 449:591–602
91. Cao X, Habibi Y, Lucia LA (2009) One-pot polymerization, surface grafting, and processing of waterborne polyurethane-cellulose nanocrystal nanocomposites. J Mater Chem 19:7137–7145
92. Shankar S, Rhim JW (2016) Preparation of nanocellulose from micro-crystalline cellulose: the effect on the performance and properties of agar-based composite films. Carbohydr Polym 135:18–26
93. Huang L, Zhao H, Yi T, Qi M, Xu H, Mo Q, Huang C, Wang S, Liu Y (2020) Reparation and properties of cassava residue cellulose nanofibril/cassava starch composite films. Nanomaterials. https://doi.org/10.3390/nano10040755
94. Vilela C, Kurek M, Hayouka Z, Röcker B, Yildirim S, Antunes MDC, Nilsen-Nygaard J, Pettersen MK, Freire CSR (2018) A concise guide to active agents for active food packaging. Trends Food Sci Technol 80:212–222
95. Yildirim S, Röcker B, Pettersen MK, Nilsen-Nygaard J, Ayhan Z, Rutkaite R, Radusin T, Suminska P, Marcos B, Coma V (2018) Active packaging applications for food. Compr Rev Food Sci Food Saf 17:165–199
96. Casalini S, Giacinti Baschetti M (2022) The use of essential oils in chitosan or cellulose-based materials for the production of active food packaging solutions: a review. J Sci Food Agric. https://doi.org/10.1002/jsfa.11918
97. Oun AA, Rhim J-W (2017) Preparation of multifunctional chitin nanowhiskers/ZnO-Ag NPs and their effect on the properties of carboxymethyl cellulose-based nanocomposite film. Carbohydr Polym 169:467–479
98. Gu R, Yun H, Chen L, Wang Q, Huang X (2020) Regenerated cellulose films with amino-terminated hyperbranched polyamic anchored nanosilver for active food packaging. ACS Appl Bio Mater 3:602–610

99. Chen S, Wu M, Lu P, Gao L, Yan S, Wang S (2020) Development of pH indicator and antimicrobial cellulose nanofibre packaging film based on purple sweet potato anthocyanin and oregano essential oil. Int J Biol Macromol 149:271–280
100. Hu Y, Guo Q, Liu P, Zhu R, Lu F, Ramaswamy S, Wu Y, Xu F, Zhang X (2021) Fabrication of novel cellulose-based antibacterial film loaded with poacic acid against staphylococcus aureus. J Polym Environ 29:745–754
101. Pawcenis D, Chlebda DK, Jędrzejczyk RJ, Leśniak M, Sitarz M, Łojewska J (2019) Preparation of silver nanoparticles using different fractions of TEMPO-oxidized nanocellulose. Eur Polym J 116:242–255
102. Shi C, An B, Zhang L, Zai Z, Shi Z, Wang Z, Ma J (2023) Contribution of surface carboxyl of cellulose in the formation mechanism and interfacial catalysis activity of ZnO/cellulose nanocomposites. Appl Surf Sci 618:156633
103. Farooq A, Patoary MK, Zhang M, Mussana H, Li M, Naeem MA, Mushtaq M, Farooq A, Liu L (2020) Cellulose from sources to nanocellulose and an overview of synthesis and properties of nanocellulose/zinc oxide nanocomposite materials. Int J Biol Macromol 154:1050–1073
104. Pirsa S, Shamusi T (2019) Intelligent and active packaging of chicken thigh meat by conducting nano structure cellulose-polypyrrole-ZnO film. Mater Sci Eng C 102:798–809
105. Liu D, Zhang C, Pu Y, Chen S, Liu L, Cui Z, Zhong Y (2022) Recent advances in pH-responsive freshness indicators using natural food colorants to monitor food freshness. Foods 11:1884
106. Roy S, Rhim J-W (2021) Anthocyanin food colorant and its application in pH-responsive color change indicator films. Crit Rev Food Sci Nutr 61:2297–2325
107. Kuswandi B, Asih NPN, Pratoko DK, Kristiningrum N, Moradi M (2020) Edible pH sensor based on immobilized red cabbage anthocyanins into bacterial cellulose membrane for intelligent food packaging. Packag Technol Sci 33:321–332
108. Moazami Goodarzi M, Moradi M, Tajik H, Forough M, Ezati P, Kuswandi B (2020) Development of an easy-to-use colorimetric pH label with starch and carrot anthocyanins for milk shelf life assessment. Int J Biol Macromol 153:240–247
109. Ma Q, Wang L (2016) Preparation of a visual pH-sensing film based on tara gum incorporating cellulose and extracts from grape skins. Sensors Actuators B Chem 235:401–407
110. Bandyopadhyay S, Saha N, Zandraa O, Pummerová M, Sáha P (2020) Essential oil based PVP-CMC-BC-GG functional hydrogel sachet for 'cheese': Its shelf life confirmed with anthocyanin (isolated from red cabbage) bio stickers. Foods 9:1–20
111. Pirayesh H, Park BD, Khanjanzadeh H, Park HJ, Cho YJ (2023) Nanocellulose-based ammonia sensitive smart colorimetric hydrogels integrated with anthocyanins to monitor pork freshness. Food Control 147:109595
112. Moradi M, Tajik H, Almasi H, Forough M, Ezati P (2019) A novel pH-sensing indicator based on bacterial cellulose nanofibers and black carrot anthocyanins for monitoring fish freshness. Carbohydr Polym 222:115030
113. Lu P, Yang Y, Liu R, Liu X, Ma J, Wu M, Wang S (2020) Preparation of sugarcane bagasse nanocellulose hydrogel as a colourimetric freshness indicator for intelligent food packaging. Carbohydr Polym 249:116831
114. Ezati P, Tajik H, Moradi M (2019) Fabrication and characterization of alizarin colorimetric indicator based on cellulose-chitosan to monitor the freshness of minced beef. Sens Actuators B Chem 285:519–528
115. Oun AA, Roy S, Shin GH, Yoo SR, Kim JT (2023) PH-sensitive smart indicators based on cellulose and different natural pigments for tracing kimchi ripening stages. Int J Biol Macromol 242:124905

Chitosan-Based Films for Food Packaging Applications

Shweta Sharma, Sahil Kohli, Divya Tripathi Bajpai, Anurag Singh, and Vivek Kumar

Abstract Chitosan is a derivative of chitin, acquired as a by-product of processing and handling of fishes and crustaceans. Due to these unique properties, it has various applications in biological and technological research areas. Chitosan is an appealing bio-polymer for the packaging and preservation of food. It possesses antibacterial activity and film-forming properties and can be applied through coating, dipping, spraying, or wrapping in the form of films. This chapter covers various properties of chitosan and its functionalities. Also, chitosan blends and nanocomposites, production applications, and techniques for extending the shelf life of meat, and fish products have also been discussed. Rapid growth is being seen in the demand for safe, wholesome food that uses as few artificial ingredients as possible. Due to the environmental hazards of plastic polymers, much effort has been made to substitute several synthetic plastic packaging materials using bio-based polymers. The current chapter relates to the knowledge of chitosan films in biodegradable and sustainable food preservation and packaging.

Keywords Chitosan · Film · Packaging · Food preservation

S. Sharma
Division of Forensic Sciences, School of Biomedical Sciences, Galgotias University, Greater Noida 203201, Uttar Pradesh, India

S. Kohli (✉)
Research and Development cell, Department of Chemistry, Manav Rachna International Institute of Research and Studies, Faridabad 121004, Haryana, India
e-mail: sahilkohli41@gmail.com; sahil.set@mriu.edu.in

A. Singh (✉) · V. Kumar
Department of Food Technology, Harcourt Butler Technical University, Kanpur 208002, Uttar Pradesh, India
e-mail: anurag.niftem@gmail.com

D. T. Bajpai
Division of Chemistry, School of Basic Sciences, Galgotias University, Uttar Pradesh 203201 Greater Noida, India

M. Alle et al. (eds.), *Functional Nanomaterials and Nanocomposites for Biodegradable Food Packaging*, Smart Nanomaterials Technology,
https://doi.org/10.1007/978-981-97-9084-5_7

1 Introduction

Packaging combines science, art, and technology to securely and affordably convey products and related information to end users. The technique of food packaging is crucial for the preservation of both fresh and processed foods. Making sure that food produced in one place is made accessible to consumers in another location even after months from the initial harvest or manufacturing is vital. The humans who lived as nomads simply collected food whenever they wanted to consume it. It became important to store food for later consumption as a result of the growing demand for homes and shelters as well as the advancement of agriculture. At that point, the demand for packing and storage arose. Leaves, shells, woven grasses, hollowed logs, animal parts, and other materials were used for this purpose [1]. These storage methods were, however, quite primitive, and as the need for advancements and inventions grew, so did the subject of food packaging research. The extensive effects of this commencement gave rise to the packaging materials that finds their place in today's world too.

An ideal material for packaging food is that which maintains food quality with time, is suitable to use, is affordable, and is biodegradable, causing no accumulation of solid waste. However, no such perfect material has yet been created, and each material has advantages and disadvantages that are relative rather than absolute based on these variables.

The most popular and greatest resources for food packaging uses are plastics and polymers. They have been heavily utilized in flexible and rigid packaging. Though plastic materials cannot decompose, it is vital to consider their contribution to the formation of municipal solid waste (MSW) in 2014, according to a report given by the US-based Environmental Protection Agency. The entire amount of MSW produced in 2014 was 258 million tonnes, of which 32.28 million tonnes, or 12.9% of the total, were plastic wastes. 14.32 million tonnes of plastic garbage, or 43% of all plastic waste, were produced by plastic packaging.

Any product that is produced industrially inevitably produces a large number of by-products, most of which are thrown away as waste. Although many byproducts have promise for other uses, they are typically not put to good use. Animal wastes, such as skins and shells of organisms, are produced in large quantities by the seafood processing business. Crustacean shells make up about 75% of the waste by-products, and because waste management techniques are ineffective, they cause major environmental concerns by being drowned in the sea, dumped on land, burned, or allowed to decompose naturally. These animal wastes are a great source of chitin and are high in proteins. It might be possible to reduce environmental risks while promoting value addition by using this trash to extract chitin.

Cell walls of fungus, yeast, and the exoskeletons of insects and crustaceans contain chitin, a biopolymer that resembles cellulose. Similar to how cellulose works as a structural component in plant cells, chitin also serves this purpose. A vast amount of 1011 metric tonnes of this polymer, which is primarily made up of (1-4)-linked 2-acetamido-2-deoxy-D-glucose monomers, is generated every year [2]. Out of this, however, only 150,000 tonnes are made accessible for commercial use [3]. With this quantity, it follows cellulose is the next most prevalent biopolymer on Earth and the most prevalent biopolymer of animal origin. Its nitrogen content sets it apart from other polysaccharides. The primary components of chitosan are (1-4)-linked 2-amino-2-deoxy-D-glucose monomers. The source of the chitin has an effect on its characteristics like the degree of deacetylation (DA), molecular size, nitrogen content, polydispersity, molecular size, and N/C ratio. Chitin usually possesses a nitrogen concentration of about 7%, resulting in an N/C ratio of 0.146, with a degree of acetylation (DA) exceeding 90%. In contrast, chitosan exhibits a nitrogen concentration higher than 7% and a DA lower than 40%. The functional attributes of chitosan stem from the presence of amino groups, imparting a spectrum of chemical, physical, and biological properties owing to their electronegative nature. Chitosan is a multipurpose polymer has been applied in many fields of research, like food packaging, due to its ion binding, film formation, solubility, viscosity, and other characteristics (Fig. 1).

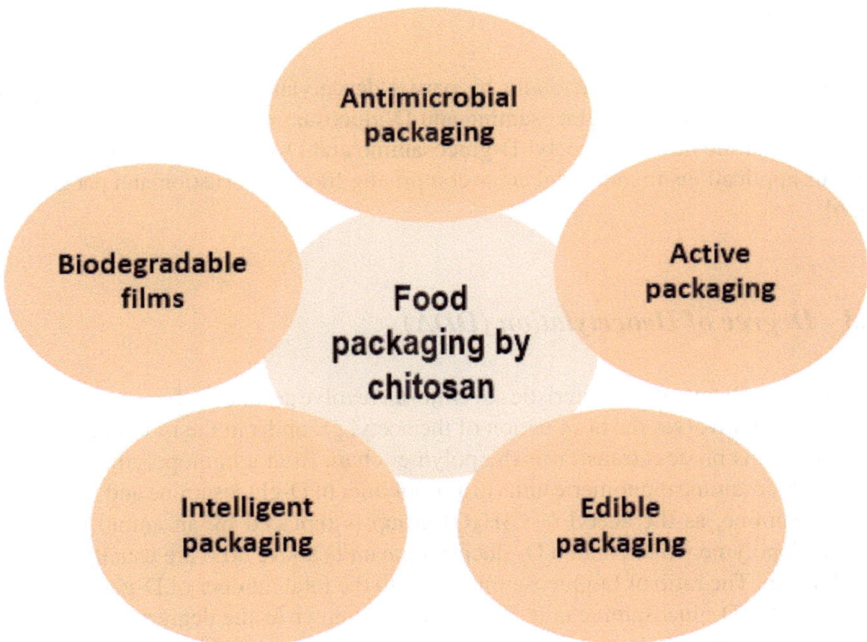

Fig. 1 Types of food packaging by chitosan

2 Properties of Chitosan

2.1 Source

After cellulose, chitin is the second most prevalent polysaccharide on the planet. It comprises of three functional groups which are reactive, i.e., primary and secondary hydroxyl groups and an amino group. As a key component of the exoskeleton and a cell wall component, chitin is mostly located in the exoskeletons of marine invertebrates, insects, along with a few fungi. Every year, people eat marine invertebrates including crabs, prawns, lobsters, and oysters because they are a good source of protein. However, these crustaceans' non-edible components, including their external shells, make up around half of their total mass and are typically discarded as waste [4].

2.2 Structure

Chitosan is acquired commercially by partial deacetylation of chitin, resulting in the formation of N-acetyl-glucosamine and D-glucosamine copolymer. Chitosan is a polysaccharide having N-acetyl D-glucosamine and D-glucosamine units and has various applications in industrial areas comprising food preservation and packaging [5, 6].

2.3 Degree of Deacetylation (DDA)

The primary chemical characteristics of chitosan revolve around its degree of deacetylation, which involves the elimination of the acetyl group from the monomeric units of chitin. This process transforms the polymer chain from a homopolymer with N-acetyl-glucosamine monomeric units to a copolymer of D-glucosamine and N-acetyl-D-glucosamine, as the acetyl ($-C_2H_3O$) group is replaced by an amino ($-NH_2$) group. Copolymers that contain D-glucosamine units above 50% are usually termed as Chitosan. The ratio of D-glucosamine units to the total number of D-glucosamine and N-acetyl-D-glucosamine units in the chain is known as the degree of deacetylation (DD) of chitosan. Hence, 70% degree of deacetylation of chitosan means that it comprises 30% of N-acetyl-glucosamine units and 70% of D-glucosamine units.

2.4 Solubility

Chitosan dissolves at pH 6 or below due to the conversion of the uncharged amino ($-NH_2$) group to the charged $-NH_3^+$ group, which results in repulsion within the polymer chain due to the formation of polycationic structure, subsequently resulting in higher solubility. The degree of deacetylation (DD) of chitosan and pH has an impact on chitosan's solubility. It is slightly more soluble in the acidic medium as its PK_a value is 6.3 [7].

2.5 Chain Conformation

Chitosan is a rigid molecule like all polysaccharides, unlike many artificial water-soluble polymers. Owing to the $^1C_4$ ring conformation in the chain, di-equatorial glycosidic connections exist in the chitosan structure, which accounts for the stiffness. As a result, the chain's rotation is constrained, extending the chain's length and increasing its intrinsic viscosity. In addition, the repulsion between the charged groups on the polymer chain causes an extra ionic strength dependent chain extension to occur when chitosan is in the protonated state [8].

2.6 Chemical Stability

Glycosidic bonds are found within the structure of chitosan, susceptible to oxidation by free radicals and degradation by both acidic and basic conditions. Chitosan's resistance to oxidation is notably lower compared to its resistance to pH changes. The degree of acetylation (DA) in chitosan seems to influence its degradation under acidic conditions, with higher DA values correlating with increased degradation rate constants. Acid hydrolysis is particularly effective in breaking A-A and A-D glycosidic bonds, being three times more efficient compared to D-A and D-D linkages. This discrepancy is attributed to the presence of nearby positively charged amino groups, which hinder the hydrolysis process of the glycosidic bonds [9].

2.7 Film-Forming Properties

The most significant chitosan property, as seen from the standpoint of food packaging in particular, is film formation, which has received much research. Due to chitosan's inherent antibacterial activity as well as its barrier qualities, it has been investigated if the films have the potential to improve food quality. An in-depth discussion of pure chitosan or chitosan membranes, edible coatings, and films will be covered from the perspective of food packaging in later sections.

2.8 Antimicrobial Activity

Chitosan acts against a variety of microorganisms, like fungi, yeast, and many bacteria. Many studies claimed that chitosan has antibacterial capabilities, while the underlying mechanisms are still unclear. Several pathways have been hypothesized as the cause of antimicrobial effects. The most common is the alteration of the membrane barrier characteristics caused by the interaction of negatively charged microbial cell membranes with positively charged amine groups in chitosan, which ultimately results in cell death [1]. Another possible mechanism is based on the properties of chitosan related to chelation. By selectively binding with metals, chitosan slows down the growth of microorganisms by impeding the active centers of several metabolic enzymes found in microbial cells. Chitosan's molecular weight can also affect its efficacy against microbes. While low-molecular-weight chitosan could enter the cytosol and bind with cell DNA, interfering with the synthesis of mRNA and proteins and ultimately leading to cell death, high-molecular-weight chitosan could form an impervious polymeric layer on the surface of the microbial cell, altering the cell permeability and blocking the entry of nutrients [10].

2.9 Antioxidant Activity

Food products are regularly subjected to various oxidative degradation methods. It is generally known that chitin and chitosan have antioxidant properties. These biopolymers' antioxidant properties come from their ability of chelating metal ions and scavenge-free radicals. Various studies have found that as chitin deacetylation levels rise, antioxidant activities noticeably improve, including the elimination of free radicals and metal ion chelation. Chitosan has emerged as a viable alternative antioxidant in food formulations, addressing increasing public apprehensions regarding the use of synthetic antioxidants as food additives [11] (Fig. 2).

3 Chitosan Production and Processing

It takes two steps to produce chitosan from animal waste. First, chitin is extracted from biological waste, which is further modified to chitosan in the second step.

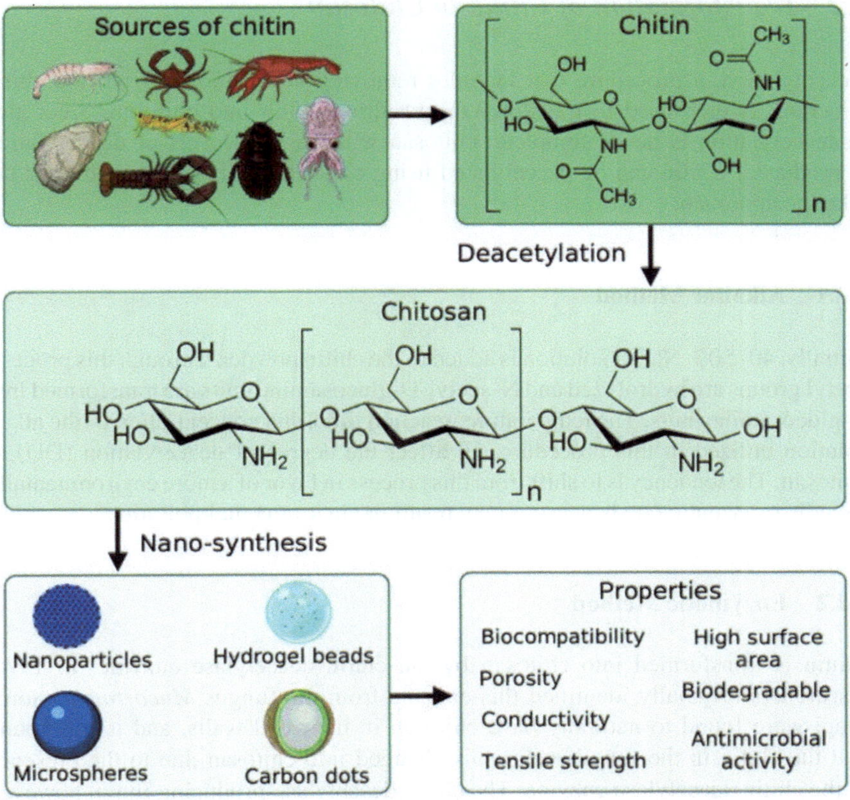

Fig. 2 Sources of chitosan for the preparation of chitosan-based nanomaterials with enhanced properties. Adapted from [12]

3.1 Chitin Extraction

Biological and chemical methods are used for the extraction of chitin from animal waste. The most often used approach is the chemical approach as it consumes the least time, despite its numerous disadvantages such as being wasteful and unfriendly to the environment. However, it appeared that the biological method produced a larger quantity of chitin as compared to the chemical method. Although the use of biological extraction is still only used in laboratory research despite being both safe and affordable, it is a rapidly expanding topic of interest.

The fundamental procedure for chitin extraction includes washing, drying, and crushing raw materials to make a fine powder. The subsequent stage comprises a three-step process known as demineralization, deproteinization, and decolorization of this powder. These three stages are employed in both chemical and biological procedures [2].

3.2 Transformation of Chitin to Chitosan

Deacetylation, a procedure that includes removing acetyl moiety from the chitin structure, is then used to transform the resulting chitin into chitosan. Since this N-deacetylation is never complete, chitosan with varying levels of deacetylation is produced. Chitin can be deacetylated using either the enzymatic method or the alkaline approach.

3.2.1 Alkaline Method

Initially, 40–50% NaOH solution is added to the chitin powder. Through this process, acetyl groups are hydrolyzed and N-acetyl-D-glucosamine units are transformed into D-glucosamine units. The temperature, reaction time, and concentration of the alkali solution utilized in this procedure, all affect the degree of deacetylation (DD) of chitosan. The tendency is to shift from this process in favor of a more environmentally friendly enzymatic one because it may result in environmental pollution.

3.2.2 Enzymatic Method

Chitin is transformed into chitosan by the chitin deacetylase enzyme. In 1974, researchers originally identified this enzyme from the fungus *Mucorrouxii*. Some fungi were found to naturally have chitosan in their cell walls, and it was found that the chitin in the fungi's cell walls changed into chitosan due to the presence of the chitin deacetylase enzyme. The chitin deacetylase-producing fungi, however, yield very little enzyme and have difficult fermentation requirements. Additionally, enzymes derived from fungal sources are less active than those from bacterial strains.

4 Industrial Applications of Chitosan

Chitin and chitosan, the biopolymers have several special qualities and characteristics, which have made it a versatile material have varied uses. The number of chitosan producers manufacturing different ranges of chitosan suitable for numerous utilities has increased during the past few years. Chitosan-based skincare and water purification products are marketed by Iceland-based Primex under the names Chito-Clear® and SeaKlear®, respectively. According to Primex EHF, the products are non-toxic and antibacterial. KitoFlokk®, a chitosan-based product made by Norwegian Chitosan, is used in wastewater and water treatment plants. For usage in food, drinks, cosmetics, pharmaceuticals, and agriculture, Chibio Biotech, established in China, produces chitosan from mushrooms and fungi like *Aspergillus*. Meron

Biopolymers, India, produces chitosan from abundant crabs and prawns for use in cosmetic, biomedical, and food applications [7].

5 Chitosan Films

Chitosan can be used to make packaging films or to apply it directly to food items for use in food packaging. Numerous researchers from all over the world have investigated both strategies. In various research investigations on formulations and uses, chitosan films that have been combined with other polymers, such as polysaccharides and synthetic plastics, have been studied as well.

5.1 Methods of Fabrication

Different techniques for film production have been developed to produce films for food packaging applications. Chitosan films can be produced through coating, direct casting, extrusion, and layer-by-layer assembly. The techniques can be applied to pure chitosan films or chitosan films that have been combined with other polymers.

5.1.1 Direct Casting

The easiest way to create biopolymer films is the direct casting approach. The steps in the preparation are as follows: (1) chitosan is dissolved in a weak acidic solution; (2) active or functional substances, such as cross-linkers and fillers are added; (3) mixing with other polymers, if required; (4) stirring to obtain a viscous and homogeneous solution; (5) filtration to eliminate insoluble particles and air bubbles; (6) casting the solution to a level, flat surface; (7) allowing the solution to dry for a predetermined interval under predetermined circumstances, such as temperature and humidity; (8) Peeling off; (9) curing and storage. For drying, it may occasionally be necessary to use specialized apparatus such as a hot air oven, humidity controller, vacuum oven, and climate chamber. As a material for food packaging, solution cast films have a lot of possibilities. They have been used on a variety of food products, and it has been found that by retaining quality and prolonging shelf life, they helped to assure food safety [13].

5.1.2 Coating

Chitosan-based thin films have frequently been applied to food products (mainly vegetables, fruits, and meat products) either directly or indirectly. The goal of a direct coating is to immobilize the polymer on the food surface in order to ensure safety from microorganisms. Additionally, the coating may alter the rate of respiration of vegetables and fruits and, thus, extend their shelf life. These coatings are generally edible, biodegradable, and environmentally friendly [14, 15].

5.1.3 Layer-By-Layer Assembly

An electrostatic deposition process known as layer-by-layer (LBL) assembly efficiently regulates the material characteristics and functionality of biomaterial films. Without the need for complicated equipment, this method can be used to fabricate multi-component films. Additionally, in this procedure, the created films are independent of the geometry of the substrate [16].

When LBL films are formed, pH plays a critical role. The pH controls the electrical charge on each layer, which has a substantial impact on how much biopolymer is deposited on the layer below. The quantity of biopolymer molecules needed to neutralize the first layer will depend on its charge. The roughness, thickness, and modulus of the films were found to increase with the drop in pH and rise in layer count. In addition to this, the films showed a strong photo-orientation [17, 18].

5.1.4 Extrusion

Another frequently used technique for processing chitosan-based biodegradable active-packaging films is melt extrusion. It entails (1) formulating materials with a variety of compositions; (2) mixing the ingredients; (3) extruding the mixture under defined processing conditions; (4) using a pelletizer to cut the extrudates into pellets; (5) drying the pellets; (6) using a twin-screw extruder attached with a flat die to extrude the pellets into flat sheets, or using a blown film extruder with an annular die to make a blown film [19].

A useful method for creating films with excellent thermal and antibacterial properties is extrusion. However, handling temperature and the interactions between the polymers must be considered in order to achieve the best performance.

6 Chitosan Blends with Other Polymers

Numerous biopolymers have been researched in order to create edible, biodegradable, and antimicrobial films for use in food packaging.

In order to be utilized effectively as films and coatings, as well as to enhance its functionality as a material for food packaging, chitosan can be combined with other biopolymers such as polysaccharides, proteins, and lipids [20]. In general, polysaccharide blends are more advantageous as compared to blends having proteins and/or lipids, like lower material costs, resource abundance, relative stability, and improved water solubility.

6.1 Chitosan Blends with Other Proteins

Protein biopolymers like collagen, gelatin, and caseinate have been combined with chitosan to create blended films. Both chitosan–gelatin bi-layer and chitosan–gelatin blend films, have been investigated, and they are hydrophilic biopolymers with compatibility and good affinity [21]. Both methods of blending produce films with significantly better elongation at break and increased barrier characteristics against water vapor penetration. Gelatin of fish, bovine or seafood origin, has been utilized most frequently for blending with chitosan [22]. A straightforward method for creating the blended film is solution casting using a mixed aqueous solution of chitosan and gelatin.

Collagen is used in the pharmaceutical, cosmetic, biomedical industries, and tissue engineering because of its exceptional capacity to produce insoluble fibers with great tensile strength and stability [23]. Meat and fish processing wastes are generally used to produce collagen [24]. Chitosan-collagen blend film can also be created by using a 1:1 combination of the two biopolymers, using the solution casting method. The hydrogen-bond network affects the structural characteristics of the collagen to completely interact with the chitosan, resulting in uniform mixing [25].

Due to electrostatic interactions between the positively charged amine groups of chitosan, and negatively charged carboxylic groups of casein, a stable blend is formed [26]. Chitosan-sodium caseinate can also be prepared by means of a straightforward technique of solution-casting.

The tensile and impact strength of blended films, among other mechanical characteristics, are greatly improved by the addition of glycerol as a plasticizer in the mixture [27]. Films made of chitosan and casein have been frequently studied for food packaging applications [22]. This blend also has exceptional thermoformability, film-forming qualities, and an enhanced moisture barrier [28].

6.2 Chitosan Blends with Other Carbohydrates

A variety of polysaccharides, including alginates, carrageenan, starch, pectin, and even bacterial cellulose, are mixed with chitosan. The aquatic polysaccharides chitosan and alginate are compatible with one another, and their blends are better suited for use in food packaging than those combining chitosan with protein or lipid polymers [29]. Alginates are polyanionic, linear, binary copolymers of (1-4) linked L-guluronic acid and (1-4) linked D-mannuronic acid. At pH levels below the alginate's pKa (acid gelation), alginates gel by ionotropic gelation [30]. Numerous studies made use of the polyelectrolyte complexation that takes place when both polyelectrolytes are charged, resulting in an electrostatic interaction between chitosan and alginates. However, mixing is necessary for manufacturing and applications on a large scale.

On the other hand, starch is also seen to blend well with chitosan. Chitosan and starch with increased amylose contents combine much better and form a stable hydrogen bond. Plasticizers, such as glycerine or sorbitol, were mixed with the blended polymers to improve the mechanical properties of the resulting films. Research was also done on barrier properties, such as the water vapor transmission rate (WVTR) [22].

Carrageenan is another anionic linear polysaccharide extracted from red seaweed. To reduce oxidative deterioration and moisture losses, fresh fruits and vegetables have already been coated with this polysaccharide [31]. Carrageenan and chitosan interact to create a polyelectrolyte complex, just like alginates do [32]. The majority of research on this chitosan mixture has been concentrated on food packaging applications.

Studies on films made with the carrageenan-chitosan complex have revealed an inferior moisture barrier and poor thermal stability characteristics [33]. Perhaps because of this, there is a scarcity of studies on using these mixed films for food packaging applications. However, mixing carrageenan and chitosan in a system of organic acids (such as acetic, lactic, citric, and ascorbic acids) because both these polysaccharides behave as polyelectrolytes with opposing ionic charges, and acids increases the tensile strength and barrier properties of the blended film [34].

Pectin and chitosan have both been thoroughly studied with regard to film formation and their application in food packaging. The two natural polymers have been blended as a way to tackle the difficulties of the individual biopolymer films' weak mechanical and barrier qualities [35]. According to reports, chitosan-pectin blended films possess superior properties compared to individual polymers. However, the blending has no effect on the mechanical properties or transparency.

The third generation of carbohydrate polymers, such as microbial cellulose and polylactic acid, have also been used in combination with chitosan. Chitosan, cyclodextrin, and carvacrol are deposited as a surface layer on the PLA matrix blended with ZnO nanoparticles [36]. Other researchers also combined chitosan and tea polyphenols with PLA film [37]. The blend films claimed improved heat-sealing of the product, while the former reported higher anti-microbial action.

7 Chitosan-Based Polymeric Antimicrobial Composites

Chitosan possesses potent antibacterial properties against a variety of microbes. However, chitosan's antibacterial characteristics could be enhanced, particularly for applications related to food packaging and preservation [38]. As a way to improve the antimicrobial characteristics of chitosan films, antimicrobial additives, such as metal or metal oxide nanoparticles or natural antimicrobial agents, have been used [39].

7.1 *Natural Antimicrobials*

Most naturally occurring antimicrobials come from animal, plant, and microbiological sources [40]. They demonstrate antimicrobial efficacy against various bacteria, including spoilage microbes, and have low toxicity, making them equivalent to their synthetic counterparts in this regard. Natural antimicrobials are utilized in food to preserve or improve its quality by preventing the growth of germs, and inhibiting natural processes that lead to food degradation or reduce its nutritional value [41]. The most widely used natural antimicrobials that are commercially available contain antimicrobials made from either animal sources (such as egg cell lysozyme), microbiological sources (such as bacteriocin), or those extracted from plants (such as essential oils) [40]. Natural anti-microbials, when used as additives in food packaging, can help maintain the appearance, flavor, and quality of the food while enhancing its shelf life.

Plant extracts and essential oils have appreciable antioxidant and antibacterial characteristics, and thus, are used as additives in composite films applied for packaging and food preservation [42]. Low-molecular-weight aliphatic and aromatic secondary metabolites, such as pinene, limonene, p-cymene, and terpenoids (such as thymol) make up the majority of essential oils. The main ingredients are phenylpropenes, such as eugenol, vanillin, or cinnamaldehyde, and carvacrol, or menthol. According to Yuan et al. [43], adding essential oils from clove, cinnamon, oregano, or eucalyptus to chitosan (CS) increased antibacterial efficacy in food packaging applications.

The recent emergence of multidrug-resistant microorganisms has directed the research on active packaging materials based on chitosan, that possess the ability to decrease, inhibit, or delay the growth of microorganisms [44]. According to Franssen and Krochta [45], adding polypeptides such as lysozyme, lactoferrin, peroxidase, and nisin to chitosan-based edible formulations, and their application to food products in the form of films, spray coating, or dip coating, may help enhance their safety and make them last longer.

7.2 *Antimicrobials with Metal and Metal Oxide Nanostructures*

Due to their strong efficacy against bacteria, fungi, protozoa, and some viruses, metal nanoparticles such as copper (Cu) and silver (Ag) nanoparticles are well-known wide-ranging antibacterial substances. To modify membrane potential and trigger injury to the membrane, antimicrobial nanomaterials can electrostatically adhere to the cell walls of microorganisms. This can lead to alterations in the normal metabolic functions of the microbial cells, such as hampering respiration, upsetting transport balance, interrupting energy transfer, and causing cell lysis, which can ultimately result in cell death [46].

Reactive oxygen species (ROS), which are formed by nanomaterials, are the foundation of another mechanism for the antimicrobial activity of the nanostructures. Toxic oxidants like hydrogen peroxide are transformed into more reactive free radicals, like hydroxyl radicals, by the catalytic action of nanostructures. Many of the semiconductor nanomaterials can also produce ROS when they interact with light, as a result of the reaction between excited electron holes with water molecules, which are hazardous to bacteria. Significant oxidative stress brought on by ROS can result in DNA/RNA damage, enzyme inhibition, lipid peroxidation, and protein modification, all of which can induce microbial cell death [47, 48].

Additionally, key microbial enzymes are directly inhibited, nitrogen reactive species (NRS) are produced, and programmed cell death is induced by antimicrobial nanomaterials [49]. The ROS production and antibacterial activity of nanostructures are primarily influenced by their surface groups, surface charges, shape, and particle size. Smaller nanostructures can easily pass through biological barriers like cell membranes. As particle size increases, the amount of uptake decreases [50].

Also, when metal ions from nanomaterials, such as Cu^+, Ag^+, and Zn^{2+} ions, are released, they may combine with molecular oxygen to produce superoxide radicals and other ROS, which have antimicrobial effects [51]. Silver nanoparticles (AgNPs) are synthesized from the bottom up using chemical and biological techniques, including green synthesis [52]. Additionally, AgNPs can be produced in situ in a chitosan matrix. Due to the presence of amine and hydroxyl groups, chitosan has a high affinity for Ag^+ ions and can reduce Ag^+ ions into AgNPs in an alkaline environment [53]. For a wide range of industrial applications, metal oxide nanoparticles are appealing. Metal oxide nanoparticles are frequently utilized as nano-fillers in biopolymer matrixes to demonstrate the antibacterial activity, mechanical, and barrier properties of the films [49].

Due to the production of reactive oxygen species (ROS), metal oxide nanoparticles such as titanium dioxide (TiO_2), copper oxide (CuO), and zinc oxide (ZnO) have also been reported to have suitable antibacterial properties [54]. As antibiofouling agents, ZnO and TiO_2 have been widely employed as photocatalytic materials [55]. They exert a photocatalytic action on the phospholipids in the microbial cell membranes, leading to their peroxidation, which results in a loss of membrane integrity.

8 Application in Packaging

8.1 *Shelf-Life of Vegetables and Fruits*

The majority of horticultural commodities, in particular, have a short postharvest life, making them perishable. Even after harvest, fruits and vegetables continue to respire and ripen, which reduces their shelf life and degrades their quality over time. Fruit ripening is a complex process that involves significant changes in physiological and biochemical parameters such as chlorophyll degradation, enzymatic cell wall degradation, changes in sugar content, respiratory activity, ethylene production, and aromatic compound levels [56, 57]. Furthermore, reactive oxygen species are formed during electron transfer in the mitochondria, which can reduce the quality of fruits and vegetables. As a result, numerous researchers have looked for safe, natural compounds that can delay fruit and vegetable ripening or lessen degradation after harvest. One such study focused on the use of an edible chitosan coating added with *Thymus capitatus* essential oil on strawberry fruit kept in cold storage for up to 15 days [58].

Coating is an inexpensive, easy, and effective way to wrap or cover fresh produce using edible film made of biopolymers to prevent moisture loss, which lowers the rate of deterioration and respiration. Due to its intrinsic antibacterial activity, capacity to form films and biodegradability, chitosan is particularly gaining importance in the manufacture of films and coatings for use in food packaging [59]

Chitosan has the potential to be employed in a wide range of industrial applications today, including those related to food, cosmetics, textile, agriculture, biomedicine, environmental protection, etc. [4]. In order to extend the shelf life and maintain the quality of postharvest products, chitosan-based composites combined with nanomaterials and/or natural antimicrobials have been proven to be highly helpful [60].

The disadvantages of the biopolymer's weak mechanical and barrier qualities, thermal stability, and limited transparency can be effectively addressed by adding nanofillers to the chitosan matrix. The physical, chemical, and biological properties of natural biopolymers are improved by the inclusion of nano-fillers.

Chitosan coating has been found to decrease weight loss, increase firmness, increase antioxidant activity, and increase the amount of soluble solids in fresh fruits and vegetables [61]. In a study conducted on strawberries coated with 1% and 2% chitosan, significantly decreased water loss, losses in titratable acidity, color loss, and ascorbic acid content, changes in the phenolic, flavonoid, and anthocyanin contents were observed in the coated fruits [62].

Further, the activity of some antioxidant enzymes increased on the chitosan coating, which stopped the oxidation of the flesh and minimized membrane damage. In a related work, the shelf life of whole bananas was extended using quaternized chitosan films combined with carboxymethyl cellulose (CMC) [63]. The findings demonstrated that the fruit degradation was slowed down by films with a high percentage of quaternized chitosan.

8.2 Shelf-Life of Meat and Fish Products

Between 1998 and 2015, 500–1500 foodborne outbreaks each year were reported in the United States, and a sizable fraction of those outbreaks were linked to meat or muscle food [64]. Only 37 foodborne outbreaks have been documented in India over the past 30 years (1980–2009), and the vast majority (24 out of 37) were brought on by bacterial pathogens [65]. In these 24 outbreaks, *Staphylococcus aureus, Vibrio* spp., and *Salmonella* spp. have been the main etiological agents. Because they are mechanically minced, meats and fish are highly perishable items that could include surface pollutants in the inner layers of the muscle [66].

Foodborne infections are substantially more resistant to cooking once they have penetrated the inner layers [67]. Microbial spoilage and food oxidation are two main issues that affect the shelf life and food quality of meat-based food products. Instead of adding synthetic food additives and preservatives in food during processing, packaging materials added with antioxidants or antimicrobials are progressively being developed. Many studies have concluded that active compounds or generally accepted as safe nanomaterials can replace antimicrobials and synthetic antioxidants, which can mitigate the possible health-related risks arising from the long-term intake of some synthetic preservatives.

The development of chitosan-based films and their use in the preservation of fish and meat products including pork, poultry, and red meat have been the subject of numerous reports [37]. Green tea extract has been included in a chitosan-based film developed as an active package for extending the shelf life of pig sausages [68]. The results displayed that compared to samples of pork sausages wrapped in films containing chitosan and control films, those wrapped in films containing green tea extract had reduced changes in texture, sensory qualities, color, and microbiological development. They concluded that adding green tea extract to chitosan film could increase the antioxidant and antimicrobial qualities of the film, which helped preserve the quality and extend the shelf life of the pork sausages.

Other researchers also studied a chitosan-based film added with acetic acid and investigated its effectiveness in extending the shelf life of meat [69]. They discovered that chitosan films could maintain the freshness and quality of chilled meat for longer periods of time than those of LDPE films. The chitosan films effectively suppressed the microbial growth in chilled meat during storage. The outcomes showed that the antibacterial properties of the chitosan films on meat were generally liable for controlling the inhibition of microbial growth. In another study, Li et al. reported that *Cynoglossus semilaevis* Günther fillets coated with chitosan added with polylysine and rosmarinic acid retained their quality throughout chilled storage [29].

9 Conclusion

Recently, the suitability of numerous biopolymers for usage in food packaging has been investigated. A widely researched biopolymer, chitosan, can be utilized to extend the shelf life of seafood, fresh vegetables, and meat products. Nanomaterials and phytochemicals added to chitosan enhance the mechanical, antibacterial, and barrier qualities of the films and coatings, maintain the quality of packed food, and extend its shelf life. Other biopolymers are blended with chitosan to create composite films and coatings. Active substances like nanomaterials, essential oils, natural extracts, and other phytochemicals have also been added to chitosan matrices to create active packaging films. However, several products based on chitosan have already been commercialized for use in various industries, including cosmetics, drugs, agriculture, water management, and food additives. The market for these goods is expanding steadily and quickly. There is no doubt that chitosan will eventually displace non-biodegradable plastics and establish itself as a natural alternative in the near future.

References

1. Berger KR (2002) A brief history of packaging: University of Florida cooperative extension service. Institute of Food and Agricultural Sciences, EDIS
2. Elieh-Ali-Komi D, Hamblin MR (2016) Chitin and chitosan: production and application of versatile biomedical nanomaterials. Int J Adv Res 4(3):411
3. Guan G, Azad MAK, Lin Y, Kim SW, Tian Y, Liu G, Wang H (2019) Biological effects and applications of chitosan and chito-oligosaccharides. Front Physiol 10:516
4. Hamed I, Özogul F, Regenstein JM (2016) Industrial applications of crustacean by-products (chitin, chitosan, and chitooligosaccharides): a review. Trends Food Sci Technol 48:40–50
5. Kausar A (2017) Scientific potential of chitosan blending with different polymeric materials: a review. J Plast Film Sheeting 33(4):384–412
6. Heras M, Huang CC, Chang CW, Lu KH (2024) Trends in chitosan-based films and coatings: a systematic review of the incorporated biopreservatives, biological properties, and nanotechnology applications in meat preservation. Food Packag Shelf Life 42:101259
7. Szymańska E, Winnicka K (2015) Stability of chitosan—a challenge for pharmaceutical and biomedical applications. Mar Drugs 13(4):1819–1846
8. Anthonsen MW, Vårum KM, Smidsrød O (1993) Solution properties of chitosans: conformation and chain stiffness of chitosans with different degrees of N-acetylation. Carbohyd Polym 22(3):193–201
9. Vårum KM, Ottøy MH, Smidsrød O (2001) Acid hydrolysis of chitosans. Carbohyd Polym 46(1):89–98
10. Hosseinnejad M, Jafari SM (2016) Evaluation of different factors affecting antimicrobial properties of chitosan. Int J Biol Macromol 85:467–475
11. Sabaghi M, Maghsoudlou Y, Habibi P (2015) Enhancing structural properties and antioxidant activity of kefiran films by chitosan addition. Food Struct 5:66–71
12. Villalba-Rodríguez AM, González-González RB, Martínez-Ruiz M, Flores-Contreras EA, Cárdenas-Alcaide MF, Iqbal HM, Parra-Saldívar R (2022) Chitosan-based carbon dots with applied aspects: new frontiers of international interest in a material of marine origin. Mar Drugs 20(12):782
13. Shankar S, Rhim JW (2018) Preparation of sulfur nanoparticle-incorporated antimicrobial chitosan films. Food Hydrocolloids 82:116–123
14. Singh S, Kumar S, Koh J (2024) Physicochemical and optical activity of chitosan based ternary nanocomposites for food packaging applications. J Mol Struct:138210
15. Ortiz-Duarte G, Pérez-Cabrera LE, Artés-Hernández F, Martínez-Hernández GB (2019) Ag-chitosan nanocomposites in edible coatings affect the quality of fresh-cut melon. Postharvest Biol Technol 147:174–184
16. Li F, Biagioni P, Finazzi M, Tavazzi S, Piergiovanni L (2013) Tunable green oxygen barrier through layer-by-layer self-assembly of chitosan and cellulose nanocrystals. Carbohyd Polym 92(2):2128–2134
17. Fernández R, Ocando C, Fernandes SC, Eceiza A, Tercjak A (2014) Optically active multilayer films based on chitosan and an azopolymer. Biomacromol 15(4):1399–1407
18. Acevedo-Fani A, Salvia-Trujillo L, Soliva-Fortuny R, Martín-Belloso O (2015) Modulating biopolymer electrical charge to optimize the assembly of edible multilayer nanofilms by the layer-by-layer technique. Biomacromol 16(9):2895–2903

19. Goskonda SR, Upadrashta SM (1993) Avicel RC-591/chitosan beads by extrusion-spheronization technology. Drug Dev Ind Pharm 19(8):915–927
20. Zhu J, Wu H, Sun Q (2019) Preparation of crosslinked active bilayer film based on chitosan and alginate for regulating ascorbate-glutathione cycle of postharvest cherry tomato (*Lycopersicon esculentum*). Int J Biol Macromol 130:584–594
21. Pereda M, Ponce AG, Marcovich NE, Ruseckaite RA, Martucci JF (2011) Chitosan-gelatin composites and bi-layer films with potential antimicrobial activity. Food Hydrocolloids 25(5):1372–1381
22. Elsabee MZ, Abdou ES (2013) Chitosan based edible films and coatings: a review. Mater Sci Eng, C 33(4):1819–1841
23. Bhuimbar MV, Bhagwat PK, Dandge PB (2019) Extraction and characterization of acid soluble collagen from fish waste: development of collagen-chitosan blend as food packaging film. J Environ Chem Eng 7(2):102983
24. Silva TH, Moreira-Silva J, Marques ALP, Domingues A, Bayon Y, Reis RL (2014) Marine origin collagens and its potential applications. Mar Drugs 12(12):5881–5901
25. Sionkowska A, Wisniewski M, Skopinska J, Poggi GF, Marsano E, Maxwell CA et al (2006) Thermal and mechanical properties of UV irradiated collagen/chitosan thin films. Polym Degrad Stab 91(12):3026–3032
26. Celli GB, Ravanfar R, Kaliappan S, Kapoor R, Abbaspourrad A (2018) Annatto-entrapped casein-chitosan complexes improve whey color quality after acid coagulation of milk. Food Chem 255:268–274
27. Pereda M, Aranguren MI, Marcovich NE (2007) Characterization of chitosan/caseinate films. J Appl Polym Sci 107(2):1080–1090
28. Huang T, Qian Y, Wei J, Zhou C (2019) Polymeric antimicrobial food packaging and its applications. Polymers 11(3)
29. Li N, Liu W, Shen Y, Mei J, Xie J (2019) Coating effects of ε-polylysine and rosmarinic acid combined with chitosan on the storage quality of fresh half-smooth tongue sole (*Cynoglossus semilaevis Günther*) fillets. Coatings 9(4):273
30. Dumont M, Villet R, Guirand M, Montembault A, Delair T, Lack S et al (2018) Processing and antibacterial properties of chitosan-coated alginate fibers. Carbohyd Polym 190:31–42
31. Lin D, Zhao Y (2007) Innovations in the development and application of edible coatings for fresh and minimally processed fruits and vegetables. Compr Rev Food Sci Food Saf 6(3):60–75
32. Spagnuolo P, Dalgleish D, Goff H, Morris E (2005) Kappa-carrageenan interactions in systems containing casein micelles and polysaccharide stabilizers. Food Hydrocolloids 19(3):371–377
33. Shahbazi M, Rajabzadeh G, Ettelaie R, Rafe A (2016) Kinetic study of κ-carrageenan degradation and its impact on mechanical and structural properties of chitosan/κ-carrageenan film. Carbohyd Polym 142:167–176
34. El-Hefian EA, Nasef MM, Yahaya AH (2010) Preparation and characterization of chitosan/polyvinyl alcohol blends-A rheological study. E-J Chem 7(s1):S349–S357
35. Younis HGR, Zhao G (2019) Physicochemical properties of the edible films from the blends of high methoxyl apple pectin and chitosan. Int J Biol Macromol 131:1057–1066
36. Andrade-Del Olmo J, Pérez-Álvarez L, Hernáez E, Ruiz-Rubio L, Vilas-Vilela JL (2019) Antibacterial multilayer of chitosan and (2-carboxyethyl)-β-cyclodextrin onto polylactic acid (PLLA). Food Hydrocolloids 88:228–236
37. Ye J, Wang S, Lan W, Qin W, Liu Y (2018) Preparation and properties of polylactic acid-tea polyphenol-chitosan composite membranes. Int J Biol Macromol 117:632–639
38. Verlee A, Mincke S, Stevens CV (2017) Recent developments in antibacterial an antifungal chitosan and its derivatives. Carbohyd Polym 164:268–283
39. Matharu RK, Ciric L, Edirisinghe M (2018) Nanocomposites: suitable alternatives as antimicrobial agents:29282001
40. Tu H, Wu G, Yi Y, Huang M, Liu R, Shi X et al (2019) Layer-by-layer im- mobilization of amphoteric carboxymethyl chitosan onto biocompatible silk fibroin nanofibrous mats. Carbohyd Polym 210:9–16

41. Davidson PM, Cekmer HB, Monu EA, Techathuvanan C (2015) The use of natural antimicrobials in food: an overview. In: Taylor TM (ed) Handbook of natural antimicrobials for food safety and quality. Woodhead Publishing, Oxford, pp 1–27
42. Pisoschi AM, Pop A, Georgescu C, Turcuş V, Olah NK, Mathe E (2018) An overview of natural antimicrobials role in food. Eur J Med Chem 143:922–935
43. Yuan G, Chen X, Li D (2016) Chitosan films and coatings containing essential oils: the antioxidant and antimicrobial activity, and application in food systems. Food Res Int 89:117–128
44. Ponce A, Roura SI, Moreira MR (2016) Casein and chitosan polymers. In: Antimicrobial food packaging. Elsevier, pp 455–466
45. Franssen LR, Krochta JM (2003) Edible coatings containing natural antimicrobials for processed foods. In: Roller S (ed) Natural antimicrobials for the minimal processing of foods, vol 12. Natural antimicrobials for the minimal processing of foods. Woodhead Publishing, pp 250–262
46. Pelgrift RY, Friedman AJ (2013) Nanotechnology as a therapeutic tool to combat microbial resistance. Adv Drug Deliv Rev 65(13–14):1803–1815
47. Pan X, Redding JE, Wiley PA, Wen L, McConnell JS, Zhang B (2010) Mutagenicity evaluation of metal oxide nanoparticles by the bacterial reverse mutation assay. Chemosphere 79(1):113–116
48. Sathe P, Richter J, Myint MTZ, Dobretsov S, Dutta J (2016) Self-decontaminating photocatalytic zinc oxide nanorod coatings for prevention of marine micro-fouling: a mesocosm study. Biofouling 32(4):383–395
49. Dizaj SM, Lotfipour F, Barzegar-Jalali M, Zarrintan MH, Adibkia K (2014) Antimicrobial activity of the metals and metal oxide nanoparticles. Mater Sci Eng C 44:278–284
50. Sathe P, Laxman K, Myint MTZ, Dobretsov S, Richter J, Dutta J (2017) Bioinspired nanocoatings for biofouling prevention by photocatalytic redox reactions. Sci Rep 7(1)
51. Joe A, Park S-H, Kim D-J, Lee Y-J, Jhee K-H, Sohn Y et al (2018) Antimicrobial activity of ZnO nanoplates and its Ag nanocomposites: Insight into an ROS-mediated antibacterial mechanism under UV light. J Solid State Chem 267:124–133
52. Roy N, Gaur A, Jain A, Bhattacharya S, Rani V (2013) Green synthesis of silver nanoparticles: an approach to overcome toxicity. Environ Toxicol Pharmacol 36(3):807–812
53. Boufi S, Vilar MR, Ferraria AM, do Rego AMB (2013) In situ photo-chemical generation of silver and gold nanoparticles on chitosan. Colloids Surf A 439:151–158
54. André RS, Zamperini CA, Mima EG, Longo VM, Albuquerque AR, Sambrano JR et al (2015) Antimicrobial activity of TiO_2: Ag nanocrystalline hetero-structures: experimental and theoretical insights. Chem Phys 459:87–95
55. Al-Naamani L, Dobretsov S, Dutta J, Burgess JG (2017) Chitosan-zinc oxide nanocomposite coatings for the prevention of marine biofouling. Chemosphere 168:408–417
56. Lee Y, Hwang KT (2017) Changes in physicochemical properties of mulberry fruits (Morus alba L.) during ripening. Sci Hortic 217:189–196
57. Perotti VE, Moreno AS, Podestá FE (2014) Physiological aspects of fruit ripening: the mitochondrial connection. Mitochondrion 17:1–6
58. Martínez O, Salmerón J, Epelde L, Vicente MS, de Vega C (2018) Quality enhancement of smoked sea bass (Dicentrarchus labrax) fillets by adding resveratrol and coating with chitosan and alginate edible films. Food Control 85:168–176
59. Harish Prashanth KV, Tharanathan RN (2007) Chitin/chitosan: modifications and their unlimited application potential—an overview. Trends Food Sci Technol 18(3):117–131
60. Xing Y, Li W, Wang Q, Li X, Xu Q, Guo X et al (2019) Antimicrobial nano- particles incorporated in edible coatings and films for the preservation of fruits and vegetables. Molecules 24(9):1695
61. Adiletta G, Zampella L, Coletta C, Petriccione M (2019) Chitosan coating to preserve the qualitative traits and improve antioxidant system in fresh figs (Ficus carica L.). Agriculture 9(4):84

62. Petriccione M, Mastrobuoni F, Pasquariello MS, Zampella L, Nobis E, Capriolo G et al (2015) Effect of chitosan coating on the postharvest quality and antioxidant enzyme system response of strawberry fruit during cold storage. Foods 4(4):501–523
63. Hu D, Wang H, Wang L (2016) Physical properties and antibacterial activity of quaternized chitosan/carboxymethyl cellulose blend films. Lebensmittel-Wissenschaft und-Technologie-Food Sci Technol 65:398–405
64. Clink K (2017) CDC WONDERRR 2017/036 CDC WONDER centers for disease control and prevention Atlanta, GAURL. Ref Rev 31(2):14–14. https://wonder.cdc.gov/LastvisitedAugust2016Gratis
65. Rao Vemula S, Naveen Kumar R, Polasa K (2012) Foodborne diseases in India—a review. Br Food J 114(5):661–680
66. Li X-F, Feng X-Q, Yang S, Fu G-Q, Wang T-P, Su Z-X (2010) Chitosan kills Escherichia coli through damage to be of cell membrane mechanism. Carbohyd Polym 79(3):493–499
67. Laine ES, Scheftel JM, Boxrud DJ, Vought KJ, Danila RN, Elfering KM et al (2005) Outbreak of Escherichia coli O157:H7 infections associated with non- intact blade-tenderized frozen steaks sold by door-to-door vendors. J Food Prot 68(6):1198–1202
68. Siripatrawan U, Noipha S (2012) Active film from chitosan incorporating green tea extract for shelf life extension of pork sausages. Food Hydrocolloids 27(1):102–108
69. Li K, Zhu J, Guan G, Wu H (2019) Preparation of chitosan-sodium alginate films through layer-by-layer assembly and ferulic acid crosslinking: film properties, characterization, and formation mechanism. Int J Biol Macromol 122:485–492

Nanoemulsions in Food Packaging

Adeeba Parveen, Vivek Kumar, Sumaiya Fatima, and Anurag Singh

Abstract With the growth of technology and globalization, people are increasingly concentrating more on healthy food items that are high in nutrients and have a long shelf life. However, notably, bacterial growth and oxidation are two variables that affect the quality and safety of food products, especially, fresh produce. To address the issue, novel food packaging solutions, that are smart, programmable, and multipurpose are being investigated. Recently a lot of work has been developed in using innovative nanoscience-based technologies to improve the quality of food items. The usage of nanoemulsions in the food packaging sector is one of the key applications in this subject. Emulsions with particle sizes 5–100 nm are comprehended as nanoemulsions. It can be created as a delivery system as a water-in-oil and water-in-oil system using a variety of ways, including low-energy and high-energy procedures. The application of nanoemulsion in the food industry has expanded because of the high optical translucency, good physical resilience against gravitational separation, and enhanced bioavailability. Moreover, nanoemulsion-based delivery systems, based on destined application, may also include colors, flavorings, preservatives, disinfectants, or nutrients. When compared to samples with standard packaging, this technology may prevent microbial development, changes in food color and appearance, reduction in moisture content, weight loss, bad flavor, and taste, and also reduce the rate of oxidation and browning. Despite the high potential of nanoemulsions, several specific difficulties must be addressed. Some food quality parameters, namely spoilage, texture, taste, flavor, color and stability, can be modified using these potential nanotechnologies. Hence, this chapter discusses the preparation of nanoemulsions, their evaluation for physical characteristics, stability and microstructure; their effect on food quality, and their applications in food packaging.

Keywords Nanoemulsions · Food packaging · Edible coating · Coating techniques · Nanoemulsion preparation

A. Parveen · V. Kumar (✉) · S. Fatima · A. Singh
Department of Food Technology, Harcourt Butler Technical University, Kanpur 208002, Uttar Pradesh, India
e-mail: viveksachan99@rediffmail.com

M. Alle et al. (eds.), *Functional Nanomaterials and Nanocomposites for Biodegradable Food Packaging*, Smart Nanomaterials Technology,
https://doi.org/10.1007/978-981-97-9084-5_8

1 Introduction

Growing consumer demand for foods with fresh-like properties, fewer preservatives and synthetic additives, and minimal environmental footprint while remaining safe to consume has prompted researchers to develop milder processing technologies and more environment-friendly packaging solutions. As the importance of sustainability in food packaging grows, bio-based and/or biodegradable materials stand out as viable alternatives to their synthetic equivalents, particularly in terms of successfully incorporating novel compounds and functionalities into existing films and coatings [1]. Nanotechnology, a growing technology possessing the capacity to alter the food industry, has shown ways to create food-grade edible packaging due to its distinct quality, as it allows modification of various micro-scale properties including texture, thermal stability, water solubility, mechanical strength, color and sensory attributes. [2]. Nanoemulsions, an important aspect of nanotechnology, have attracted significant attention in the food packaging sector, as they possess unique properties such as tiny droplet size, translucent appearance, adjustable rheology and remarkable stability [3]. They are also useful for entrapping lipophilic substances and increasing their functional characteristics [4]. Food-grade nanoemulsions are being employed to increase targeted administration, bioavailability, digestibility and effective and efficient encapsulation. Moreover, owing to their antioxidant and antibacterial properties, nanoemulsions are embedded in the edible films/coatings, which in turn prolong the shelf life of fruits and vegetables, meat, and other foods [5].

Nanoemulsions, basically, are dispersion of nanoscale droplets formed by mixing two immiscible liquids into a kinetically stable form. Water and oil are the two immiscible liquids mostly employed in the commercial environment. Nanoemulsions are of two types, oil-in-water (O/W) nanoemulsion specifying the dispersion of small droplets of oil in an aqueous medium, water-in-oil (W/O) nanoemulsion which is the dispersion of small water droplets in oil [6]. Oil phase, aqueous phase, and stabilizers are the main components in nanoemulsion formulation. Stabilizers are used to prevent aqueous and oil phase disintegration in nanoemulsions, hence extending their longevity. Nanoemulsion can be prepared using either a low-energy approach, such as spontaneous emulsification, membrane emulsification, and phase inversion, or a high-energy approach, which comprises various methods such as homogenization, microfluidic homogenization, and ultrasonic homogenization. Change in environmental factors or the composition of nanoemulsions that result in the preparation of oil nanoemulsion droplets within the mixture that has surfactants, oil, and water is known as the low-energy method [7–10]. High energy method involves breaking down and blending of oil and water using the shear, powerful cavitational, and turbulent flow profiles made through specifically designed devices. The small droplets are made after consuming a significant amount of energy which ranges from ~108–1010 W/kg [1, 11].

With the help of the nanoemulsion technique, it is feasible to incorporate a variety of food products and additives, such as antioxidants, vitamins, bioactive lipids, acidulants, flavorings, preservatives, colorings, and so on. There have been several researches on the use of nanoemulsions in food packaging. They can be either utilized directly in films/coatings or encapsulated in a polymeric carrier before being employed in films/coatings [12]. Among the most common ways to include nanoemulsions in films/coatings is to scatter the nanoemulsion in the film-forming solution under high shear stress before casting the mixture on the product [13]. The use of nanoemulsion has shown effective performance in food packaging. As a result, this chapter examines the preparation of nanoemulsions, their evaluation for physical properties, stability, and characterization, as well as their influence on food quality and applications in food packaging.

2 Nanoemulsion Formation

A nanoemulsion is a typical emulsion with a particle having a radius less than 100 nm, more commonly between 1 and 100 nm. An emulsion comprises a dispersed phase and a continuous phase, wherein the liquid that creates the droplets is the dispersed phase while the liquid that forms the surrounding solvent is termed the continuous phase. As previously stated, nanoemulsions can be classified into oil in water (o/w) emulsions, water in oil (w/o) emulsion, and bicontinuous nanoemulsion, where oil is in the dispersed phase, water in the dispersed phase, and both oil and water are interspersed within the system, respectively. Coating of o/w nanoemulsion is typically done by hydrophilic emulsifier and w/o nanoemulsion is coated by lipophilic emulsifier. The addition of surface-active compounds improves nanoemulsion kinetic stability and delays its destabilization. Emulsifiers play a significant role in stabilizing the nanoemulsion. The oil phase, aqueous phase, and stabilizers are the main components in the nanoemulsion (Fig. 1).

2.1 Oil Phase

Free fatty acids (FFA), waxes, triacylglycerols (TAG), diacylglycerols (DAG), monoacylglycerols (MAG), mineral oils, or different lipophilic functional compounds can be used to create the oil phase in nanoemulsions [14]. The physicochemical parameters of the oil phase component (such as density, viscosity, polarity, water solubility, chemical stability, refractive index, and interfacial tension) greatly affect the nanoemulsion characteristics, such as its formation, stability, and functional properties. It also frequently constrains the type of homogenization method(s) that may be utilized to create a nanoemulsion [15]. Figure 2 summarizes the influence of different physicochemical parameters of aqueous and oil phases on the characteristics of nanoemulsions. Generally, long-chain triacylglycerols are preferred in the food industry, owing to their high availability, functional properties, and low cost [16].

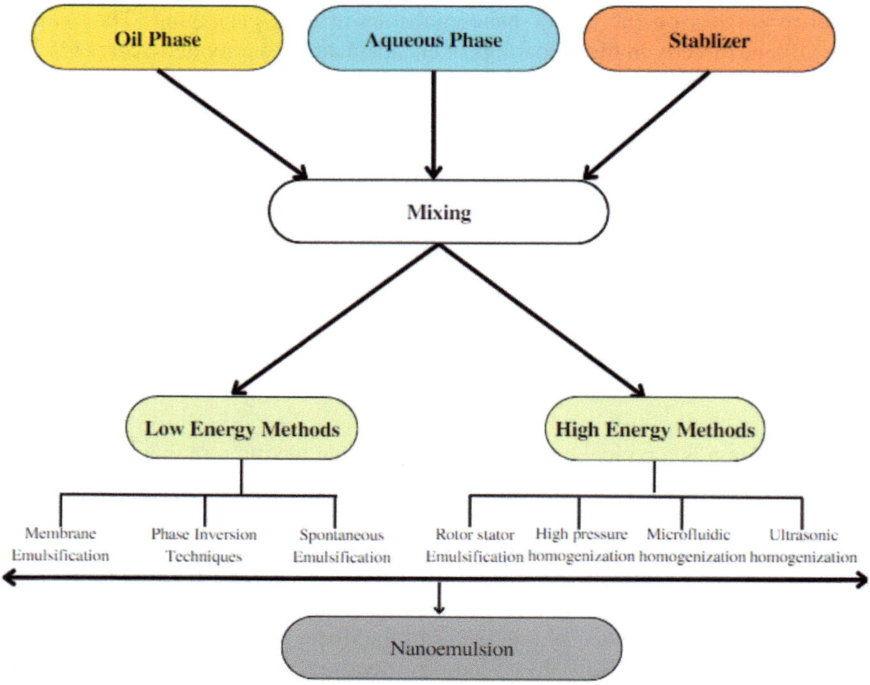

Fig. 1 Flowchart of formation of nanoemulsions

2.2 *Aqueous Phase*

The aqueous phase is made up of a polar solvent (water, most often) and some cosolvents, such as carbohydrates, proteins, alcohol, and polyols. The aqueous phase affects both the stability of nanoemulsion and its droplet size. The properties of nanoemulsion such as its density, ionic strength, interfacial tension, pH, polarity, rheology, and refractive index are all governed by the type and concentration of the aqueous phase (Fig. 2). Hence, the composition of the aqueous phase may be adjusted to optimize the preparation and increase the stability of nanoemulsions [17].

2.3 *Stabilizers*

Stabilizers are employed to prevent the aqueous and oil phase breakdown in the nanoemulsion in order to extend its lifespan. The homogenized nanoemulsion when allowed to stand will quickly breakdown as a result of gravity separation, flocculation, and Ostwald ripening, thus stabilizers need to be added. Emulsifiers, weighing

Fig. 2 Influence of different physicochemical parameters of oil and aqueous phases on the performance of nanoemulsions

agents, ripening retarders, and texture modifiers are some examples of these stabilizers. Emulsifiers are most often used stabilizers for nanoemulsions. They shield tiny droplets because they are surface-active molecules that lessen interfacial tension and facilitate the production of compact and stable nano-molecules. Emulsifiers can be categorized as anionic, cationic, non-ionic, or zwitterionic based on their charge. Surfactants, polysaccharides, proteins, and phospholipids are a few examples of emulsifiers [18]. Additionally, in order to prevent gravitational forces from acting on the oil phase of an oil-in-water emulsion, weighting agents are typically used, as they balance the density of the oil phase with the surrounding aqueous phase. This slows the sedimentation or creaming process in the emulsion. Ester gum and margarine gum are the frequently used weighing agents in nanoemulsions. Besides, by acting hydrophobically, ripening agents like long-chain triacylglycerols and mineral oils, slow down the ripening process [19]. Texture modifiers work with the aqueous phase to alter its texture by thickening it like gel; as a result, they retard the migration of the oil droplets and impart a thick, cream-like consistency to the nanoemulsion. Gums, polysaccharides, and vegetable proteins are the most often employed texture modifiers in nanoemulsions [20].

3 Nanoemulsion Preparation Technique

The nanoemulsions include a large number of minute droplets that lead to increased surface area. As a result, a significant quantity of energy is required to produce nanoemulsion. The energy necessary for the generation of nanoemulsions (ΔG) is calculated as

$$\Delta G = \Delta A\gamma - T\Delta S$$

where ΔA is the increase in interfacial area, γ is the surface tension, and $T\Delta S$ is the entropy of dispersion [21]. Either high or low-energy methods can be used to prepare nanoemulsions. Nanoemulsion formation method relies upon the selection of the composition of surfactants and the amount of energy to be applied. Additionally, the size of the nanoemulsion is determined by the ingredients, operating circumstances, and techniques of preparation.

3.1 Low-Energy Methods

The low-energy techniques result in the formation of oil droplets spontaneously, within the oil–water emulsion by altering the composition of the solution or the ambient conditions. These are cost-effective methods as they use less amount of energy for the preparation of nanoemulsion, however, formation of nanoemulsion particles of the desired size is dependent primarily on the type of oil used, surfactants, and cosurfactants. Low energy techniques used for the formation of nanoemulsions involve nominal heat generation, and are therefore suitable for heat labile constituents [22]. Membrane emulsification, spontaneous emulsification, and phase inversion techniques are the low-energy methods used for the development of nanoemulsions (Fig. 3).

3.1.1 Membrane Emulsification

The concept behind this emulsification method is to force a scattered phase through the pores of a microporous membrane into a continuous phase. A drop-by-drop technique is used to form and separate the droplets that have been emulsified at the end of the pores. It may be used to create emulsions with regulated and narrow droplet

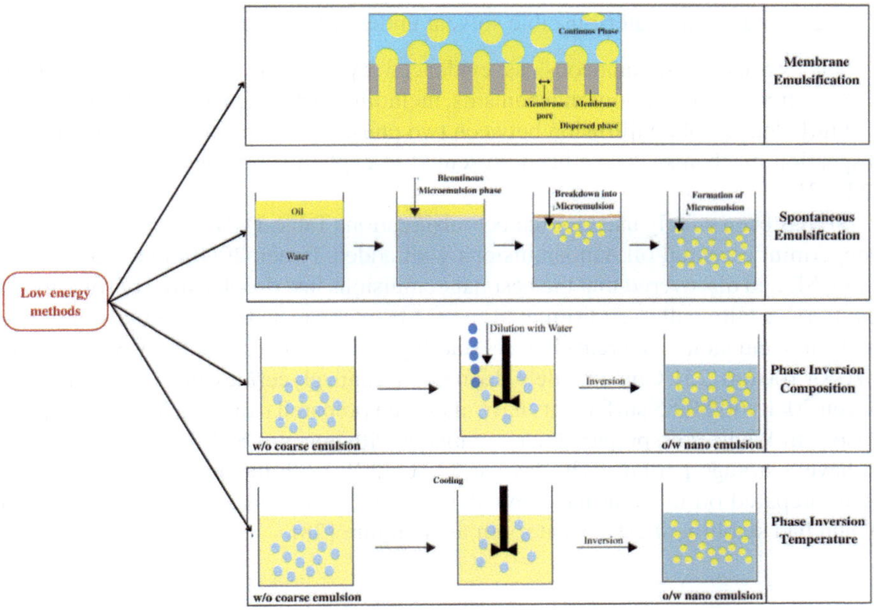

Fig. 3 Schematic diagram to represent various low-energy approaches to produce food-grade nanoemulsions

sizes. This type of emulsification method requires less surfactant when compared to high-energy methods. This low-energy approach has a restricted size distribution range varying with the membrane type, porosity, average pore size, cross-flow velocity, emulsifier, and trans membrane pressure. Moreover, some shear-sensitive components, like proteins and starch, are utilized in the transit of a dispersed phase across a membrane into a continuous phase, resulting in reduced shear stress impact. However, this approach can't be used on a large-scale production due to the low flow of droplets through the membrane [17].

3.1.2 Spontaneous Emulsification (SE)

Aqueous and oil phases are mixed to create a spontaneous emulsion. The aqueous phase is a mixture of water and a surfactant, whereas the organic phase constitutes oil and a hydrophobic surfactant [22]. When the organic and aqueous phases are mixed without first stirring, an initial nonequilibrium condition results, however, spontaneous emulsification takes place under specific conditions. Some of the steps involved in the process of spontaneous emulsification are listed below.

(1) Mixing of oil with hydrophobic surfactants and water with hydrophilic surfactants, to prepare organic phase and aqueous phase, respectively.
(2) Inoculation of the organic phase into the aqueous phase with constant stirring.

(3) Evaporation of water-miscible solvent under reduced pressure.

The process is spontaneous, as evidenced by the rise in entropy and fall in Gibbs's free energy. Various mechanisms, including surface tension gradient, interfacial turbulence, solute diffusion between two phases, condensation mechanism, and dispersion mechanism, have been presented to explain spontaneous emulsification [22, 23].

SE has been widely used for nanoemulsification (Table 1). Barzegar et al. created peppermint essential oil nanoemulsions with added Tween 20/40/80 as surfactants using SE and discovered that the best nanoemulsions had droplet sizes of roughly 50 nm [24]. Additionally, Yildirim et al. used SE to create stable nanoemulsions. The best nanoemulsions had great physical stability and antibacterial activity with particle sizes of 100 nm [25]. Cinnamaldehyde/medium chain glycerides nanoemulsions with tween 20/40/60/80/85 surfactants have also been prepared using the SE method [26]. Also, Lui Wang.et.al. prepared nanoemulsion with eugenol by SE which was found to have a storage period of 30 days at 22 °C [27]. With the same process, Mariss et al. prepared oil in water nanoemulsion with vitamin D as a bioactive compound which had stability of 30 days at room temperature [28].

3.1.3 Phase Inversion Techniques

Phase inversion techniques, also referred to as emulsion phase inversion (EPI), are commonly used to describe the production of w/o nanoemulsions utilizing conventional high-speed mixers followed by subsequent conversion into o/w nanoemulsions induced by changing the composition or temperature. Phase inversion induced by changing the composition is termed as phase inversion composition, while if temperature induces phase change it is called phase inversion temperature [1]. The phase inversion techniques, in particular, identify transitory inversion caused by changing elements that alter the lipophilic-hydrophilic balance in the system [29].

3.1.3.1 Phase Inversion Composition (PIC)

The PIC-based nano-emulsification technique includes the gradual incorporation of the aqueous phase into the oil phase, which results in a continuous rise in the water volume percentage. A phase inversion occurs at a certain level, resulting in the formation of a bicontinuous phase that can eventually capture oil phases into the water phase and create o/w nanoemulsions. The PIC approach has several benefits, such as minimal cost and minimal requirement for basic equipment. Nevertheless, due to the weaker driving pressures of the PIC approach, the preparation time is found to be greater than that of the spontaneous emulsion technique [30].

Using the PIC approach, Ostertag et al. created stable nanoemulsions (Table 1). The produced nanoemulsions had particle diameters of 200 nm [31]. Borrin et al. also created curcumin-loaded nanoemulsions using the PIC approach and used them

Table 1 Application of low-energy methods for the preparation of nanoemulsions

Emulsification method	Dispersed phase composition	Bioactive compounds encapsulated	Optimal processing condition	Droplet size (nm)	Remarks	References
Spontaneous emulsification	Peppermint essential oil/Tween 20, 40, 80	Peppermint essential oil	• Magnetic stirring at 600 rpm for 10 min • At room temperature	30–75	The speed of mixing, type of surfactant, surfactant-to-oil ratio, and temperature affected the droplet size and the storage stability	[24]
	Cinnamon oil and coconut oil/Tween 80	Cinnamon oil	• Stirring constantly at 750 rpm • Ambient temperature	<200	Antimicrobial against *E. coli* was enhanced	[25]
	Cinnamaldehyde/ medium chain triglycerides/Tween 80	Cinnamon oil	• Constant stirring at 400 rpm • 15 min, room temperature	8–300	Significant antimicrobial activity against. *S. aureus, E. coli, C. albicans*	[26]
	Eugenol oil	Eugenol oil	• Deprotonated eugenol in hot alkaline added to surfactant mixtures • Mixtures acidified to pH 7.0, • Stirring at 600 rpm, room temperature	≈109–139	Stable for 30 days at 22 degree Celsius Good entrapment efficiency with NaCaS and zein mass ratio 1:1	[27]
	Vitamin D/Medium chain triglycerides 10% wt/Tween 20, 40, 60, 80, 85	Vitamin D	• Constant stirring at 600 rpm • 15 min, 25 degree Celsius	160	Enhanced stability (1 month) Possibility of increase in growth of droplets at high temperature	[28]

(continued)

Table 1 (continued)

Emulsification method	Dispersed phase composition	Bioactive compounds encapsulated	Optimal processing condition	Droplet size (nm)	Remarks	References
Phase inversion composition	Limonene mineral oil, olive oil, canola oil, orange oil, sesame oil, grape seed oils, medium chain glycerides	Limonene	• Titration of aqueous phase • Constant stirring at 750 rpm for 1 h, 25 °C	<200	Disadvantage of high usage of surfactant as it produces small droplets	[31]
	Sunflower oil, curcumin/polysorbate 80	Curcumin	• Organic phase was mixed using magnetic stirring for 15 min • Organic phase and aqueous phase were mixed together using a magnetic stirrer for 30 min	–	Successful incorporation of curcumin-loaded nanoemulsions in pineapple ice creams to replace artificial yellow dyes	[32, 33]
	Virgin coconut oil, essential oils blend/ polyethylene Glycol, sorbitan monooleate	Limonene and linalool	• Slow titration of the aqueous phase into the oil phase with continuous stirring at room temperature	29.55–37.2	Highly stable at various storage conditions	[34]
	Canola oil, EPA, DHA/ Tween 20, 40, 80, 85, Span 20, 80	EPA, DHA	• The organic phase was mixed using a magnetic stirrer for 30 min after being added with an aqueous phase (deionized water) and the system was stirred for another 30 min at a flow rate of 2 drops per second • Constant stirring at 300 rpm	<150	Application of DHA- and EPA-based nanoemulsion in liquid food, and to improve the physicochemical stability	[35]

(continued)

Table 1 (continued)

Emulsification method	Dispersed phase composition	Bioactive compounds encapsulated	Optimal processing condition	Droplet size (nm)	Remarks	References
	Medium chain triglyceride, orange oil/ Tween 20, 40, 60, 80, 85	Orange and limonene oil	• Organic phase and surfactant stirring at 300 rpm for 30 min • Aqueous phase was titrated into the organic phase, along with constant stirring (500 rpm) at a flow rate of 20 drops per 10 s using a burette	41	Smallest droplets were produced when 8% vitamin E acetate 2% Medium chain triglycerides 2 wt% and 20 wt% Tween 80	[36]
Phase inversion temperature	Cinnamaldehyde/ benzyl alcohol	Cinnamon oil	• Mechanical stirring for 30 min • Heated above PIT till 15 degree Celsius • temperature reduced to PIT	101	Enhanced antimicrobial activity on adding nanoemulsions	[37, 38]
	Sodium caseinate	Lemon oil	• Emulsion heated till 21–98 degree Celsius • Quenched in ice water • Shaking	≈100	2%(w/v) NaCs short-term stability	[39]
	Hydrogenated castor oil	Curcuminoids	• Mixing the components • Temperature combination range (90–60, 60–90, 90–60,90–75)	20–100	Nanoemulsions were stable for 2 months	[40]

to replace artificial yellow pigments in pineapple ice creams [32, 33]. PIC was also used by Nantarat et al. to create nanoemulsions with average droplet sizes ranging from 29.55 to 37.12 nm. High stability was found in nanoemulsions made with incorporated limonene and linalool as bioactive compounds in different oils (canola oil, olive oil, grape seed oil, sesame oil) [34]. Also, nano-emulsified EPA/DHA with carrier canola oil and Tween 20/40/60/80/85 and Span 20/80 as surfactants have been formulated using PIC for application in liquid food to improve the physiochemical stability [35]. Moreover, Mayer et al., also reported the use of the PIC method to prepare nanoemulsions with medium chain triglyceride and orange oil as dispersed phase and Tween 20/40/60/80/85 as surfactants [36].

3.1.3.2 Phase Inversion Temperature (PIT)

The PIT approach makes advantage of the phase inversion property of the molecules, wherein the lipophilicity and hydrophilicity of the emulsifiers at a given composition, vary with the temperature (Fig. 3). The spontaneous curvature of non-ionic surfactants changes depending on the hydration of their polar heads. Oil-in-water emulsions are developed at low temperatures, whereas water-in-oil emulsions are formed at elevated temperatures because the solubility of emulsifiers in water diminishes with increasing temperature. The temperature at which an emulsion transforms from o/w to w/o is referred to as phase inversion temperature [17].

Chuesiang et al. created nanoemulsions utilizing the PIT technique, in which a composition of oil, water, and surfactants was heated above the phase inversion temperature followed by cooling abruptly to fabricate nanoemulsions with <120 nm particle size [37, 38] (Table 1). Su et al. also used the PIT approach to create nanoemulsions. The nanoemulsions had droplet sizes of ~100 nm and were stable for more than 15 days [39]. Besides, Jintapatanakit et al. used the PIT approach to create nanoemulsions with droplet size of 20–100 nm, and polydispersity index less than 0.2 [40].

3.2 *High-Energy Methods*

High-energy methods are associated with the use of high mechanical energy for the disruption of the droplets. Application of a high amount of mechanical energy is done by using mechanical devices such as homogenisers, microfluidizers, or high-shear stirring devices. As high-energy methods are permitted to use natural emulsifiers at a lower concentration, hence they tend to be more felicitous for the preparation of nanoemulsion. This method is also suitable for industrial-scale production. Generally, there are only two steps involved during oil-in-water nanoemulsion production with a high-energy method. The first step is the creation of coarse oil-in-water emulsion by combining the two components using a high-speed mixer or stirrer. Secondly, the coarse emulsion is subjected to disruptive forces to reduce the droplet diameter to

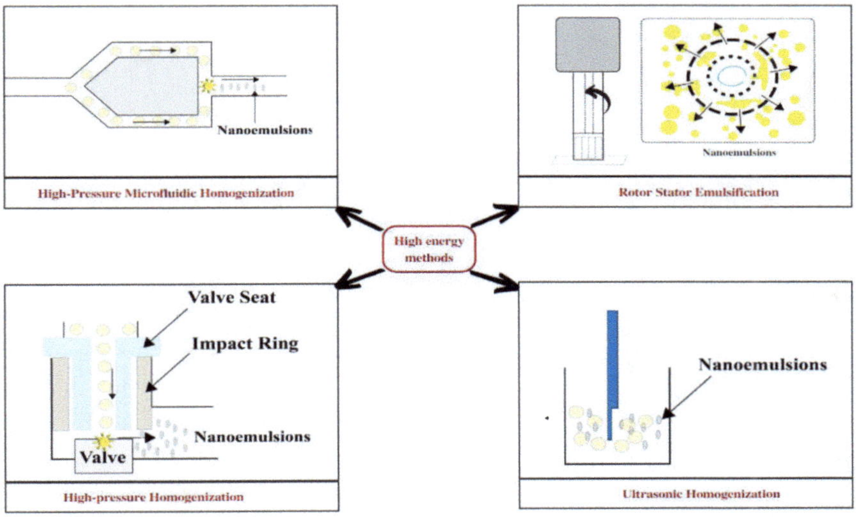

Fig. 4 Schematic diagram to represent various high-energy approaches to produce food-grade nanoemulsions

200 ~500 nm to form the nanoemulsions. High-energy technique devices that are used include rotor stator emulsification, high-pressure homogenization, microfluidic homogenization, and ultrasonic homogenization (Fig. 4) [41].

3.2.1 Rotor Stator Emulsification (RSE)

RSE or high-speed homogenization (HSH) has been commonly utilized in food, cosmetics, and pharmaceutical industries for emulsion production [42]. This method is generally regarded as standard for preparing emulsion with respect to the intermediate-to-high viscosity of the dispersed phase and its volume fractions [43, 44]. The rotor of the RSE head is attached to a rotor shaft and a perforated stator screen exterior to the rotor shaft which works on the same principle as that of the impeller in a centrifugal pump, wherein the fluid enters axially in the rotor along the rotor shaft and the rotor accelerates the fluid tangentially into the outer clearance region of the rotor. As the mixture is passed through the rotor–stator region, the particle size reduces. The size and the hydrodynamic intensity of the subsequent emulsion are greatly dependent on the rotor speed. For industrial applications, the rotor speed ranges from 10–30 m/s [11]. Karthik et al. reported the application of rotor–stator emulsification to produce nanoemulsion incorporated with DHA (docosahexaenoic acid), and particle size was reduced to 87 nm at 800 bar pressure [45] (Table 2).

Table 2 Application of high-energy methods for the preparation of nanoemulsions

Emulsification method	Dispersed phase	Bioactive compound	Processing conditions	Droplet diameter (nm)	Evaluated remarks	References
Rotor stator emulsification	Docosahexaenoic acid	–	24,000 rpm stirring for 25 min 800 bar 8 process cycles; 4 °C	87	Increase in storage duration (100 days) antimicrobial activity	[45]
High-pressure valve homogenization	Soybean oil, pepper extract/tween 20, 40, 60, and 80; Span 20, 40, 60, and 80)	Capsaicin	6.9–13.8 *10^7 Pa for 5 to 20 cycles	132–145	The type and concentration of surfactants, hydrophilic-lipophilic balance, lipid/ aqueous phase ratio, surfactant/oil ratio, pepper extract composition in nanoemulsion, and processing conditions influenced the prepared nanoemulsions	[48]
	Canola oil, sunflower oil, linseed oil, corn oil, and medium chain glycerides/lecithin, Tween 80	Curcumin	60 MPa for three cycles	203.6–260.6	The increase in oil phase concentration increased the curcumin content, particle size, and viscosity of the emulsions, but decreased their stability	[49]
	Carvacrol, bergamot, mandarin, and lemon essential oils; sunflower oil/ glycerol monooleate, Tween 20	Carvacrol	200 MPa; 80°C; 5 processing cycles	133–180	Emulsion exhibited antibacterial activity	[50]

(continued)

Table 2 (continued)

Emulsification method	Dispersed phase	Bioactive compound	Processing conditions	Droplet diameter (nm)	Evaluated remarks	References
	Sunflower and olive oil	–	200 MPa; 25 °C; 1 processing cycle	110–112	-	[51]
High-pressure microfluidic homogenization	Thyme, lemongrass, and sage essential oil/Tween 80, glycerol	Thyme, lemongrass and sage essential oil	150 MPa Temperature less than 15 °C 3 process cycles	50–85	Edible films containing thyme essential oil had the highest antibacterial efficacy against inoculated *E. coli,* attaining up to 4.71 log reductions after 12 h	[55]
	Canola and ginger oil	Ginger essential oil	69 MPa 3 Process cycles	133–138	Antioxidant property, Structural roughness found in film with incorporated ginger essential oil-based nanoemulsions	[56]
	Rutin, soybean oil/ Tween 80	–	100 MPa 3 process cycles	150	Antioxidant activity, Rutin declined to 22–35% after storage for 30 days at 25 degree Celsius	[57]
	Lemon grass oil/ Tween 80	Lemon grass oil	150 MPa 3 process cycles > 20 °C	7–10	Increased processing pressure and cycles through the interaction chamber of the microfluidizer device reduced the average droplet size, viscosity, and whiteness index of nanoemulsions	[58]

(continued)

Table 2 (continued)

Emulsification method	Dispersed phase	Bioactive compound	Processing conditions	Droplet diameter (nm)	Evaluated remarks	References
Ultrasonic Homogenization	Resveratrol and cyclodextrin	–	1000 rpm Power sonication at 350 W for 5 min 50 °C	19–30 (resveratrol and incorporated cyclodextrin)	Nanoemulsions had optimal stability	[62]
	Eucalyptus oil	–	750 W/20 kHz 10 min	9	Nanoemulsion-incorporated chitosan films had better antibacterial properties	[61]

3.2.2 High-Pressure Homogenization (HPH)

HPH reduces the coarser pre-emulsion droplet size, prepared by the rotor–stator mixer, directly into nano-size particles [11]. Macroemulsions are pushed with the help of a piston pump through a narrow valve downstream, which turns the macro-droplets into much smaller size droplets as the macroemulsion droplet breaks apart due to the extreme mechanical forces, i.e., hydraulic shear and turbulence. Moreover, after repetitive several processing cycles ideal size nanoemulsions are formed. Three primary variables recreate the significance of optimization for HPH: system temperature, working pressure, and number of pass cycles [46]. Although HPH has great efficiency, it is highly exothermic and requires very high energy for the formation of nanoemulsions [47].

Galvao et al. prepared nanoemulsion by using HPH, preceded by the formation of coarse emulsion by mixing soybean oil/pepper extract with surfactants and aqueous phase using high-speed homogenizer (Table 2). The mean droplet size of nanoemulsions ranged from 132 to 145 nm, depending on the working pressure enforced and the number of times the mixture is passed through the machine [48]. A similar study was conducted by Ma et al., wherein the formed nanoemulsions had a particle size of 203.6–260.6 nm [49]. Besides, Severino and Ferrari stated that antibacterial property was enhanced when Carvacrol was used in the mixture of sunflower and other essential oils. The particle size of the nanoemulsion was reduced to 133–180 nm by passing it through the high-pressure homogenizer at 200 MPa for 5 processing cycles [50]. In another study, the droplet size of sunflower oil-olive oil nanoemulsions was reduced to 110–112 nm when a high pressure of 200MPa was applied by Hebishy et al. [51].

3.2.3 High-Pressure Microfluidic Homogenization (HPMH)

Emulsification by microfluidic homogenization works on the same principle as high-pressure valve homogenization, except that the microfluidization technique uses a particular microchannel with diameters ranging from 50 to 300 m to create droplets [1, 52]. In a microfluidizer, the product is pumped using a high-pressure positive displacement pump which works on 500–20,000 psi, and then is passed through the interaction chamber made of rigid stainless steel microchannels to create sturdy coherent flow. Firstly, a coarse emulsion is formed by a homogenizer by processing the two phases with a surfactant. The piston amplifies the high-pressure air (~100 psi) in a pulse mode to create liquid pressure that can get as high nearly 30,000 psi. The premixed coarse emulsion has a droplet size of less than 10μm. The microscale droplets are routed through the microfluid channel and transmitted to the interaction area where extreme coherent shear flow brings out stable emulsion. Moreover, the coarse emulsion is cycled many times through the interaction chamber in a microfluidizer to achieve the desired particle size [53]. Finally, the large droplets are eliminated by passing the bulk emulsion through a filter to obtain the nanoemulsion. Moreover, microfluidic flow generates temporal and spatial inhomogeneities. As a result, the produced nanoemulsion must be recirculated through a high shear zone [54].

Acevedo et al. used high-pressure microfluidic homogenization to produce nanoemulsion based on thyme, lemon grass, and sage essential oils with Tween 80/glycerol as surfactants (Table 2). Coarse emulsion was first prepared by mixing the aqueous phase and organic phase in a high-speed mixer at 17,500 rpm for 2 min followed by microfluidization at 150 MPa for three cycles (15 °C) [55]. Alexandre et al. (2016) also employed high-pressure microfluidization to prepare ginger essential oil and canola oil nanoemulsions. The average droplet size of obtained nanoemulsions ranged between 133 and 138 nm [56]. Besides, a similar process was utilized to produce nanoemulsions of soybean oil/rutin and lemongrass oil with Tween 80 as surfactants as reported by researchers [57, 58].

3.2.4 Ultrasonic Homogenization (USH)

Ultrasonic emulsification is the process of creating emulsion droplets from two immiscible liquids in the presence of a surfactant by using sound waves of low or high frequencies. Two processes are used to create the nanoemulsion for low-frequency sound waves. First, the formation of interfacial waves, which are caused by an acoustic field that directs the breakup of an oil phase into an aqueous phase as droplets. Second, by employing acoustic cavitation, which results in the generation and dissolution of microbubbles as a result of pressure fluxes from sound waves, extremely high levels of localized turbulence are generated, and the original droplets are broken down into submicron size. Moreover, in the case of the application of high-intensity sound waves, the surrounding liquid is subjected to intense shock waves, that result in the production of emulsion droplets. However, subsequent recirculation of the emulsion is essential, owing to the inhomogeneity cast by the sound field [59]. This technology was discovered to be extremely successful for small-batch manufacture, however, is not appropriate for large industrial-scale applications [54]. The laboratory-scale ultrasonic homogenizer has numerous advantages, including affordability, ease of use, and excellent energy efficiency. Further, under optimal conditions added benefits are feasible, such as less emulsifier content requirements, superior dispersion stability, and lowered risk of microbial contamination at the entrance of processing equipment [60].

Ultrasonic homogenization is reported to have been employed to prepare stable nanoemulsions (Table 2). Sugumar reported the formation of coarse emulsion by mixing eucalyptus oil, Tween 20/80, and water using a magnetic stirrer, followed by ultrasonic emulsification using a 20 kHz sonicator with a maximum power output of 750 W [61]. In another study, Kumar et al. also used the USH approach to create nanoemulsion. Two separate nanoemulsions prepared with revestrol and cyclodextrin along with lecithin and Tween 80 had average diameters of 24.48 nm and 20.41 nm, respectively [62].

4 Evaluation for Stability in Nanoemulsions

The undesirable interaction of molecules in the emulsion is due to its hydrophobic nature, which in turn leads to instability in nanoemulsion thermodynamics. Thus, eventually, nanoemulsions degrade and destabilize following various causes such as Ostwald ripening, droplet aggregation, and gravitational separation [63, 64]. Understanding the fundamental principles of nanoemulsion instability is critical for building emulsion systems with acceptable stability properties (Fig. 5).

4.1 Ostwald Ripening

The Ostwald ripening happens when the droplet mean size in a nanoemulsion increases over the span of time by oil droplet diffusion from small to larger droplets through a continuous phase. [65]. Small oil droplets in freshly prepared nanoemulsions have high aqueous-solubility, resulting in a huge quantity of dissolved oil molecules. However, with the passage of time, these solubilized oil molecules migrate to bigger droplets, causing the droplet size to gradually expand resulting in Ostwald ripening [66, 67]. The ripening can be decreased by using surfactant at a higher concentration as it leads to the formation of a large number of micelles, and a rise in the number of micelles in the continuous phase prevents the dispersion of the oil particles into the continuous phase. When the concentration of surfactant is below critical micelle concentration, small droplets with low interfacial tension have a mono-dispersed distribution which decreases the Oswald ripening process. Polymeric surfactant in o/w nanoemulsion results in a decrease in the Ostwald ripening process as it gets absorbed at the oil/water interface and modifies the interfacial tension. Ostwald ripening can occur more quickly in nanoemulsions prepared with comparatively higher water solubility oils (essential oils, short-chain triglycerides,

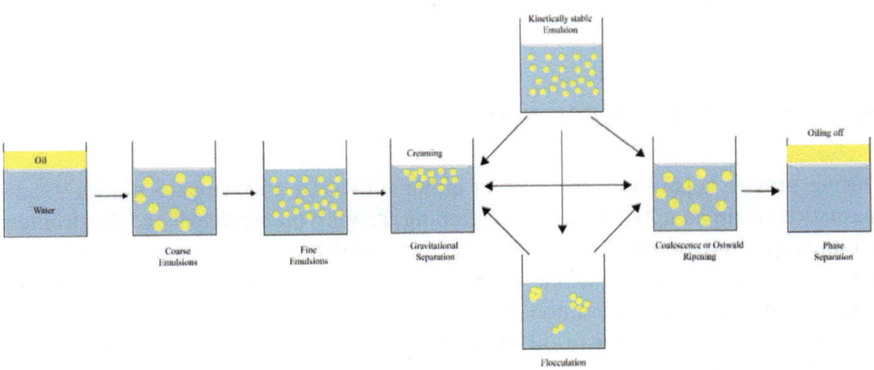

Fig. 5 Schematic diagram to represent various mechanisms responsible for physical instability in nanoemulsions

and so on) when compared to nanoemulsions prepared with oils that have lower water solubility like long-chain triglycerides (fish oil, corn oil, or sunflower) [68]. Hence, it is essential to know the water-solubility of the oil phase. Moreover, a ripening inhibitor can be used for retardation of the Ostwald ripening. The ripening inhibitor is a non-polar molecule, insoluble in the aqueous phase but has good solubility in the oil phase. These inhibitors (such as long-chain triacylglycerol) prevent the small oil droplets from moving toward large droplets because they are less soluble in the aqueous phase [69].

4.2 Droplet Aggregation

Colloidal interaction between droplets causes two forms of droplet aggregation, namely (a) coalescence and (b) flocculation. Flocculation is the mechanism through which two or more droplets are drawn together to form clusters [8]. Using amphiphilic polysaccharides as emulsifiers, over a particular level, results in flocculation which might cause emulsion instability, and subsequently would reduce the long-term stability of various nanoemulsion products [70]. While coalescence is a process in which a large droplet is formed by collision and amalgamation. This mechanism is caused due to low electrostatic repulsion among the droplets and due to the low pH of the emulsion due to the presence of salt (200–500 mM). Destabilization of nanoemulsion by coalescence can also be due to the addition of n-alcohol in excess amounts. The characteristics of the colloidal interactions used in a specific nanoemulsion depend on the physical and chemical characteristics of the phases, the particle-core characteristics, (such as radius); the shell characteristics, (including charge, hydrophobicity, packing, rheology, and thickness), as well as the characteristics of the continuous fluid (such as ionic strength, pH, temperature, and osmotic pressure) [70]. In a nanoemulsion, the amount the van der Waals, electrostatic, steric, and hydrophobic interactions between two droplets can be used to illustrate the whole range of colloidal interaction[41, 71].

4.3 Gravitational Separation

During gravity separation, the droplet either moves in a downward motion (sedimentation) or in an upward motion (creaming) relying on the density of the droplets and the liquid surrounding it. Water in the nanoemulsion tends to separate and move downward due to gravity, whereas oil being less dense moves in an upward direction, hence creaming is more common in oil-in-water nanoemulsions and sedimentation generally happens in water-in-oil nanoemulsions. Droplet size, density difference, and a decrease in aqueous phase viscosity all impact the motion speed of droplets caused by gravitational separation [41].

5 Characterization of Nanoemulsions

Characterization of nanoemulsions based on their physiochemical properties like physical properties, stability, rheological effects, and microstructure, is important to decide their application in food as they affect the final flavor, texture, taste, and stability of foods. Particle size, charge, and lipid crystallinity are the fundamental physical characteristics of nanoemulsions. A few characterization techniques are discussed in this section.

5.1 Dynamic Light Scattering

The particle size distribution and its structure can be measured by dynamic light scattering (DLS). The proportion of different-sized suspended tiny particles or droplets in suspension, with the range of 3nm-5μm, can be measured through DLS. It measures the fluctuations in intensity which is a result of relative spatial location as a consequence of Brownian motion. In case of nanoemulsions, DLS measurements employ the "Mie theory" mathematical tool to estimate the scattering pattern of emulsion droplets. This technique also measures the polydispersity index, which is the broadness of size distribution. The polydispersity index describes the homogeneity of the nanoemulsion system. Additionally, DLS may also be used to assess the stability of nanoemulsions under different environmental conditions [72, 73].

5.2 Zeta Potential

The distributed particles between the low-charged surface and the high-charged surfaces bind the stationary layer fluid. The potential distinction between the stationary and dispersion medium is called the zeta potential. The zeta potential value corresponds to the stability of the colloids. It also tells about the degree of repulsion between adjacent, equal-charged particles within the dispersion. Nanoemulsions are considered electrically stable colloidal dispersions if the zeta potential value is ± 30 mV, while emulsions with low zeta potential coagulate. A low zeta potential value causes flocculation in the system as the attraction between the molecules becomes higher than usual. Hence, from the zeta potential calculations, the stability of the system and surface charge properties of nanoemulsions can be known. Zeta potential value is affected by many factors including type of surfactants, electrolyte concentration, source of particles, particle morphology and size, pH of the solution, and state of hydration [74].

5.3 *Conductivity*

Conductivity determines the traits of the continuous phase and phase inversion phenomena in the nanoemulsion. Electrical conductivity can be measured through a digital conductivity meter at ambient temperature. Water-in-oil nanoemulsion has a low conductivity value, whereas oil-in-water nanoemulsion has a higher conductivity value [75].

5.4 *Rheology*

Shear device is used to assess the apparent viscosity and the interrelation between apparent viscosity and shear stress in nanoemulsions. The viscoelastic characteristics of emulsions are evaluated using dynamic oscillation techniques. At low shear rates(~$0.01 s^{-1}$), the flow curves of stable emulsions have a constant apparent viscosity and substantial shear thinning at high shear rates [76]. Surfactants, water, and oil concentrations and their nature affect the viscosity of nanoemulsions. Viscosity is inversely proportional to the amount of water in the solution. With an increase in the amount of surfactant and cosurfactant, the viscosity reduces due to a reduction in the interfacial tension between the aqueous and organic phases. The value of viscosity gathers the knowledge of the rheological properties and stability of the nanoemulsions [77].

5.5 *Interfacial Tension*

The degree of interfacial tension that exists between two liquids can be used to determine the mechanism of nanoemulsion creation and its properties. The addition of surfactants greatly affects the interfacial tension in the emulsion, as it would lower interfacial tension. The value of interfacial tension is measured using a spinning drop instrument. This can be done by rotating a low-density phase drop in a high-density phase-filled cylindrical tube and analysing the drop shape [78].

5.6 *Differential Scanning Calorimetry (DSC)*

By using a thermo-analytical technique, differential scanning calorimetry measures the lipid crystallinity of nanoemulsions. This technique gives knowledge about the impact of oil crystallization on the stability of nanoemulsion by measuring the difference in the crystallization temperature of pure oil and oil in the emulsion. Nanoemulsion stability is altered due to the formation of crystals and nucleation which is the result of impurities in the oil. DSC gives information on melting, crystallization, and phase transition, and hence, can be used to investigate the structure and stability of the nanoemulsions. Moreover, the DSC instrument can be used to calculate the percentage of the destabilized fat in the emulsion and the effect of droplet size in the oil phase by analyzing its effect on crystallization [79, 80].

5.7 Small-Angle X-Ray Scattering

Small-Angle X-Ray Scattering (SAXS) is employed to characterize the colloidal particles on the basis of their size, shape, and nanostructure. The photon-electron interaction is the fundamental mechanism by which materials scatter X-rays, wherein, the sample and X-ray collide, which leads to the scattering and subsequent change in intensity and scattering angle of the X-rays. As a result, the electron density function directly affects the characteristics of this function. The obtained SAXS pattern is a function of intensity and scattering vector produced at scattering angles of less than 10 degrees. The scattering intensity is concentrated within a very tiny angle domain, near the direct X-ray beam, which is covered by the narrow direct beam stopper placed close to the detector, if the objects are very large relative to the X-ray wavelength (sizes exceeding 1 nm). Thence, SAXS is used for examining material structures from a great distance or at acute angles. In the case of nanoemulsions, this technique is used to examine the structure, and shape-size of macromolecules in emulsion, characteristic distances of partially ordered materials, and pore diameters in the emulsions [81].

5.8 Nuclear Magnetic Resonance

Nuclear magnetic resonance (NMR) methods have also been utilized to investigate the kinds, structure, and diffusion properties of nanoemulsion components [77, 82]. The Fourier transform pulsed gradient spin echo (FT-PGSE) technique, one of the NMR techniques, evaluates the self-diffusion coefficients of the molecules of oil, water, and surfactant all at once. It is also important in assessing emulsion connectivity, and the PGSE aids in determining the phase transition between the droplet emulsion phase and bicontinuous emulsion [83].

5.9 Imaging Techniques

Microscopic imaging approaches are used to identify the shape, size, and orientation of the components inside the nanoemulsion sample. The characterization of nanoemulsions using microscopic methods is piquing the attention of researchers.

It might be regarded as the most useful and dependable tool for characterization of nanoemulsions. Particle sizes in nanoemulsion formulations range from submicrometer to nanoscale and as standard light microscopy cannot view submicrometer particles, electron microscopy is utilized, which provides adequate resolution for the imaging of nano-sized objects (<5 nm). Electrons have shorter wavelengths than visible-light photons used in optical microscopes resulting in the increased spatial resolution in electron microscopy [2, 77, 84]. The size, form, and internal structure of nanoemulsions have all been studied using scanning electron microscopy (SEM), transmission electron microscopy (TEM), and atomic force microscopy (AFM) techniques.

SEM is used for the determination of microstructure morphology and chemical content at the nanoscale or micrometer scale, as it employs a narrow electron beam to the surface structures [85]. As a result, SEM is utilized to determine the surface structure of nanoemulsion oil droplets [84]. SEM can process vast volumes of material at low magnification and provide high-resolution pictures at greater magnification. The disadvantages of SEM include the expensive cost, the usage of high vacuum, and the high sample conductivity. Moreover, surfactants present in the sample can also coat the particle surfaces, which can interfere with SEM imaging [81]. Besides, TEM is reported to have been used to investigate the shape and structure of nanoemulsions (resolution ~0.2 nm) [53]. Nevertheless, high-energy electron beams are used in sample preparation for TEM, which can cause structural damage to the materials [81], thus cryogenic preparation (cryo-TEM) and freeze-fracture (FFTEM) procedures can be used [77]. Additionally, AFM microscopy is used to characterize the structural properties of nanoemulsions. It senses force changes between a form probe and an immobilized sample in order to generate a high-resolution 3-D profile of the sample's surface [74].

6 Nanoemulsions in Food Packaging

A food packaging material should primarily prevent contamination, protect the food contents, and communicate its nutritional properties to the consumers. Additionally, it should establish the conditions essential to preserve the products' quality, safety, and shelf life while they are being transported and stored. Moreover, the packing material should be able to offer the fundamental qualities of mechanical, optical, and thermal properties. Choosing the right food packaging should also ensure that there are no unfavorable changes in food quality (such as color and flavor), microbial growth, lipid oxidation, or loss of nutrients in the food [86]. Active packaging, which includes the addition of bioactive components, has been introduced as it aims to extend the shelf life and preserve the product quality. Spray drying, spray cooling, spray chilling, fluidized bed coating, coacervation, interfacial polymerization liposome entrapment, nanoemulsion, air suspension, co-crystallization, freeze-drying, molecular inclusion, and sol–gel are all methods that can be used to incorporate bioactive chemicals into environmentally friendly food packaging. A further step is required in the creation of films, coatings, and membranes containing nanoemulsions for applications relevant to food packaging: the "solidification" of the already-formed nanoemulsion. A supporting polymer/matrix is often spread in the continuous phase to achieve this. In the continuous phase, the addition of (bio)polymer can be essential to the droplet mono dispersity and stability, even before the act of solidification occurs, due to the thickening effect, which occurs when the viscosity increases during the continuous phase, the rate of inter-droplet coalescence reduces [87, 88]. Table 3 represents (bio)polymers that have been referred to as continuous phases in food packaging materials made of nanoemulsions, as well as the fundamental compositional features of the dispersed phase and the processing parameters for the preparation of nanoemulsion-based films/coatings. These methods mostly blend the biopolymers with organic antibacterial elements present in plant essential oils, that are made to stay encapsulated within the oil droplets of an o/w nanoemulsion that has already been created.

6.1 Production of Nanoemulsion Based Films/Coating

The production of nanoemulsion-based food packaging requires the combination of nanoemulsions (liquid phase) with polymeric film (solid phase). It is essential to create a film with uniform dispersion of nanoemulsions inside its matrix while retaining the true properties of the nanoemulsions. The most prevalent procedures for making films are solvent casting (SC), extrusion (EX), electrospinning (ESP), melt mixing (MM), and film spreading (FS). For the integration of nanoemulsions in polymeric films, SC has been widely used in the literature, and ESP and FS [89–91] have seen moderate applications [92], while EX and MM have not been considered appropriate because of the high operating temperature, the inapplicability of some materials, and the control challenges, moreover, high temperatures cause the active components to evaporate and get denaturized since the heating process is inappropriate for the liquid phase. [93].

As mentioned before, SC is primarily used in the manufacturing of various sorts of films, particularly food packaging films, that comprise a variety of nanomaterials, including nanoparticles and nanoemulsions (Fig. 6). This approach is widely used since it requires less preparation time, is less expensive, and is simple to use and modify. The most popular technique for creating films with nanoemulsions is to first disperse the active ingredients within a continuous phase containing the film-forming matrix, followed by the addition of an emulsifier, homogenization, or another suitable mixing process, and finally to create an evenly dried layer, cast the film-forming compositions at a predetermined thickness [94]. However, it is expected that the initial structural conformation will change when nanoemulsions are added to polymer matrixes, demanding structural, morphological, thermal, mechanical, and barrier characterizations [1].

Table 3 Nanoemulsions used in food packaging

Food product	Film/coating	Essential oil/bioactive compound in nanoemulsion	Polymer film material	Method of nanoemulsion preparation	Beneficial effects	References
Beef	Film	Eugenol	Poly vinyl pyrrolidone	Electrospinning	Nanofibrous membranes demonstrated antioxidant action	[97]
Fish sausage	Coating	Tocopherol	–	High shear mixer	Delay lipid oxidation and improved fish sausage quality	[98]
Chicken fillets	Coating	Curcumin-cinnamon essential oil, Curcumin-garlic essential oil, Curcumin-sunflower oil	Pectin	High-speed stirring and homogenization	Shelf life enhancement	[99]
Tomato	Coating	Sweet orange essential oil	Sodium alginate	Ultrasonication homogenization	Delayed ripening, improved post-harvest quality	[100]
Chinese bayberry fruit	Film	Cinnamon	Polyvinyl alcohol	High shear mixer	Active packaging to control the decay of fruit	[101]
Strawberry	Coating	Cinnamon essential oil	Pullulan	Ultrasound treatment	Prolong shelf life of fruit, during room storage	[102]
Papaya	Coating	Sunflower oil	Psyllium gum	Mixing	Improved shelf life	[103]
Okra	Coating	Basil oil	Sodium alginate coating	Blending and Ultrasonication	Shelf life extension	[104]

(continued)

Table 3 (continued)

Food product	Film/coating	Essential oil/bioactive compound in nanoemulsion	Polymer film material	Method of nanoemulsion preparation	Beneficial effects	References
Cantaloupe (fresh cut)	Coating	Cinnamaldehyde	Chitosan	Mixing	Shelf life up to 7–9 days. effective against mesophilic microorganisms	[105]
Cucumber slices	Coating	Carvacrol	Chitosan	High-pressure homogenization	Increased microbial reduction	[106]
Grape berry	Coating	Lemongrass oil nanoemulsions	Carnauba wax	Dynamic high-pressure processing	Reduction in microbial effect by inhibition of *Salmonella typhimurium* and Escherichia coli	[107]
Green beans	Coating	Mandarin essential oil	Coating of modified chitosan containing a nanoemulsion	High hydrostatic pressure	Reduction in the growth of *L. innocua* with the storage time anti-microbial effect	[108]
	Coating	Carvacrol, bergamot, mandarin and lemon essential oils	Chitosan	High-pressure homogenization	Reduced growth of *E. coli and S. typhimurium*, during storage	[50]
Plum	Coating	Lemongrass oil	Carnauba wax	Dynamic high-pressure processing	Enhanced storage quality microbial growth depletion	[109]
Fish fillet	Coating	Curcumin	–	Phase inversion	Retarded growth of spoilage bacteria improves the storability	[110]
Sliced bread	Film	Clove bud and oregano essential oil	Methyl cellulose	Low-speed mixing and ultrasonication	Improved antimicrobial properties	[111]

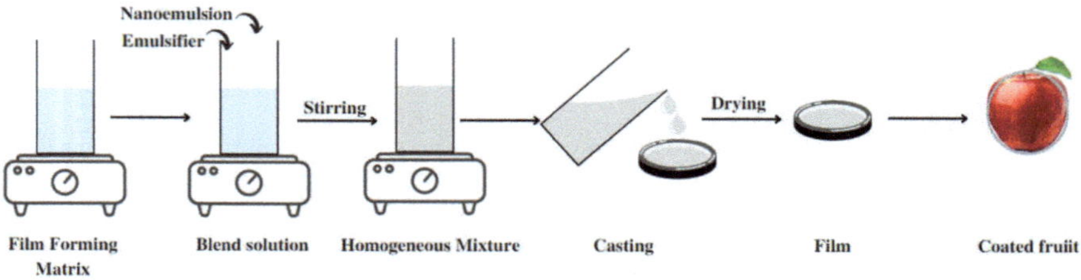

Fig. 6 Schematic representation of the preparation of nanoemulsions-based food film/coating

6.2 Application of Nanoemulsions-Based Food Film/Coating on Food Materials

Nanoemulsion-based food coating and films have been identified owing to their effect on diminishing food quantity and quality loss, nutritional loss, and hence market value loss [95]. Essential oil nanodroplets, which have been widely used in nanoemulsions formulated for food packaging, contain a range of ingredients, including antibacterial compounds. Table 3 summarizes the different nanoemulsions based film coatings and their effect on coated food products. Eugenol the major compound of clove oil, nutmeg oil, and cinnamon oil, has been widely known for its antibacterial, antioxidant, anti-inflammatory, and local anesthetic effects [96]. Eugenol adsorbed in silica nanoparticles were utilized in liposome encapsulation, followed by immobilization into electrospun nanofibrous membranes in polyvinyl pyrrolidone. The prepared edible film was applied to beef to study the prevention of oxidation of meat produce, expressed in terms of thiobarbituric acid reactive substances (TBARS) values. TBARS values of the system increased with the storage time, indicating lipid oxidation, however, the increase was very less in case of nanoemulsion-based film when compared with the control sample [97]. Similarly, delayed lipid oxidation and improved fish sausage quality were observed in samples coated with tocopherols nanoemuslions [98]. A similar study on the effect of pectin coating with incorporated curcumin-cinnamon essential oil, curcumin-garlic essential oil, and curcumin sunflower oil nanoemulsions, on the shelf life enhancement of chicken fillets has been conducted [99].

Besides, nanoemulsion-impregnated edible coatings have been reported to induce delayed ripening resulting in subsequent extended shelf life of fruits. Sweet orange essential oil nanoemulsion prepared using ultrasonication homogenizer was infused in sodium alginate to coat fresh tomatoes which resulted in delayed ripening and improved post-harvest quality [100]. In another study, an active packaging film made of polyvinyl alcohol with cinnamon essential oil nanoemulsion has been found to control the decay of Chinese bayberry fruits [101]. Additionally, strawberries with coated pullulan and cinnamon essential oil based nanoemulsion depicted prolonged shelf life during storage under ambient conditions [102]. Synonymous results were recorded in papaya and okra coated in basil oil nanoemulison-sodium alginate coating and sunflower oil nanoemulsion-psyllium gum coating, respectively [103, 104].

Moreover, edible films/coating with incorporated nanoemulsions have been found to possess antimicrobial properties. Edible coating of chitosan with incorporated nanoemulsion containing cinnamaldehyde has been found to be effective against mesophilic microorganisms and lead to an increase in food shelf life up to 7–9 days [105]. Additionally, chitosan coating with carvacrol nanoemulsion made by high-pressure homogenization has been reported to cause increased microbial reduction in coated cucumber slices [106]. In another study, coating grape berries with carnauba wax and adding lemongrass oil nanoemulsions hindered the growth of *Salmonella typhimurium* and *Escherichia coli*, resulting in increased shelf life [107]. In two other research studies, researchers found reduced growth of *L. innocua, E. coli,* and *S. typhimurium* in green beans coated with chitosan enriched with mandarin essential oil, and with a combination of carvacrol, bergamot, mandarin, and& lemon essential oil nanoemulsions [50, 108]. Additionally, lemon grass essential oil, curcumin, and clove bud-oregano essential oil nanoemulsions imparted antimicrobial properties on coated plum, fish fillet, and sliced bread, respectively [109–111].

6.3 *Influence of Nanoemulsions-Based Films/Coatings on Food Properties*

Nanoemulsions have shown great potential for integrating bioactive ingredients into foods. Scientists are interested in developing a multifunctional film/coating that can preserve food quality while inhibiting bacterial and fungal development. Nanoemulsions have the potential to provide several benefits in food packaging, including environmental stress protection, release control, solubility enhancement, and natural antimicrobial absorption [112]. Bearing this in mind, nanoemulsions demonstrated their impact on the following features.

6.3.1 Appearance

Food appearance is considered a critical factor influencing its quality perception and marketability because customers are drawn to the exterior look of the food matrix. Freshness, color, and the absence of flaws are all aspects of the appearance of the food. It is critical that the food seems fresh and free of rotting [113, 114]. Food products throughout their shelf life, are susceptible to many negative effects such as enzymatic browning and oxidation reactions, which may prohibit them from preserving their original look and result in undesirable changes. Additionally, microbial, mechanical, and insect damage can all have an unfavorable effect on food surfaces (fruit skin) and generate blemishes. Application of edible coating on the food material results in retardation of various physiological activities and hence delay the negative changes that affect the overall appearance of the food. Besides, coating the produce with nanoemulsion incorporated edible coating can retard the changes further. A study reported the retarded decrease in L* values of fresh-cut pineapple coated in citral nanoemulsion when compared to control samples during a storage period of 12 days [115]. Similar observations were recorded in the case of apricots coated with nano-chitosan emulsion coating infused with pomegranate peel extract (PPE) [116]. Additionally, Rudra et al. coated okra with a nanoemulsion-based coating (alginate coatings with basil oil nano-emulsified with Tween 20 and *Sapindus* extract) that resulted in delayed discoloration [104]. Furthermore, in another study, it was observed that adding a nanoemulsion of α -tocopherol and natamycin increased the a*, b*, L*, and opacity values of the produced films, which further impacts the appearance of the food material coated [117]. Ghadetaj et al. observed similar findings [118].

6.3.2 Moisture Loss

Fruits and vegetables lose weight and lose quality owing to the migration of water molecules from the fruit to the environment during storage. This moisture loss associated with subsequent uncoordinated shrinkage may result in negative changes in food structure and diminished product transparency. As a result, it is critical to provide a coating/film that acts as a barrier to decrease water integrity, texture, and firmness of food [119]. Rudra et al. developed a nanoemulsion-based coating (Alginate coatings with basil oil nano-emulsified with Tween 20 and *Sapindus* extract) to extend the shelf life of okra [104]. It has been reported that coatings can slow down the processes of moisture loss, discoloration, and lower strength during storage. In another study, avocado fruits were covered with orange oil based nanoemulsion coating, which depicted nanoemulsion as a covering was a practical alternative for extending the fruit's postharvest storage period. [120]. Similar observations were reported in the case of strawberries coated in quinoa protein/chitosan edible coating and thymol nanoemulsion incorporated quinoa protein/chitosan edible coating during a 16-day storage time, wherein the weight loss for coated samples were similar (nearly 20%) and significantly lower than that of uncoated samples (36% weight loss) [121]. Similar findings were reported in the case of kiwi slices coated in the sodium alginate and carboxymethylcellulose-based nanoemulsion coating with enriched antioxidants, wherein it resulted in significantly delayed physiological weight loss by 8–14% [122].

6.3.3 Shelf-Life

Fresh produce and various other edible commodities are prone to a short span of shelf-life owing to moisture loss and solute migration, exchange of gases, respiration, and oxidative reaction rates, and physiological disorders. Nanoemulsion-based edible coatings contribute to extending the shelf-life attributed to barrier properties and strength. The majority of food products are oxygen-sensitive; as oils and fats along with enzymes are most vulnerable to oxygen exposure. When oxygen comes into contact with free fatty acids (FFA), it induces chemical changes that result in flavor modifications [123, 124]. Moreover, food pigments as well as various food additives might react with the oxygen exhibiting a negative impact on the food color [125]. Fresh fruits and vegetables, on the other hand, require controlled oxygen to prevent accelerated aging and senescence of the matrix because of their respiratory metabolism [126]. Oxygen permeates through the film and reaches the food resulting in subsequent changes. Thus, edible coating imparts oxygen barrier properties, subsequently resulting in increased shelf-life of the food materials, such as in the case of carnauba wax-lemongrass oil nanoemulsions coating on grapes [127]. Besides, it was observed that the barrier properties of the film/coating are impacted by the addition of plasticizer, however, the presence of nanoemulsion reduces it [128]. To summarize, embedded nanoemulsions prevent oxygen from diffusing into the film matrix and reacting with the food components [129].

Furthermore, foods supplemented with a large amount of polyunsaturated lipids, such as fish containing omega-3 fatty acids can get oxidized, leading to food spoilage [124]. A typical and powerful coating/film has the capacity to suppress oxidation reactions. Nanoemulsions, despite their intrinsic antioxidant capabilities [130, 131], have shown significant potential in the encapsulation and delivery of antioxidant compounds [132]. It has been observed that essential oils, whether in the form of nanoparticles, combined with biomaterial, or pure administration, reduce the oxidation process, thus dispersing essential oils in nanoemulsion-based food packaging would offer a great technique to lower oxidation levels in oxidative sensitive, high-fat foods [133].

Besides, nanoemulsion coatings also impart microbial safety to the coated food samples, thereby prolonging their shelf-life further. The antimicrobial agents added to the nanoemulsion being prepared impart favorable effects. Alginate and carboxylmethylcellulose-based nanoemulsion coatings with vanillin (0.5 and 1%) were reported to improve microbial stability of fresh-cut kiwi slices during storage for 7 days, as the antimicrobial agent in combination with emulsion coatings successfully slowed bacteria, yeast, and mold growth [122]. In another study, coating with quinoa-chitosan-based nanoemulsion with added thymol was reported to reduce the fungal decay of strawberries to 41.7% during a period of 16 days when stored at 5 °C [121]. Similar observations were recorded in case of fresh-cut pineapples coated in citral nanoemulsion-based coating [115].

6.3.4 Textural Properties

The firmness of food commodities, specifically fresh produce, is one of the primary textural parameters particularly important in terms of its overall desirability and acceptance. However, its texture is expected to lose its firmness and soften during storage due to the loss of water as a result of cell breakdown and other physiological changes [134]. Coating the food commodity would act as a water transfer barrier, resulting in slower dehydration, and would serve as a tool for extending the firmness of it. Besides, the coating might also retard their respiration rate, creating a low oxygen environment, that in turn result in a decrease in the enzymatic activity of polygalacturonase and pectin esterase, which hasten the depolymerization of pectin and other pectic materials that give stiffness to their structures [135, 136]. A similar observation has been recorded in the case of grapes coated with the carnauba wax-lemongrass oil nanoemulsions coating, wherein grape berries with coatings had more firmness right away (day 0), and coated samples usually had more firmness while stored at 4 and 25 °C [107]. In another study, the retention of firmness of 'Delicious' apples coated with carnauba wax based coating has also been reported [137]. Also, in the case of apricots, samples coated with nano-chitosan emulsion coating with incorporated pomegranate peel extract (PPE) had higher retention of firmness during 30 days of refrigerated storage, when compared to control uncoated samples. This might be attributed to the fact that the coating inhibits the action of an enzyme that breaks down cell structures, coating the cuticle and lenticels and decreasing respiration and other ripening processes during storage [116].

Furthermore, food products may be subjected to extreme tension and pressure, during the transit or storage procedures. This increases the possibility of damaging the film or coating layer, which will have an impact on the toughness, crispness, juiciness, and hardness product quality indices. The mechanical behavior of films is primarily described by two parameters: elongation at break (EAB) and tensile strength (TS) [138]. Poor mechanical qualities are often caused by moisture loss, microbial degradation, and oxidation, which would ultimately affect the properties of the food material coated. Adjusting the above-stated factors will enhance and promote mechanical characteristics. The addition of cinnamaldehyde nanoemulsion in the gelatin-based film has been reported to boost EAB while decreasing TS. The presence of cinnamaldehyde nanoemulsion interacts with the gelatin polymer chain to produce flexible regions within the matrix, which may be the source of the improved EAB [90]. Additionally, it has been reported, that sunflower, cinnamon, or garlic oil-loaded nanoemulsion enhanced the quality, nutrition, and textural profile of chicken fillets [99]. Moreover, it has been observed that when nanoemulsions were added to films, the films were significantly, less stretchy, softer, and less break-resistant than films without nanoemulsions [126].

7 Conclusion and Future Prospects

In recent years, consumers' desire for safe natural goods has driven the hunt for mild preservation strategies to increase the quality and safety of bacteria without inflicting food or organoleptic harm. The purpose of this chapter was to offer a brief explanation of the function of nanoemulsions and their multiple benefits in the food packaging business. Different techniques can be used to create nanoemulsions for various applications, but they may result in varying efficiencies. The primary ways for producing nanoemulsions are classified as low- and high-energy procedures.

Low-energy nanoemulsions are generally well-suited for the development of antimicrobial essential oil delivery systems. Moreover, mechanisms such as gravitational separation, droplet aggregation, and Oswald ripening are associated with emulsion instability. Therefore, the incorporation of surfactants shows significant efficacy in producing a stable nanoemulsion. Surfactant combinations (multicomponent) have been found to outperform pure (single-component) surfactants. As well it was also discovered that the type of surfactant used can influence the size of the nanodroplets. Additionally, a food-grade essential oil with a low viscosity is advised for creating a very stable nanoemulsion. Both water-in-oil and oil-in-water, industries offer a lot of opportunities for nanoemulsions, but in food packaging systems, oil-in-water systems are mostly used as the oil phase is inexpensive and has high antimicrobial properties. Nanoemulsions have been found to result in powerful coatings and films. They demonstrated their effectiveness in preserving the high quality of meals, fruits, and vegetables by inhibiting bacterial and fungal development, administering antibacterial agents, increasing mechanical characteristics, controlling the pH, and delivering agents. Coatings made of nanoemulsions performed better than polymer coatings or oil by itself. Also, it is essential to create nanoemulsions that are more homogeneous, stable, and transparent, which means that all of their dimensions must be lower than an order of magnitude of the wavelength of visible light. As a result, nanoemulsions must be chemically and physically evaluated before use. Yet, there are constraints and hurdles to consider while preparing a nanoemulsion-based coating/film. Moreover, further research on essential oil based nanoemulsion into the toxicity is required and the introduction of this new packaging method will require the permission of the legal organization and expert committees.

References

1. Espitia PJ, Fuenmayor CA, Otoni CG (2019) Nanoemulsions: synthesis, characterization, and application in bio-based active food packaging. Compr Rev Food Sci Food Saf 18(1):264–285
2. Saxena A, Maity T, Paliwal A, Wadhwa S (2017) Technological aspects of nanoemulsions and their applications in the food sector. In: Nanotechnology applications in food. Academic Press, pp 129–152
3. Harwansh RK, Deshmukh R, Rahman MA (2019) Nanoemulsion: promising nanocarrier system for delivery of herbal bioactives. J Drug Deliv Sci Technol 51:224–233
4. Acevedo-Fani A, Soliva-Fortuny R, Martín-Belloso O (2017) Nanoemulsions as edible coatings. Curr Opin Food Sci 15:43–49
5. Al-Tayyar NA, Youssef AM, Al-Hindi RR (2020). Edible coatings and antimicrobial nanoemulsions for enhancing shelf life and reducing foodborne pathogens of fruits and vegetables: a review. Sustain Mater Technol
6. McClements DJ (2012) Advances in fabrication of emulsions with enhanced functionality using structural design principles. Curr Opin Colloid Interface Sci 17(5):235–245
7. Jafari SM, McClements DJ (eds) (2018) Nanoemulsions: formulation, applications, and characterization. Academic Press
8. Zhang Z, McClements DJ (2018) Overview of nanoemulsion properties: stability, rheology, and appearance in nanoemulsions. Academic Press, pp 21–49
9. Gupta A, Eral HB, Hatton TA, Doyle PS (2016) Nanoemulsions: formation, properties and applications. Soft Matter 12(11):2826–2841

10. Komaiko JS, McClements DJ (2016) Formation of food-grade nanoemulsions using low-energy preparation methods. Compr Rev Food Sci Food Saf (A review of available methods)15(2):331–352
11. Håkansson A, Rayner M (2018) General principles of nanoemulsion formation by high-energy mechanical methods. In: Nanoemulsions. Academic Press, pp 103–139
12. Rehman A, Jafari SM, Aadil RM, Assadpour E, Randhawa MA, Mahmood S (2020) Development of active food packaging via incorporation of biopolymeric nanocarriers containing essential oils. Trends Food Sci Technol 101:106–112
13. Frank K, Garcia CV, Shin GH, Kim JT (2018) Alginate biocomposite films incorporated with cinnamon essential oil nanoemulsions: physical, mechanical, and antibacterial properties. Int J Polym Sci:1–8
14. McClements DJ (2011) Edible nanoemulsions: fabrication, properties, and functional performance. Soft Matter 7(6):2297–2316
15. Salem MA, Ezzat SM (2019) Nanoemulsions in food industry. In: Some new aspects of colloidal systems in foods
16. Witthayapanyanon A, Acosta EJ, Harwell JH, Sabatini DA (2006) Formulation of ultralow interfacial tension systems using extended surfactants. J Surf Deterg 9(4):331–339
17. McClements DJ, Rao J (2011) Food-grade nanoemulsions: formulation, fabrication, properties, performance, biological fate, and potential toxicity. Crit Rev Food Sci Nutr 51(4):285–330
18. McClements DJ (2004) Food emulsions: principles, practices, and techniques. CRC Press
19. Sonneville-Aubrun O, Simonnet JT, L'alloret F (2004) Nanoemulsions: a new vehicle for skincare products. Adv Coll Interface Sci 108:145–149
20. Imeson A (2011) Food stabilisers, thickeners and gelling agents. John Wiley & Sons
21. Tadros T, Izquierdo P, Esquena J, Solans C (2004) Formation and stability of nano-emulsions. Adv Coll Interface Sci 108:303–318
22. Anton N, Vandamme TF (2009) The universality of low-energy nano-emulsification. Int J Pharm 377(1–2):142–147
23. Sonneville-Aubrun O, Babayan D, Bordeaux D, Lindner P, Rata G, Cabane B (2009) Phase transition pathways for the production of 100 nm oil-in-water emulsions. Phys Chem Chem Phys 11(1):101–110
24. Barzegar H, Mehrnia MA, Nasehi B, Alipour M (2018) Fabrication of peppermint essential oil nanoemulsions by spontaneous method: effect of preparing conditions on droplet size. Flavour Fragr J 33(5):351–356
25. Yildirim ST, Oztop MH, Soyer Y (2017). Cinnamon oil nanoemulsions by spontaneous emulsification: formulation, characterization and antimicrobial activity 84:122–128
26. Chen H, Hu X, Chen E, Wu S, McClements DJ, Liu S, Li B, Li Y (2016) Preparation, characterization, and properties of chitosan films with cinnamaldehyde nanoemulsions. Food Hydrocolloids 61:662–671
27. Wang L, Zhang Y (2017) Eugenol nanoemulsion stabilized with zein and sodium caseinate by self-assembly. J Agric Food Chem 65(14):2990–2998
28. Guttoff M, Saberi AH, McClements DJ (2015) Formation of vitamin D nanoemulsion-based delivery systems by spontaneous emulsification: factors affecting particle size and stability. Food Chem 171:117–122
29. Li Y, Zhang Z, Yuan Q, Liang H, Vriesekoop F (2013) Process optimization and stability of D-limonene nanoemulsions prepared by catastrophic phase inversion method. J Food Eng 119(3):419–424
30. Perazzo A, Preziosi V, Guido S (2015) Phase inversion emulsification: current understanding and applications. Adv Coll Interface Sci 222:581–599
31. Ostertag F, Weiss J, McClements DJ (2012) Low-energy formation of edible nanoemulsions: factors influencing droplet size produced by emulsion phase inversion. J Colloid Interface Sci 388(1):95–102
32. Borrin TR, Georges EL, Brito-Oliveira TC, Moraes IC, Pinho SC (2018) Technological and sensory evaluation of pineapple ice creams incorporating curcumin-loaded nanoemulsions obtained by the emulsion inversion point method. Int J Dairy Technol 71(2):491–500

33. Borrin TR, Georges EL, Moraes IC, Pinho SC (2016) Curcumin-loaded nanoemulsions produced by the emulsion inversion point (EIP) method: an evaluation of process parameters and physico-chemical stability. J Food Eng 169:1–9
34. Nantarat T, Chansakaow S, Leelapornpisid P (2015) Optimization, characterization and stability of essential oils blend loaded nanoemulsions by PIC technique for anti-tyrosinase activity. Int J Pharm Pharm Sci:308–312
35. Zhang L, Zhang F, Fan Z, Liu B, Liu C, Meng X (2019) DHA and EPA nanoemulsions prepared by the low-energy emulsification method: process factors influencing droplet size and physicochemical stability. Food Res Int 121:359–366
36. Mayer S, Weiss J, McClements DJ (2013) Vitamin E-enriched nanoemulsions formed by emulsion phase inversion: factors influencing droplet size and stability. J Colloid Interface Sci 402:122–130
37. Chuesiang P, Siripatrawan U, Sanguandeekul R, Yang JS, McClements DJ, McLandsborough L (2019) Antimicrobial activity and chemical stability of cinnamon oil in oil-in-water nanoemulsions fabricated using the phase inversion temperature method 110:190–196
38. Chuesiang P, Siripatrawan U, Sanguandeekul R, McLandsborough L, McClements DJ (2018) Optimization of cinnamon oil nanoemulsions using phase inversion temperature method: impact of oil phase composition and surfactant concentration. J Colloid Interface Sci 514:208–216
39. Su D, Zhong Q (2016) Lemon oil nanoemulsions fabricated with sodium caseinate and tween 20 using phase inversion temperature method. J Food Eng 171:214–221
40. Jintapattanakit A, Hasan HM, Junyaprasert VB (2018) Vegetable oil-based nanoemulsions containing curcuminoids: formation optimization by phase inversion temperature method. J Drug Deliv Sci Technol 44:289–297
41. Liu Q, Huang H, Chen H, Lin J, Wang Q (2019) Food-grade nanoemulsions: preparation, stability and application in encapsulation of bioactive compounds. Molecules 24(23):4242
42. Utomo A, Baker M, Pacek AW (2009) The effect of stator geometry on the flow pattern and energy dissipation rate in a rotor–stator mixer. Chem Eng Res Des 87(4):533–542
43. Jafari SM, Assadpoor E, He Y, Bhandari B (2008) Re-coalescence of emulsion droplets during high-energy emulsification. Food Hydrocolloids 22(7):1191–1202
44. Van der Schaaf US, Karbstein HP (2018) Fabrication of nanoemulsions by rotor-stator emulsification. In: Nanoemulsions. Academic Press, pp 141–174
45. Karthik P, Anandharamakrishnan C (2016) Fabrication of a nutrient delivery system of docosahexaenoic acid nanoemulsions via high energy techniques. RSC Adv 6(5):3501–3513
46. Shakeel F, Baboota S, Ahuja A, Ali J, Faisal MS, Shafiq S (2008) Stability evaluation of celecoxib nanoemulsion containing tween 80. Thai J Pharm Sci 32:4–9
47. Bhatt P, Madhav S (2011) A detailed review on nanoemulsion drug delivery system. Int J Pharm Sci Res 2(10):2482–2489
48. Galvão KC, Vicente AA, Sobral PJ (2018) Development, characterization, and stability of O/W pepper nanoemulsions produced by high-pressure homogenization. Food Bioprocess Technol 11:355–367
49. Ma P, Zeng Q, Tai K, He X, Yao Y, Hong X, Yuan F (2017) Preparation of curcumin-loaded emulsion using high pressure homogenization: impact of oil phase and concentration on physicochemical stability 84:34–46
50. Severino R, Ferrari G, Vu KD, Donsì F, Salmieri S, Lacroix M (2015) Antimicrobial effects of modified chitosan based coating containing nanoemulsion of essential oils, modified atmosphere packaging and gamma irradiation against escherichia coli O157: H7 and salmonella typhimurium on green beans. Food Control 50:215–222
51. Hebishy E, Buffa M, Guamis B, Blasco-Moreno A, Trujillo AJ (2015) Physical and oxidative stability of whey protein oil-in-water emulsions produced by conventional and ultra high-pressure homogenization: effects of pressure and protein concentration on emulsion characteristics. Innov Food Sci Emerg Technol 32:79–90
52. Qadir A, Faiyazuddin MD, Hussain MT, Alshammari TM, Shakeel F (2016) Critical steps and energetics involved in a successful development of a stable nanoemulsion. J Mol Liq 214:7–18

53. Bouchemal K, Briançon S, Perrier E, Fessi H (2004) Nano-emulsion formulation using spontaneous emulsification: solvent, oil and surfactant optimisation. Int J Pharm 280(1–2):241–251
54. Ramesh M, Sakthishobana K, Suriya SB (2022) Microbial polysaccharides as cell/drug delivery systems. Polysacch Microb Orig:1025–1043
55. Acevedo-Fani A, Salvia-Trujillo L, Rojas-Graü MA, Martín-Belloso O (2015) Edible films from essential-oil-loaded nanoemulsions: physicochemical characterization and antimicrobial properties. Food Hydrocolloid 47:168–177
56. Alexandre EM, Lourenço RV, Bittante AM, Moraes IC, do Amaral Sobral PJ (2016) Gelatin-based films reinforced with montmorillonite and activated with nanoemulsion of ginger essential oil for food packaging applications. Food Packag Shelf Life 10:87–96
57. Dammak I, de Carvalho RA, Trindade CS, Lourenco RV, do Amaral Sobral PJ (2017) Properties of active gelatin films incorporated with rutin-loaded nanoemulsions. Int J Biol Macromol 98:39–49
58. Salvia-Trujillo L, Rojas-Graü MA, Soliva-Fortuny R, Martín-Belloso O (2013) Effect of processing parameters on physicochemical characteristics of microfluidized lemongrass essential oil-alginate nanoemulsions. Food Hydrocolloids 30(1):401–407
59. Mason TG, Wilking JN, Meleson K, Chang CB, Graves SM (2006) Nanoemulsions: formation, structure, and physical properties. J Phys Condens Matter 18(41):R635
60. Abbas S, Hayat K, Karangwa E, Bashari M, Zhang X (2013) An overview of ultrasound-assisted food-grade nanoemulsions. Food Eng Rev:139–157
61. Sugumar S, Mukherjee A, Chandrasekaran N (2015) Eucalyptus oil nanoemulsion-impregnated chitosan film: antibacterial effects against a clinical pathogen, staphylococcus aureus, in vitro. Int J Nanomed 10(sup2):67–75
62. Kumar R, Kaur K, Uppal S, Mehta SK (2017) Ultrasound processed nanoemulsion: a comparative approach between resveratrol and resveratrol cyclodextrin inclusion complex to study its binding interactions, antioxidant activity and UV light stability. Ultrason Sonochem 37:478–489
63. Friberg S, Larsson K, Sjoblom J (2004) Food emulsions. Marcel Dekker, New York
64. McClements DJ (2012) Nanoemulsions versus microemulsions: terminology, differences, and similarities. Soft Matter 8(6):1719–1729
65. Kabalnov AS, Pertzov AV, Shchukin ED (1987) Ostwald ripening in emulsions: i direct observations of ostwald ripening in emulsions. J Colloid Interface Sci 118(2):590–597
66. Kabalnov AS, Shchukin ED (1992) Ostwald ripening theory: applications to fluorocarbon emulsion stability. Adv Colloid Interface Sci 38:69–97
67. Aoki T, Decker EA, McClements DJ (2005) Influence of environmental stresses on stability of O/W emulsions containing droplets stabilized by multilayered membranes produced by a layer-by-layer electrostatic deposition technique. Food Hydrocolloids 19(2):209–220
68. Wooster TJ, Golding M, Sanguansri P (2008) Impact of oil type on nanoemulsion formation and ostwald ripening stability. Langmuir 24(22):12758–12765
69. Zhang L, Yang H, Xie X, Zhang F, Li L (2009) Preparation and photocatalytic activity of hollow ZnSe microspheres via ostwald ripening. J Alloy Compd 473(1–2):65–70
70. McClements DJ (2005) Food emulsions in practice. In: Food emulsions: principles, practices and techniques. CRC Press, Boca Raton, pp 515–544
71. Bai L, McClements DJ (2016) Formation and stabilization of nanoemulsions using biosurfactants: rhamnolipids. J Colloid Interface Sci 479:71–79
72. Li X, Anton N, Ta TM, Zhao M, Messaddeq N, Vandamme TF (2011) Microencapsulation of nanoemulsions: novel trojan particles for bioactive lipid molecule delivery. Int J Nanomed:1313–1325
73. Saint Ruth H, Attwood D, Ktistis G, Taylor CJ (1995) Phase studies and particle size analysis of oil-in-water phospholipid microemulsions. Int J Pharm 116(2):253–261
74. Preetz C, Hauser A, Hause G, Kramer A, Mäder K (2010) Application of atomic force microscopy and ultrasonic resonator technology on nanoscale: distinction of nanoemulsions from nanocapsules. Eur J Pharm Sci 39(1–3):141–151

75. Rao SV, Shao J (2008) Self-nanoemulsifying drug delivery systems (SNEDDS) for oral delivery of protein drugs: I formulation development. Int J Pharm 362(1–2):2–9
76. Sharma G, Wilson K, Van der Walle CF, Sattar N, Petrie JR, Kumar MR (2010) Microemulsions for oral delivery of insulin: design, development and evaluation in streptozotocin induced diabetic rats. Eur J Pharm Biopharm 76(2):159–169
77. Acharya DP, Hartley PG (2012) Progress in microemulsion characterization. Curr Opin Colloid Interface Sci 17(5):274–280
78. Leong TS, Wooster TJ, Kentish SE, Ashokkumar M (2009) Minimising oil droplet size using ultrasonic emulsification. Ultrason Sonochem 16(6):721–727
79. Mcclements DJ (2007) Critical review of techniques and methodologies for characterization of emulsion stability. Crit Rev Food Sci Nutr 47(7):611–649
80. Thanasukarn P, Pongsawatmanit R, McClements DJ (2004) Influence of emulsifier type on freeze-thaw stability of hydrogenated palm oil-in-water emulsions. Food Hydrocolloids 18(6):1033–1043
81. Luykx DM, Peters RJ, van Ruth SM, Bouwmeester H (2008) A review of analytical methods for the identification and characterization of nano delivery systems in food. J Agric Food Chem 56(18):8231–8247
82. Hathout RM, Woodman TJ (2012) Applications of NMR in the characterization of pharmaceutical microemulsions. J Control Release 161(1):62–72
83. Gradzielski M (2008) Recent developments in the characterisation of microemulsions. Curr Opin Colloid Interface Sci 13(4):263–269
84. Silva HD, Cerqueira MÂ, Vicente AA (2012) Nanoemulsions for food applications: development and characterization. Food Bioprocess Technol:854–867
85. Ferreira LM, Cervi VF, Gehrcke M, da Silveira EF, Azambuja JH, Braganhol E, Sari MH, Zborowski VA, Nogueira CW, Cruz L (2015) Ketoprofen-loaded pomegranate seed oil nanoemulsion stabilized by pullulan: selective antiglioma formulation for intravenous administration. Colloids Surf B 130:272–277
86. Velloso MS, Otoni A, de Paula SA, de Castro WV, Pinto SW, Marinho MA, Rios DR (2014) Peritoneal dialysis and inflammation. Clin Chim Acta 430:109–114
87. Artiga-Artigas M, Acevedo-Fani A, Martín-Belloso O (2017) Effect of sodium alginate incorporation procedure on the physicochemical properties of nanoemulsions. Food Hydrocolloids 70:191–200
88. Artiga-Artigas M, Acevedo-Fani A, Martín-Belloso O (2017) Improving the shelf life of low-fat cut cheese using nanoemulsion-based edible coatings containing oregano essential oil and mandarin fiber. Food Control 76:1–2
89. Li X, Yang X, Deng H, Guo Y, Xue J (2020) Gelatin films incorporated with thymol nanoemulsions: physical properties and antimicrobial activities. Int J Biol Macromol 150:161–168
90. Ji M, Wu J, Sun X, Guo X, Zhu W, Li Q, Shi X, Tian Y, Wang S (2021) Physical properties and bioactivities of fish gelatin films incorporated with cinnamaldehyde-loaded nanoemulsions and vitamin C
91. Shao P, Yan Z, Chen H, Xiao J (2018) Electrospun poly (vinyl alcohol)/permutite fibrous film loaded with cinnamaldehyde for active food packaging. J Appl Polym Sci 135(16)
92. Rieger KA, Schiffman JD (2014) Electrospinning an essential oil: cinnamaldehyde enhances the antimicrobial efficacy of chitosan/poly (ethylene oxide) nanofibers. Carbohyd Polym 113:561–568
93. Ahari H, Naeimabadi M (2021) Employing nanoemulsions in food packaging: Shelf life enhancement. Food Eng Rev:858–883
94. Otoni CG, de Moura MR, Aouada FA, Camilloto GP, Cruz RS, Lorevice MV, de FF Soares N, Mattoso LH (2014) Antimicrobial and physical-mechanical properties of pectin/papaya puree/ cinnamaldehyde nanoemulsion edible composite films. Food Hydrocolloids 41:188–194
95. Loi M, Paciolla C, Logrieco AF, Mulè G (2020) Plant bioactive compounds in pre-and postharvest management for aflatoxins reduction. Front Microbiol

96. Phunpee S, Saesoo S, Sramala I, Jarussophon S, Sajomsang W, Puttipipatkhajorn S, Soottitantawat A, Ruktanonchai UR (2016) A comparison of eugenol and menthol on encapsulation characteristics with water-soluble quaternized β-cyclodextrin grafted chitosan. Int J Biol Macromol 84:472–480
97. Cui H, Yuan L, Li W, Lin L (2017) Antioxidant property of SiO_2-eugenol liposome loaded nanofibrous membranes on beef. Food Packag Shelf Life 11:49–57
98. Feng X, Tjia JY, Zhou Y, Liu Q, Fu C, Yang H (2020) Effects of tocopherol nanoemulsion addition on fish sausage properties and fatty acid oxidation. https://doi.org/10.1016/j.lwt.2019.108737
99. Abdou ES, Galhoum GF, Mohamed EN (2018) Curcumin loaded nanoemulsions/pectin coatings for refrigerated chicken fillets. Food Hydrocolloids 83:445–453
100. Das S, Vishakha K, Banerjee S, Mondal S, Ganguli A (2020) Sodium alginate-based edible coating containing nanoemulsion of citrus sinensis essential oil eradicates planktonic and sessile cells of food-borne pathogens and increased quality attributes of tomatoes. Int J Biol Macromol 162:1770–1779
101. Wen P, Zhu DH, Wu H, Zong MH, Jing YR, Han SY (2016) Encapsulation of cinnamon essential oil in electrospun nanofibrous film for active food packaging. Food Control 59:366–376
102. Chu Y, Gao C, Liu X, Zhang N, Xu T, Feng X, Yang Y, Shen X, Tang X (2020) Improvement of storage quality of strawberries by pullulan coatings incorporated with cinnamon essential oil nanoemulsion. https://doi.org/10.1016/j.lwt.2020.109054
103. Yousuf B, Srivastava AK (2015) Psyllium (plantago) gum as an effective edible coating to improve quality and shelf life of fresh-cut papaya (carica papaya). Int J Bioeng Life Sci 9(7):765–770
104. Gundewadi G, Rudra SG, Sarkar DJ, Singh D (2018) Nanoemulsion based alginate organic coating for shelf life extension of okra. Food Packag Shelf Life 18:1–2
105. Martiñon ME, Moreira RG, Castell-Perez ME, Gomes C (2014) Development of a multilayered antimicrobial edible coating for shelf-life extension of fresh-cut cantaloupe (cucumis melo L.) stored at 4 C. LWT-Food Sci Technol 56(2)341–350
106. Taştan Ö, Pataro G, Donsì F, Ferrari G, Baysal T (2017) Decontamination of fresh-cut cucumber slices by a combination of a modified chitosan coating containing carvacrol nanoemulsions and pulsed light. Int J Food Microbiol 260:75–80
107. Kim IH, Oh YA, Lee H, Song KB, Min SC (2014) Grape berry coatings of lemongrass oil-incorporating nanoemulsion. LWT-Food Sci Technol 58(1):1
108. Donsì F, Marchese E, Maresca P, Pataro G, Vu KD, Salmieri S, Lacroix M, Ferrari G (2015) Green beans preservation by combination of a modified chitosan based-coating containing nanoemulsion of mandarin essential oil with high pressure or pulsed light processing. Postharvest Biol Technol 106:21–32
109. Kim IH, Lee H, Kim JE, Song KB, Lee YS, Chung DS, Min SC (2013) Plum coatings of lemongrass oil-incorporating carnauba wax-based nanoemulsion. J Food Sci 78(10):E1551–E1559
110. KhoshbouyLahidjani L, Ahari H, Sharifan A (2020) Influence of curcumin-loaded nanoemulsion fabricated through emulsion phase inversion on the shelf life of Oncorhynchus mykiss stored at 4 C. J Food Process Preserv
111. Otoni CG, Pontes SF, Medeiros EA, Soares ND (2014) Edible films from methylcellulose and nanoemulsions of clove bud (Syzygiumaromaticum) and oregano (origanum vulgare) essential oils as shelf life extenders for sliced bread. J Agric Food Chem 62(22):5214–5219
112. Bahrami F, Ahari H, Yousefi SS (2019) The effect of efficient bioactive nano-emulsion formulation based on polylophium involucratum on improving quality features of green tiger pawn fridge storage. Ann Mil Health Sci Res 17(1)
113. Ghaderi-Ghahfarokhi M, Barzegar M, Sahari MA, Gavlighi HA, Gardini F (2017) Chitosan-cinnamon essential oil nano-formulation: application as a novel additive for controlled release and shelf life extension of beef patties. Int J Biol Macromol 102:19–28

114. Falguera V, Quintero JP, Jiménez A, Muñoz JA, Ibarz A (2011) Edible films and coatings: structures, active functions and trends in their use. Trends Food Sci Technol 22(6):292–303
115. Prakash A, Baskaran R, Vadivel V (2019) Citral nanoemulsion incorporated edible coating to extend the shelf life of fresh cut pineapples. LWT:108851
116. Gull A, Bhat N, Wani SM, Masoodi FA, Amin T, Ganai SA (2021) Shelf life extension of apricot fruit by application of nanochitosan emulsion coatings containing pomegranate peel extract. Food Chem 349:129149
117. Agudelo-Cuartas C, Granda-Restrepo D, Sobral PJ, Hernandez H, Castro W (2020) Characterization of whey protein-based films incorporated with natamycin and nanoemulsion of α-tocopherol. Heliyon
118. Ghadetaj A, Almasi H, Mehryar L (2018) Development and characterization of whey protein isolate active films containing nanoemulsions of grammosciadium ptrocarpum bioss essential oil. Food Packag Shelf Life 16:31–40
119. Saberi B, Golding JB, Marques JR, Pristijono P, Chockchaisawasdee S, Scarlett CJ, Stathopoulos CE (2018) Application of biocomposite edible coatings based on pea starch and guar gum on quality, storability and shelf life of 'valencia' oranges. Postharvest Biol Technol 137:9–20
120. Cenobio-Galindo AD, Ocampo-López J, Reyes-Munguía A, Carrillo-Inungaray ML, Cawood M, Medina-Pérez G, Fernández-Luqueño F, Campos-Montiel RG (2019) Influence of bioactive compounds incorporated in a nanoemulsion as coating on avocado fruits (perseaamericana) during postharvest storage: antioxidant activity, physicochemical changes and structural evaluation. Antioxidants 8(10):500
121. Robledo N, López L, Bunger A (2018) Effects of antimicrobial edible coating of thymol nanoemulsion/quinoa protein/chitosan on the safety, sensorial properties, and quality of refrigerated strawberries (fragaria × ananassa) under commercial storage environment. Food Bioprocess Technol 11:1566–1574
122. Manzoor S, Gull A, Wani SM, Ganaie TA, Masoodi FA, Bashir K, Dar BN et al (2021) Improving the shelf life of fresh cut kiwi using nanoemulsion coatings with antioxidant and antimicrobial agents. Food Biosci 41:101015. https://doi.org/10.1016/j.fbio.2021.101015
123. Hirsch A (1991) Flexible food packaging: questions and answers. Springer Science & Business Media
124. Das AK, Nanda PK, Das A, Biswas S (2019) Hazards and safety issues of meat and meat products. In: Food safety and human health. Academic Press, pp 145–168
125. Bao Y, Puolanne E, Ertbjerg P (2016) Effect of oxygen concentration in modified atmosphere packaging on color and texture of beef patties cooked to different temperatures. Meat Sci 121:189–195
126. Dey A, Neogi S (2019) Oxygen scavengers for food packaging applications: a review. Trends Food Sci Technol 90:26–34
127. Rovera C, Ghaani M, Farris S (2020) Nano-inspired oxygen barrier coatings for food packaging applications: an overview. Trends Food Sci Technol 97:210–220
128. Otoni CG, Avena-Bustillos RJ, Olsen CW, Bilbao-Sáinz C, McHugh TH (2016) Mechanical and water barrier properties of isolated soy protein composite edible films as affected by carvacrol and cinnamaldehyde micro and nanoemulsions. Food Hydrocolloids 57:72–79
129. Imran M, Revol-Junelles AM, René N, Jamshidian M, Akhtar MJ, Arab-Tehrany E, Jacquot M, Desobry S (2012) Microstructure and physico-chemical evaluation of nano-emulsion-based antimicrobial peptides embedded in bioactive packaging films. Food Hydrocolloids 29(2):407–419
130. Jimenez-Escobar MP, Pascual-Mathey LI, Beristain CI, Flores-Andrade E, Jiménez M, Pascual-Pineda LA (2020) In vitro and in vivo antioxidant properties of paprika carotenoids nanoemulsions
131. Rehman A, Jafari SM, Tong Q, Karim A, Mahdi AA, Iqbal MW, Aadil RM, Ali A, Manzoor MF (2020) Role of peppermint oil in improving the oxidative stability and antioxidant capacity of borage seed oil-loaded nanoemulsions fabricated by modified starch. Int J Biol Macromol 153:697–707

132. Das AK, Nanda PK, Madane P, Biswas S, Das A, Zhang W, Lorenzo JM (2020) A comprehensive review on antioxidant dietary fibre enriched meat-based functional foods. Trends Food Sci Technol 99:323–336
133. Vilela C, Kurek M, Hayouka Z, Röcker B, Yildirim S, Antunes MD, Nilsen-Nygaard J, Pettersen MK, Freire CS (2018) A concise guide to active agents for active food packaging. Trends Food Sci Technol 80:212–222
134. Martínez-Ferrer M, Harper C, Pérez-Muñoz F, Chaparro M (2006) Modified atmosphere packaging of minimally processed mango and pineapple fruits. J Food Sci 67(9):3365–3371
135. Perez-Gago MB, Serra M, Del Rio MA, Alonso M, Mateos M (2003) Effect of solid content and lipid content of whey protein-beeswax edible composite coatings on color change of freshcut apple. J Food Sci 68:1–6
136. Maftoonazad N, Ramaswamy HS, Marcotte M (2008) Shelf-life extension of peaches through sodium alginate and methyl cellulose edible coatings. Int J Food Sci Technol 43(6):951–957
137. Bai J, Baldwin EA, Hagenmaier RH (2002) Alternatives to shellac coatings provide comparable gloss, internal gas modification, and quality for 'delicious' apple fruit. Hortic Sci 37(3):559–563
138. Huang T, Lin J, Fang Z, Yu W, Li Z, Xu D, Yang W, Zhang J (2020) Preparation and characterization of irradiated kafirin-quercetin film for packaging cod (gadusmorhua) during cold storage at 4 °C. Food Bioprocess Technol:522–532

Nanoclays in Food Packaging

Tejaskumar A. Suhagia, Prachitee Ayare, and Madhusudhan Alle

Abstract This chapter explores the pivotal role of nanoclays in modern food packing in response to the growing demand for minimally processed and safe food products. Nanoclays, when integrated into packing materials, enhance thermal, mechanical, and barrier properties, making them valuable for preserving food quality. They exhibit versatility by enabling controlled ingredient release, functioning as antimicrobial agents, and facilitating colorimetric indicators for food safety. Crucially, safety concerns are addressed through migration studies, affirming their compliance with safety regulations. Notably, nanoclays-based active films and coatings prove highly effective for high-moisture content foods, prolonging shelf life and preventing spoilage. In a world increasingly prioritizing sustainable practices, nanoclays emerge as contenders in the future of food packaging. This chapter offers a concise overview of their applications, underscoring their potential to revolutionize food preparation, safety, and sustainability.

Keywords Nanoclays · Food packaging · Nanocomposites · Migration · Layered silicates · Meat · Fruits and vegetables

T. A. Suhagia · M. Alle (✉)
Department of Chemistry, University of Memphis, Memphis, TN 38152, USA
e-mail: mralle@memphis.edu

T. A. Suhagia
e-mail: tejaskumarsuhagia@gmail.com

P. Ayare
Department of Pharmaceutics, VES College of Pharmacy, Mumbai, India
e-mail: prachitee.ayare@ves.ac.in

M. Alle et al. (eds.), *Functional Nanomaterials and Nanocomposites for Biodegradable Food Packaging*, Smart Nanomaterials Technology,
https://doi.org/10.1007/978-981-97-9084-5_9

1 Introduction

The packaging of food products matters a lot for maintaining their quality, safety, and shelf-life so that they reach consumers in the appropriate condition [1, 2]. In recent times, innovative packaging solutions have driven the incorporation of nanotechnology into the packaging materials consequently improving the sector with new functionalities and outstanding performance. While myriad nanomaterials have been exploited in this regard, nanoclay stands out as a promising candidate owing to its peculiar properties and feasibility [1, 2].

As the name suggests, nanoclay refers to clay particles that have nanoscale dimensions with their size ranging from a few nanometers to a few micrometers. These nanoclays have unique structural features, being constituted of layered silicate compounds with high aspect ratios and large surface areas. As for nanoclays, montmorillonite, kaolinite, and halloysite are the most frequent types. Their use in food packaging provides numerous benefits [1, 2]. These materials provide outstanding mechanical properties, and barrier characteristics to the packaging material. Their compatibility with the polymer matrices enables their use as reinforcing agents in packaging materials and enhances their performance [1, 2].

The addition of nanoclay to food packaging materials provides many major advantages mainly related to improvements in the barrier properties. Nanoclay particles act as efficient barriers against the permeation of gases such as oxygen, moisture, and volatile organic compounds causing aroma deterioration. Nanoclay-based packaging materials hinder the inward migration of the substances thereby maintaining the shelf-life of packaged food products, ensuring freshness and quality for extended periods. Nanoclay reinforcement further leads to better mechanical properties such as strength, toughness, and flexibility resulting in better integrity of packaging materials [1, 2].

Secondly, nanoclay-based food packaging (NCFP) materials have shown great potential as antimicrobials, capable of hindering the reproduction of spoilage microorganisms and pathogens, which pose a threat to human health. This antimicrobial functionality provides additional protection against contamination and thus reduces the probability of food poisoning, ensuring food safety and consumer welfare. Besides, the appearance of smart packaging applications using nanoclay technology allows real-time monitoring of food quality and safety parameters which enables the implementation of preventive measures to eliminate risks and maintain product integrity throughout the whole supply chain [3, 4].

A variety of nanoclay incorporation methods have been utilized to fabricate nanocomposites for food packaging, including direct blending with polymer matrices, in-situ polymerization, and layer-by-layer deposition. Each of these methods has its advantages and disadvantages that impact mostly the uniformity, compatibility, and performance of nanoclay-based packaging materials. Characterization techniques are X-ray diffraction (XRD), transmission electron microscopy (TEM), and scanning electron microscopy (TEM), which are crucial in analyzing the structural and morphological characteristics of NCFP materials, providing knowledge about their performance characteristics and accomplishments. Techniques such as scanning electron microscopy (SEM), Fourier-transform infrared spectroscopy (FTIR), and gas permeability testing enable the determination of the overall characteristics of nanoclay-based packaging materials, their optimal formulation, and processing parameters [2, 5].

However, some aspects need to be addressed before safe and hygienic use of NCFP. The risks associated with the translocation of nanoclay particles, and their toxicological implications require careful safety assessment and regulatory compliance. In addition, the scalability, cost-effectiveness, and environmental performance of nanoclay-based packaging materials must be evaluated meticulously owing to their practical viability and long-term sustainability [6, 7].

This chapter aims to carry out a systematic evaluation of the applications of nanoclay in food packaging from diverse aspects such as its properties, applications, incorporation techniques, characterization methods, safety aspects, challenges, and prospects. This chapter has the objective of clarifying the potential of nanoclays as versatile nanomaterials to improve the functionality, performance, and sustainability of food packaging materials.

2 Classification of Nanoclays

Nanoclay comprises a division of clay minerals referred to as clay particles with nanoscale dimensions that measure from several nanometers to several micrometers. These materials have distinctive properties and qualities caused by their tiny size and large surface area to volume ratio, making them suitable for diverse applications such as food packaging.

Nanoclay is characterized by its nano-scale dimensions which makes it different from conventional clay minerals. However, conventional clay minerals like Kaolinite, Montmorillonite, and Halloysite are naturally occurring minerals with macroscopic particle size, but the nanoclays are processed or the engineered materials on the nanoscale. The reduction in particle size gives nanoclays specific properties and features that make them suitable for advanced materials and applications [2, 8].

Nanoclays can be classified in many ways such as by their composition, structure, surface modification, or synthesis. The classification of nanoclays contributes to their characterization and applicability for certain uses such as food packaging [2, 9].

2.1 Composition-Based Classification

Montmorillonite: Montmorillonite is a type of clay mineral belonging to the smectite group, which has high cation exchange capacity and swelling properties. It consists of layers of alumina (Al_2O_3) and silica (SiO_2) octahedra alternating with tetrahedral layers of silica [10].

Kaolinite: Kaolinite is a non-swelling clay mineral with a layered structure of alumina and silica frameworks composed of stacked octahedral sheets. It shows low cation exchange capacity and is commonly used in ceramic and paper coatings [11].

Halloysite: Halloysite is a clay mineral having a tube-like structure made from rolling layers of alumina and silica sheets. Its hollow tube design makes it suitable for encapsulation and controlled release operations [12].

The structure of different nanoclays types classified based on their composition is shown in Fig. 1 (Table 1).

2.2 Structure-Based Classification

1:1 and 2:1 nanoclays can be classified based on their layer disposal and the ratio of tetrahedral to octahedral layers. 1:1 layered nanoclays, e.g., kaolinite, have a 1-to-1 ratio of tetrahedral to octahedral layers while 2:1 nanoclays such as montmorillonite have two tetrahedral layers for every octahedral layer [10, 11].

2.3 Surface Modification-Based Classification

Organically Modified Nanoclays: The nanoclays can be combined with organic molecules which promote their compatibility with polymer matrices and results in enhancement of screening and dispersion, barrier performance, and mechanical strength. Examples of common organic modifiers are quaternary ammonium salts, alkylammonium ions, and surfactants [17].

2.4 Synthesis-Based Classification

Natural vs. synthetic nanoclays: The nanoclays can be generated from natural clay minerals via mechanical or chemical processes for reducing particle size and changing properties. Alternatively, the synthesis of artificial nanoclays can be performed through top-down approaches, e.g., hydrothermal synthesis, or sol–gel methods, which provide versatility in adjusting their properties for desired applications [2, 18].

Nanoclay classification helps in choosing appropriate materials for food packaging applications by properties and performance specifications. Through the utilization of the unique qualities of nanoclays, like high surface area, aspect ratio, and surface reactivity, scientists and producers can create advanced packaging that improves food safety, shelf-life, and sustainability [2, 18, 19].

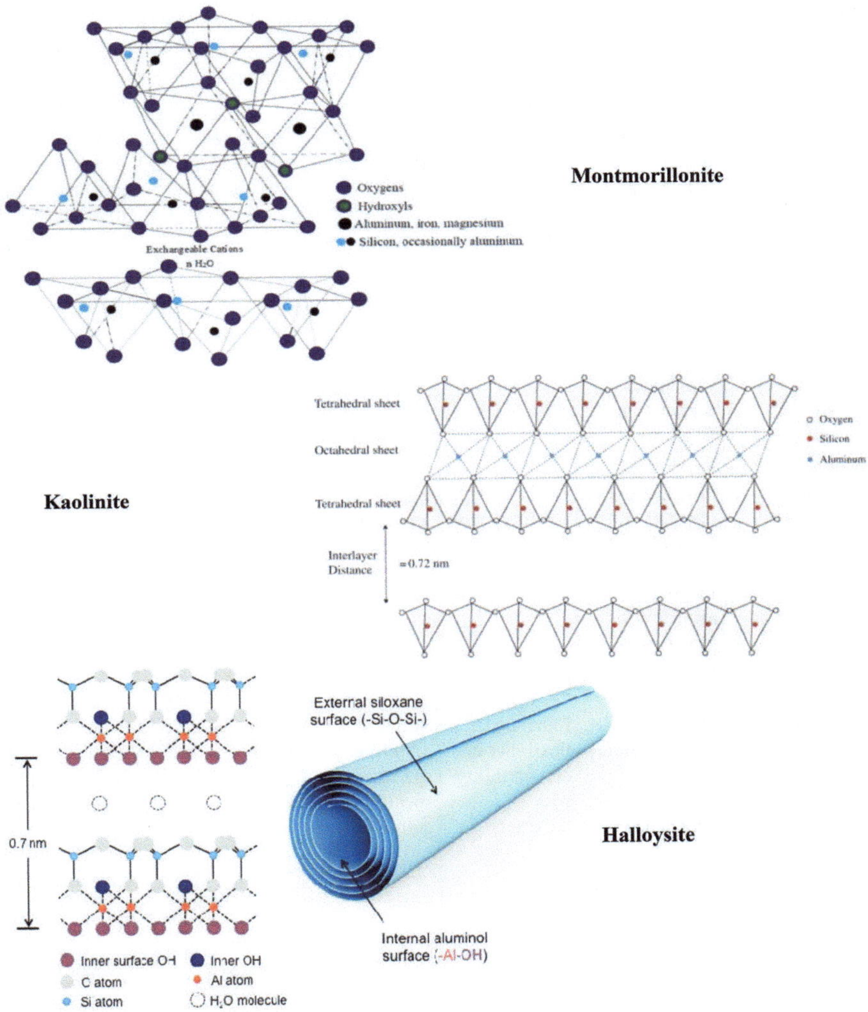

Fig. 1 Schematic representation and its chemical structure of Montmorillonite, Kaolinite, and halloysites (Adapted from [13–16])

3 Structure and Morphology of Nanoclays

Nanoclays are made from layers of silicate structures, more often stacks of individual clay platelets. The platelets are arranged in layers in a recurring style. Each layer comprises a two-dimensional network of silicon (Si) and oxygen (O) atoms forming the silicate layers. The silicate layers are surrounded by an interlayer space that can

Table 1 Properties of nanoclays in food packaging and antimicrobial effectiveness

Nanoclay type	Characteristics	Advantages in food packaging	Antimicrobial effectiveness	Microorganisms inhibited	Applications in food packaging
Montmorillonite	Layered clay minerals belonging to the smectite group with a high aspect ratio and large surface area. Exhibits high cation exchange capacity and swelling properties	Excellent barrier properties against gases and moisture	High antimicrobial activity	Spoilage microorganisms, pathogens	Barrier properties enhancement, mechanical reinforcement
Kaolinite	Non-swelling clay mineral with a layered structure and a moderate aspect ratio. Low cation exchange capacity	Good reinforcement for mechanical properties	Moderate antimicrobial properties	Bacteria and fungi	Moisture barrier, grease, and oil resistance in coatings and films
Halloysite	Tubular clay mineral with hollow interiors and a high aspect ratio. Rolled layers of alumina and silica sheets	Enhanced flexibility and compatibility with polymers	Strong inhibition against specific pathogens	Pathogenic bacteria, mold	Controlled release systems, and encapsulation for food safety enhancement
Bentonite	Plate-like clay mineral with excellent swelling properties and ion exchange capacity. Adsorbs toxins, contaminants, and volatile compounds	Enhanced adsorption of toxins, extending the shelf-life of perishable foods	High antimicrobial activity	Spoilage microorganisms, pathogens	Adsorption of toxins, extending shelf-life of perishable foods

Table 2 Morphological characteristics of nanoclays [2, 9, 22]

Morphology	Description
Platelets	Platelet-shaped nanoclays are characterized by flat, disc-like structures with nanoscale dimensions. They typically have diameters ranging from a few nanometers to several micrometers and thicknesses on the order of nanometers
Tubes	Some nanoclays, such as halloysite, exhibit a tubular shape, consisting of cylindrical structures with hollow interiors. These nanotubes have unique properties, such as tunable aspect ratio, inner diameter, and length
Aggregates	During processing, nanoclays may form aggregates due to interactions between individual platelets or external forces. Understanding and control of aggregate formation are essential to ensure uniform dispersion and properties of nanoclay-based materials

trap water molecules and various cations like Na^+ and Ca^{2+} through electrostatic interaction and hydration force.

The spatial configuration of atoms and ions within nanoclay structures results in remarkable characteristics such as high aspect ratios and large surface areas with interlayer spacing. The high aspect ratio, which is the ratio of the diameter to the thickness of the platelet, confers the reinforcement and barrier enhancement properties of nanoclays when they are incorporated into packaging material. Additionally, the large specific surface area-to-volume ratio favors easy interaction with polymer matrices and other additives thus leading to effective dispersion and compatibility of resultant composite systems.

Nanoclays display a variety of morphologies owing to their composition, fabrication techniques, and environmental conditions. The most numerous morphologies are plates, tubes, and aggregates which present a particular performance within each group [9, 20, 21] (Table 2).

4 Properties of Nanoclays

4.1 Physical Properties

Particle size and surface area: Nanoclays are characterized by the small size of their particles which is approximately at the nanometer level, hence the high surface area-to-volume ratio. This enhanced surface area permits numerous interactions with surrounding molecules and improves properties such as barrier performance and mechanical reinforcement [8, 9, 23].

Aspect ratio: Nanoclays often possess a high aspect ratio, defined by the ratio of their length to width, which is significantly larger than conventional clay minerals. The long morphology makes them capable of polymer matrix enhancement and strength reinforcement [8, 9, 23].

Porosity and pore structure: Nanoclays have inherent porosity and complex pore structure, which determine their adsorption capacity. These can be employed in wastewater treatment, controlled release systems, and adsorbent materials [8, 9, 23].

Density: The density of nanoclays governs their dispersibility and compatibility with polymer matrices. Low-density nanoclays are supposed to disperse better when introduced to polymeric matrices leading to homogeneous nanocomposites with improved properties [8, 9, 23].

4.2 Chemical Properties

Surface chemistry: The surface chemistry of nanoclays dramatically affects their interaction with polymer matrices as well as other additives. Surface modification methods such as cation exchange and functionalization with organic molecules help in tuning the surface characteristics of nanoclays to improve compatibility and adhesion in composite materials [8, 9, 23].

Ion exchange capacity: Layered nanoclays such as smectite-group clays, such as montmorillonite, have high ion exchange capacities. Such property provides nanoclays with the ability to adsorb and retain ions and molecules in the interlayer space which influences properties such as breathability and antimicrobial effect [8, 9, 23].

Hydrophilicity/Hydrophobicity: Hydrophilic and hydrophobic properties of nanoclays determine their interactions with moisture and other active components present in the packaging atmosphere Surface modification can be used to change the hydrophobicity/hydrophilicity of nanoclays for better suitable to specific applications [8, 9, 23].

Thermal stability: The nanoclays usually have a high thermal stability which allows them to be fabricated at high temperatures during the composite production. The thermal properties of those affect processing conditions and the thermal stability of nanoclay-containing packaging materials throughout the storage and application [8, 9, 23].

5 Applications of Nanoclays in Food Packaging

5.1 Types of Nanoclays Used in Food Packaging

Several types of nanoclays are commonly employed in food packaging applications, each offering unique properties and advantages. Montmorillonite is the most widely used nanoclay in food packaging due to its high aspect ratio, cation exchange capacity, and barrier properties It increases the mechanical strength and gas permeability of packaging materials, thereby ensuring long-term storage of packaged foods

[24]. Kaolinite nanoclays are preferred for low water absorption and good chemical resistance making them suitable for applications where water barrier properties are important. They are mainly used in food packaging coatings and films to enhance water resistance and prevent grease and oil migration [11]. Halloysite nanotubes offer unique properties such as tubular morphology with a high aspect ratio making them ideal for controlled release and encapsulation applications in food packaging. They can be modified with antimicrobial agents or antioxidants, for applications in food safety and shelf-life extension [12]. Bentonite nanoclays possess outstanding swelling properties and ion exchange capacity. They absorb toxins, contaminants, and volatile compounds in food packaging. They are widely incorporated into packaging materials in order to have high barrier properties and longer shelf life for perishable foods [25].

5.2 Barrier Properties Enhancement

It has been found that the addition of nanoclay additives improves the barrier characteristics of food packaging materials in particular oxygen, moisture, and aroma barriers.

5.2.1 Oxygen Barrier

Oxygen is a major reason for food spoilage, as it promotes oxidative reactions that affect flavor, color, texture, and nutritive value. Among the various types of nanoclays, montmorillonite and kaolinite show superiority in improving the oxygen barrier properties of packaging material. This unique combination of high aspect ratio and tortuous diffusion pathways creates barriers that dramatically suppress oxygen permeation through packaging films [26, 27].

For nanoclays incorporated into polymer matrices, nanocomposites with decreased oxygen permeability are formed. The nanoclay platelets lead to a tortuous path for diffusion of oxygen molecules which effectively limits oxygen ingress into packaged foods. This barrier function ensures that food freshness and quality are maintained by hindering the oxidative reactions and thereby extending the shelf-life of perishable items such as meat, dairy, and bakery items [26, 27].

5.2.2 Moisture Barrier

Moreover, moisture is one of the major factors that influence food decomposition since it leads to microbial growth, enzymatic reactions, and textural changes. The nanoclays have innate hygroscopic properties and they can form hydrogen bonds with water molecules hence they are suitable for improving the moisture barrier characteristics of food packaging [28, 29].

Packages made of nanoclay significantly lower water vapor transmission rates relative to standard materials. The limited permeability no longer allows moisture uptake and keeps adequate moisture levels in packed food. The shelf life of various food products, such as snacks, cereals, and confectionery items, is extended with the help of nanoclay-enhanced packaging materials which dodge moisture-induced problems like spoilage and texture degradation [28, 29].

5.2.3 Aroma Barrier

Food aroma preservation is vital for sensory appeal and consumer acceptance. Aroma is one of the most important factors contributing to the tasty food and sensory appeal of foods. Nanoclays participate in aroma barrier enhancement by impeding the penetration of volatile organic compounds attributed to flavor loss and undesirable odor development [30, 31].

Nanoclays incorporated into packaging materials generate intercalated structures inside polymer matrices that impede aroma compounds' diffusion. The barrier function protects the sensory qualities and organoleptic characteristics of packaged foods. Thus, consumers get products with genuine tastes and aromas. Flavor retention is very important for products that are sensitive to aroma loss, hence food products like coffee, spices, and baked goods have aroma barrier enhancement as a significant factor [30, 31].

5.3 Mechanical Reinforcement

Mechanical properties are among the important properties as they govern the packaging material strength during food handling, transportation, and storage. Nanoclays considerably improve the mechanical properties of the polymers by enhancing their strength and flexibility [32, 33]. Nanoclay addition significantly strengthens packaging materials which gives rise to higher tensile strength, modulus, and impact resistance. This is due to the high aspect ratio and surface area of nanoclays which result in strong interfacial interactions with polymer matrices.

When incorporated into the polymer matrices, nanoclays create a network of filler particles that act as load-bearing elements. The intimate morphological-interface relationship between nanoclays and polymer chains enables highly effective stress transfer and load distribution which effectively increases the overall strength and structural integrity of the packaged material. Nanoclay reinforcement enhances the strength which in turn reduces the risk of mechanical failure like tearing, puncturing, or rupture. It guarantees safe and reliable packaging for the foods. This is a very crucial property for packaging applications where mechanical strength is a key requirement like in flexible pouches, trays, and containers used for shipping and storage [32, 33].

Besides strength enhancement, nanoclays influence the increase of flexibility in packaging materials through the reduction of brittleness and enhancement of plasticity features. This flexibility reinforcement enables the production of ultra-thin, light packaging films with outstanding flexibility and conformability to food surfaces.

The nanoscale dispersion of clay platelets within polymer matrices provides flexibility without sacrificing strength leading to packaging materials that are both strong and flexible. This flexibility is beneficial in applications that need pressure-sensitive conformability, opening ease, and creasing/folding resistance, like stand-up pouches, wrappers, and lids.

Furthermore, the nanoclay-filled packaging materials show increased cracking, tearing, and delamination resistance also under dynamic loading. The enhancement of packaging materials' flexibility extends their longevity and robustness. Therefore, the risk of products being damaged during handling and distribution is reduced [34, 35].

5.4 Antimicrobial Properties

Data safety is of utmost importance in the food packaging sector, and nanoclay-based materials possess antimicrobial properties that inhibit microbial development and ensure the security of packaged foods. Nanoclays such as silver-exchanged montmorillonite and zinc oxide-modified halloysite exhibit intrinsic antibacterial activity against a wide range of microorganisms, like bacteria, fungi, and yeast. Such antimicrobial properties arise due to several mechanisms like the release of biocidal ions, surface charge interactions, and microbial cell membrane disruption upon contact with nanoclay surfaces.

The inclusion of antimicrobial nanoclays into food packaging materials is extremely efficient in terms of microbial control thereby minimizing the risk of food contamination, spoilage, and deterioration. This antimicrobial functionality lengthens the shelf-life of packaged foods by inhibiting the growth of pathogenic and spoilage microorganisms, hence, preserving freshness, safety, and quality throughout the supply chain [36].

5.5 Shelf-Life Extension

The shelf-life extension, minimizing degradation processes, and maintenance of product integrity over time is an important goal in food packaging and nanoclay-based materials contribute to the shelf-life extension. Nanoclays-based composite materials possess superior shielding parameters against oxygen, moisture, and volatile permeability, which are major causes of food degradation. By restricting oxygen intake, nanoclays inhibit oxidative reactions resulting in flavor loss, color changes, and nutrient degradation. At the same time nanoclays decrease moisture absorption, and prevent the formation and development of microorganisms, enzyme reactions, and texture decay.

Also, nanoclay-based materials possessing antimicrobial properties impede microbial multiplication, thus prolonging the shelf-life of the preservative foods by decreasing the potential for spoilage and contamination. It allows manufacturers to deliver to consumers high-quality shelf shelf-stable products minimizing food waste and increasing economic sustainability [1, 2].

5.6 Smart Packaging Applications

Advances in nanoclay technology have led to the development of new packaging systems that provide real-time monitoring and regulation of nutrition, safety, and renewal. Using nanoclay-based sensors, indicators, and nanocomposite films offered in packaging materials, manufacturers can create smart functions including gas sensing, freshness indicators, time and temperature monitoring, and damage detection. These smart packaging systems interact with the environment to detect changes in air quality, humidity, and pH levels.

For example, nanoclay-based gas sensors can detect changes in the gas composition of packaging materials, indicating the presence of corrosive gases such as ammonia or ethylene. Similarly, nanoclay-based freshness indicators can change color due to certain chemical reactions or microbial activity providing consumers with visual clues about food condition and shelf life.

Smart nanoclay-based packaging can be used to improve consumer knowledge and perceptions of food quality and safety, thereby increasing consumer confidence, comfort, and satisfaction. These innovative packaging solutions make it easier to prevent food waste, simplify inventory management, and improve supply chain efficiency [37, 38].

6 Nanoclay Incorporation Methods in Food Packaging Materials

6.1 Direct Blending with Polymer Matrices

Often used, direct blending is one of the easiest methods to introduce nanoclay into food packaging materials. In this approach, clay nanoparticles are directly mixed or blended with polymer matrices at the moment of manufacture of packaging films, coatings, or containers (Fig. 2).

Nanoclay particles are dispersed in the polymer matrix using mechanical mixing methods such as melt compounding, solution mixing, and extrusion in the direct

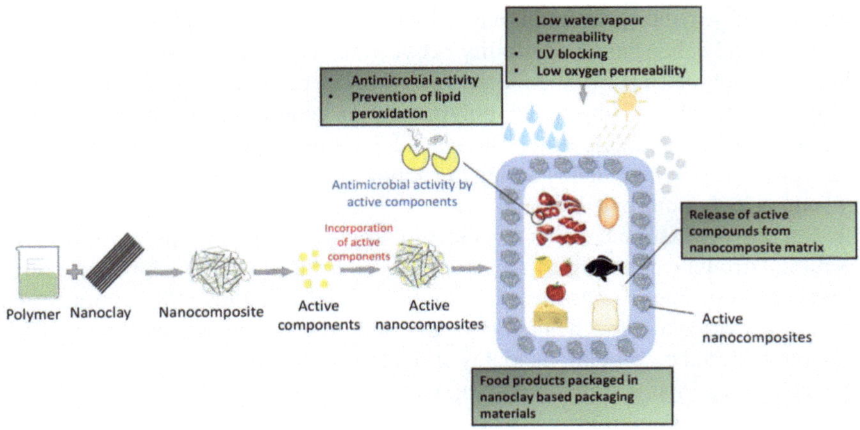

Fig. 2 Diagram illustrating the processes involved in the production of active nanocomposites (Adapted from [1, 2])

blending process. During blending shear forces and mechanical agitation are imposed to achieve homogeneous dispersion of nanoclay in the polymer matrix.

Direct blending has several advantages such as simplicity, scalability, and cost-effectiveness. It makes it possible for the facilitated inclusion of nanoclay into the original packaging formulations without specially designed apparatus and intricate filling procedures. Additionally, in situ, blending offers the possibility to tailor the nanoclay content and its distribution throughout the polymeric matrix for optimal barrier properties, mechanical strength, and other functions of the packaging material.

However, direct blending could face difficulties in achieving uniform particle dispersion and the interfacial adhesion between nanoclay particles and polymer chains. Very poor dispersion can result in agglomeration or phase separation which causes the packaging material to be heterogeneous, thus reducing its performance significantly. In order to solve the aforementioned complications, adequate processing conditions, surfactants, and compatibilizers should be exploited to promote nanoclay dispersion and good compatibility with the polymer matrix.

Though these obstacles exist, direct blending is still a choice method for infusing nanoclay into food packaging materials offering a way to improve barrier properties, mechanical performance, and functionality to meet the growing needs of the food industry [39].

6.2 *In-Situ Polymerization*

The synthesis of nano clay or incorporation of nano clay into the polymer matrix happens on-site (during the ongoing polymerization). This technique permits the co-creation of polymer chains and nanoclay dispersion, thus resulting in intensive interactions and enhanced compatibility between the two components.

In in-situ polymerization, the monomer, nanoclay particles, and polymerization initiators are mixed and subjected to polymerization conditions that include temperature, pressure, and reaction time. The nanoclay particles get embedded or grafted into the growing polymer chains during the polymerization process so a homogenous nanocomposite structure is realized.

In-situ polymerization benefits from several improvements such as increased clay dispersion, stronger adhesion at the interface, and the ability to shape structures and properties of nanocomposites. In-situ polymerization allows the optimization of barrier performance, mechanical strength, and other functional attributes of the packaging material employing nanoclay at the molecular level.

Nevertheless, in-situ polymerization may need specialized equipment, exact setting of reaction conditions, as well as monomers and polymerization methods compatibility. Furthermore, the choice of proper monomers and initiators and adjustment of reaction parameters are important factors determining the efficiency and scale-up of the process.

Taking into account these difficulties, in-situ polymerization provides possibilities for modifying nanoclay-polymer affinity and improving the characteristics of nanocomposite food packaging materials for particular applications of the food industry [40].

6.3 Coating Techniques

The coating method comprises the act of applying a film of nanoclay dispersion on the surface of food packaging material to improve the characteristics. Among these techniques are dip coating, spray coating, and roll-to-roll coating, which allow the uniform deposition of nanoclay coatings on different substrates.

In coatings techniques, nanoclay dispersions are prepared by dispersing nanoclay particles in a selected solvent or dispersion medium. Then, the substrate material, for example, paperboard, plastic film, or metal foil, is coated with the nanoclay dispersion using the selected coating method. After the application, the coated substrate is dried or cured to remove the solvent thereby forming a thin nanoclay layer on the surface.

Nanoclay incorporation into food packaging materials through coating techniques provides several benefits such as flexibility, fitting in the existing packaging formats, and easy application. The nanoclay coatings can be applied onto a wide range of substrates such as flexible films, rigid containers, and paper-based packaging without significantly affecting their mechanical or optical properties. Nanoclay coatings deliver targeted improvement in specific features of the substrate especially moisture barrier, mechanical strength, and surface functionality with no interference with inherent properties of the substrate material. Selective nanoclay coating of packaging material critical zones, such as seals, or high-stress areas, allows for targeted performance and functionality improvements.

Nevertheless, coating methods necessitate adjustment of coating formulations, drying parameters, and coating thickness to guarantee homogeneous coverage and adhesion of nanoclay coatings. Also, compatibility with substrate material should be done to avoid adhesion issues and substrate degradation. The coating methods present a diverse and efficient way of adding nanoclay into food packaging materials, having their main application in selectively improving barrier properties, mechanical performance, and functionality toward specific food industry requirements [41].

6.4 Layer-By-Layer Assembly

Layer-by-layer assembly, a versatile technique, enables uniform multilayered films with nanoclay. This method provides control over the film thickness, shape, and properties by sequentially depositing nanoclay layers and polyelectrolytes on the substrate surface through a series of immersion and extraction steps. Filler solvents are used to prepare polyelectrolyte solutions containing charged polymers or surfactants. The substrate material, usually a thin layer or film, is slowly dispersed in the nanoclay dispersion and polyelectrolyte solution followed by washing to remove excess material. Successive deposition of nanoclay polyelectrolyte layers provides precise control of the thickness structure of nanoclay-based multilayer films. Electrostatic interactions between opposing layers result in strong adhesion and interfacial bonding, resulting in long-term stability coatings made of nanocomposites [42].

6.5 Electrospinning

Electrospinning is an alternative method for incorporating nanoclays into food packaging materials, allowing for precise control of fiber size, structure, and properties. This method involves electrospinning ultrafine fibers with a polymer solution or liquid and depositing them on the substrate.

The electrospinning process starts with the formation of a polymer solution or hydrated nanoclay particles. The polymer solution is dispensed into a syringe or reservoir, and the electrical system applies high voltage to the solution. When an electric field is applied, a drop of polymer solution is formed at the tip of the spinneret, which is suspended as a micro-jet under the influence of the electric field. The nanoclay particles are uniformly dispersed in the polymer solution, ensuring uniform distribution throughout the resulting nanofibers. The electrospun nanofibers accumulate in a grounded collector, producing characteristic nanocomposite membranes or mats improved due to the addition of nanoclay [43] (Table 3).

Table 3 Comparison of nanoclay incorporation methods

Incorporation method	Description	Impact on material properties	Advantages	Challenges
Direct Blending	Nanoclay mixed with polymer matrices	Improved mechanical strength and barrier properties	Simplicity, Cost-effective	Uneven dispersion, Compatibility issues
In-situ Polymerization	Nanoclay synthesized within the polymer matrix	Enhanced compatibility and uniform dispersion	Homogeneous dispersion, Improved bonding	Complex process, High production cost
Layer-by-Layer Assembly	Alternating layers of polymer and nanoclay	Precise control over nanoclay distribution and properties	Precision in layer deposition	Time-consuming, Limited scalability

7 Characterization of Nanoclay-Based Food Packaging Materials

7.1 X-Ray Diffraction (XRD)

X-ray diffraction (XRD) is a powerful analytical technique for determining the crystal structure and morphology of the nanoclay particles in food packaging by exposing the sample to X-rays to obtain information on the interfacial phases and crystallinity of the nanoclay.

In NCFP materials, XRD can reveal the degree of nanoclay dispersion, and its bonding in the polymer matrix. Changes in the surface area, intensity, and width of the XRD patterns may indicate the presence or precipitation of nanoclay, as well as the formation of additional crystals due to interactions with the polymer matrix.

XRD is important for analyzing the process integrity and stability of nanoclay-based packaging materials under processing conditions, such as blending, extrusion, or coating. XRD helps optimize processing parameters by characterizing nanoclay morphology and its structural properties, for effective food packaging applications [44].

7.2 Transmission Electron Microscopy (TEM)

Transmission electron microscopy (TEM) enables high-resolution imaging and characterization of nanoclay particles and distribution in food packaging detailing nanoclay size, shape, morphology, and dispersion at the TEM nanoscale by transferring electrons to thin sections of the sample.

TEM of NCFP materials reveals the spatial distribution of nanoclays in the polymer matrix, including their orientation, stacking, and network interactions. TEM images can distinguish between interstitial, adsorbent, and aggregate nanoclays, providing valuable information for conversion control and optimization methods. TEM also allows observing nanoclay-polymer interfaces, elucidating mechanisms governing nanoclay dispersion and reinforcement in food packaging [45].

7.3 *Scanning Electron Microscopy (SEM)*

Scanning electron microscopy (SEM) is used to investigate the surface morphology, topography, and microstructure of NCFP materials. By scanning the sample surface with a focused electron beam, SEM produces a high-resolution image that reveals the details of the nanoclay diffusion, distribution, and interaction with the polymer matrix.

In NCFP materials, SEM provides valuable information about the surface roughness, porosity, and uniformity of nanocomposite films, coatings, or membranes. SEM images can detect nanoclay aggregates, voids, or defects. SEM is also used to investigate the cracking of nano-clay packaging under mechanical or environmental stress conditions. By investigating crack properties such as ductile deformation, crack propagation, and interface failure, SEM helps to elucidate the mechanisms governing the mechanical behavior and failure mechanisms governed by nanocomposites [46].

7.4 *Fourier Transform Infrared Spectroscopy (FTIR)*

Fourier-transform infrared spectroscopy (FTIR) is a spectroscopic technique for investigating chemical structure, functional groups, and molecular interactions in NCFP materials. FTIR radiation provides insight into the chemical composition and bond state of a nanoclay-polymer matrix through infrared absorption and transmission measured on nanoclay. The spectra of the functional groups of nanoclays, polymers, and additives depend on the peak position, intensity, and visibility of the characteristic absorption bands. Changes in band shapes and sizes may reflect changes in chemical structure and interactions. FTIR is particularly useful for studying nanoclay-polymer interactions, such as hydrogen bonding, van der Waals effects, or covalent interactions, which affect nanocomposite properties such as assembly, dispersion, and stability [47].

7.5 Gas Permeability Testing

Gas permeability testing measures how fast gas passes through NCFP materials and provides important information about barriers, shelf-life, and product stability. Samples are exposed to controlled gas conditions, and gas diffusion is monitored over time. Gas permeability tests in NCFP materials examine the ability of nanoclays to reduce oxygen, water, and odor permeation, which are major contributors to food spoilage. The permeability values of nanocomposites and conventional packaging materials can be compared and their impact on the shelf-life extension of packed food products can be ascertained. Gas permeability testing also enables analysis of the performance of nanocomposites under various environmental conditions such as temperature, humidity, and pressure. By simulating real-world storage and transport conditions, researchers can predict the stability and reliability of NCFP products and design, and operating parameters are optimized accordingly [48] (Table 4).

8 Safety and Regulatory Considerations

NCFP materials offer many advantages, but it is important to address potential toxicity concerns associated with nanoclay particles that can migrate into food. Understanding the aspects of toxicity is important to ensure consumer safety and compliance with regulatory standards [49, 50]. The nanoscale shape of nanoclays raises concerns about their ability to penetrate biological barriers and interact with cellular components. The study showed that nanoclay particles can exhibit different biological behaviors compared to their bulk counterparts due to their large surface area and reactivity [49, 50]. Although nanoclays are generally considered inert and non-toxic, studies are underway to evaluate possible health effects, especially when ingested or inhaled. Studies have examined the cytotoxicity, genotoxicity, and inflammation associated with the use of nanoclay particles, differing in such things as the size of particles, surface chemistry, and method of expression [49, 50].

Regulatory agencies around the world have developed guidelines and standards to ensure the safety and compliance of food products containing nanoclays. This regulation aims to protect public health and eliminate potential hazards associated with chemical migration from packaging materials [49, 50]. Law enforcement agencies such as the U.S. The Food and Drug Administration (FDA), the European Food Safety Authority (EFSA), and the World Health Organization (WHO) provide guidelines and regulations for the use of food additives, where requirements regarding safety inspection, migration testing, and compliance with specific migration limits are defined [49, 50]. Some regulatory agencies have developed specific guidance documents or policies to address specific issues and challenges associated with nanomaterials in food contact. These guidelines provide recommendations for risk assessment, characterization, and safety assessment of nanoclays and other nanomaterials applied to food packaging [49, 50].

Table 4 Characterization techniques for nanoclay-based packaging materials

Technique	Description	Application in nanoclay characterization
X-ray diffraction (XRD)	Analyzes crystal structure, composition, and orientation of nanoclay particles within food packaging materials. Provides information on interlayer spacing, crystallinity, and phase composition	Determines nanoclay dispersion, intercalation, and exfoliation within the polymer matrix. Essential for assessing the structural integrity and stability of nanoclay-based packaging materials under different processing conditions. Helps optimize processing parameters and enhance performance in terms of barrier properties, mechanical strength, and functionality
Transmission electron microscopy (TEM)	Provides high-resolution imaging and characterization of nanoclay particles and their distribution within food packaging materials. Offers insights into size, shape, morphology, and dispersion at the nanoscale	Reveals the spatial distribution of nanoclays within the polymer matrix. Distinguishes between intercalated, exfoliated, and agglomerated nanoclays. Elucidates mechanisms governing nanoclay dispersion and reinforcement. Correlates observations with mechanical, barrier, and antimicrobial properties to tailor morphology and distribution for desired performance outcomes
Scanning electron microscopy (SEM)	Examines surface morphology, topography, and microstructure of NCFP materials. Generates high-resolution images revealing details of nanoclay dispersion, distribution, and interaction	Provides information on surface roughness, porosity, and homogeneity of nanocomposite films, coatings, or membranes. Identifies nanoclay agglomerates, voids, or defects affecting barrier properties, mechanical strength, or processing performance. Analyzes fracture surfaces to elucidate mechanical behavior and failure modes

(continued)

Table 4 (continued)

Technique	Description	Application in nanoclay characterization
Fourier-transform infrared spectroscopy (FTIR)	Analyzes chemical composition, functional groups, and molecular interactions in NCFP materials. Provides insights into chemical structure and bonding environment	Reveals absorption bands associated with functional groups in nanoclays, polymers, and additives. Indicates chemical interactions, bonding, or modifications induced by processing or environmental exposure. Useful for studying nanoclay-polymer interactions influencing properties such as compatibility, dispersion, and stability. Assess the impact of processing on nanoclay-based packaging materials and optimize conditions accordingly
Gas permeability testing	Measures gas transmission rate through NCFP materials. Evaluate barrier properties, shelf-life, and product stability. Quantifies permeability coefficient, solubility, and diffusion rate	Evaluate the effectiveness of nanoclays in reducing oxygen, moisture, and aroma permeation. Assesses barrier enhancement and impact on shelf-life extension and product quality. Predicts long-term stability and reliability under different environmental conditions. Optimizes formulation and processing parameters for improved performance

Manufacturers of NCFP products should conduct safety assessments to demonstrate compliance with regulatory requirements. These studies typically include toxicity studies, migration tests, and risk assessments to assess the safety of nanoclay particles and their potential impact on food safety and human health [49, 50]. Risk assessment and mitigation strategies are necessary to manage the potential risks associated with NCFP products and ensure consumer safety [49, 50]. Detailed characterization of nanoclay particles and nanocomposite materials requires understanding their physicochemical properties, behavior, and potential hazards. This includes evaluating particle size, shape, surface area, and stability under normal food contact conditions. Migration studies are required to monitor the release of nanoclay particles from packaging materials to estimate consumer exposure and potential health risks. This includes measurement of migration rates and analysis of the toxicity of migrating nanoparticles [49, 50].

It is important to ensure consumer protection through the implementation of risk management measures to reduce the potential risk of nanoclay migration. This includes selecting fewer mobile materials, optimizing operating conditions, applying a barrier coating or coating, or using a surface treatment to reduce nanoparticle release. Continuous monitoring of NCFP resources is essential to identify emerging threats, address security concerns, and ensure compliance.

9 Challenges and Future Perspectives

One of the main challenges in the application of NCFP materials is achieving perfect dispersion and homogeneity in polymer matrices. Poor dispersion can lead to aggregation, separation, and heterogeneity, compromising packaging functionality and performance. To overcome dispersion and compatibility issues, alternative methods of processing, surface modification, and material compatibility are necessary to maximize nanoclay-polymer interactions and ensure uniform distribution throughout the case [51].

Although NCFP materials offer significant advantages in barrier properties, mechanical properties, and performance, cost and scalability remain key considerations for widespread adoption. Production costs associated with nanoclay synthesis, processing, and incorporation into packaging materials may be higher than conventional alternatives. It requires advances in nano-clay manufacturing, processing technologies, and industrial processes to reduce costs while maintaining quality and efficiency [52].

Environmental concerns and sustainability are increasingly important trends shaping the future of food packaging including nanoclay-based solutions. Although nanoclays offer opportunities for product delivery, improve the shelf life of raw materials, and maximize recyclability through food packaging, they must meet environmental considerations, such as energy consumption, and waste generation management. Sustainable nano-clay manufacturing processes, biodegradable polymer matrices, and environmentally friendly packaging material design are needed to reduce environmental impact and promote circular economic principles [53].

Continued research in new manufacturing processes such as nanocomposite extrusion, injection molding, and 3D printing will enable the development of durable nanoclay-based packaging materials with tailored properties and functionalities [54]. The incorporation of functional additives, nanoparticles, and bioactive compounds into NCFP materials will enable the development of multifunctional nanocomposites with improved barrier properties, antimicrobial activity, and intelligent functionalities to improve food safety and quality [55]. The use of bio-based polymers, biodegradable matrices, and renewable nano-clay sources will contribute to the development of sustainable NCFP products with reduced environmental footprint and improved end-of-life processes [56]. The integration of sensors, indicators, and reactive materials into nanoclay-based packaging will enable smart packaging technology to monitor food freshness, quality, and safety in real time, increase consumer confidence, and reduce food waste [38]. Harmonization of regulatory requirements, standardized testing protocols, and collaboration between stakeholders, regulatory agencies, and research institutions will facilitate the commercialization and regulatory approval of NCFP products worldwide (Fig. 3).

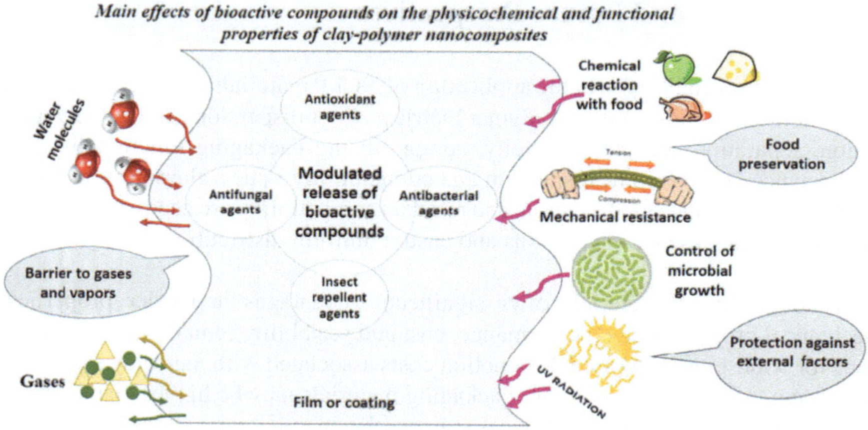

Fig. 3 Transforming nanoclay composites for enhanced performance in diverse environments (Adapted from [57])

10 Conclusion

In conclusion, nanoclay-based food packaging (NCFP) represents a breakthrough innovation that could revolutionize the food packaging industry. By harnessing the unique properties of nanoclay and increasing research and development efforts, we can open new opportunities for safe, innovative, and sustainable packaged foods to overcome existing challenges.

As we move forward, collaboration, innovation, and responsible stewardship will be essential in realizing the full potential of NCFP products. Industry stakeholders, researchers, policymakers, and consumers must work together to address safety concerns, regulatory requirements, and sustainability considerations, and exploit the benefits of nanoclay to develop solutions used in food packaging that enhance food safety, quality, and sustainability.

The prospects for the NCFP are bright, with continued advances in materials science, process technology, and regulatory frameworks that encourage innovation and adoption. By seizing these opportunities and all food packaging atomically we can build a safe, resilient, and sustainable food system for future generations.

Abbreviations

NCFP	NanoClay-based Food Packaging
PET	Polyethylene Terephthalate
FDA	U.S. Food and Drug Administration
EFSA	European Food Safety Authority

WHO	World Health Organization
LbL	Layer-by-Layer
XRD	X-ray Diffraction
TEM	Transmission Electron Microscopy
SEM	Scanning Electron Microscopy
FTIR	Fourier-Transform Infrared Spectroscopy

References

1. Nath D, Santhosh R, Pal K, Sarkar P (2022) Nanoclay-based active food packaging systems: a review. Food Packag Shelf Life 31:100803
2. Perera KY, Hopkins M, Jaiswal AK, Jaiswal S (2023) Nanoclays-containing bio-based packaging materials: properties, applications, safety, and regulatory issues. J Nanostruct Chem [Internet]:1 [cited 2024 Feb 11]. (Articles/PMC9893189)
3. Zorraquín-Peña I, Cueva C, Bartolomé B, Moreno-Arribas MV (2020) Silver nanoparticles against foodborne bacteria: effects at intestinal level and health limitations. Microorganisms [Internet] 8(1) [cited 2024 Feb 11]. (Articles PMC7022296)
4. Mostafa AA, Al-Askar AA, Almaary KS, Dawoud TM, Sholkamy EN, Bakri MM (2018) Antimicrobial activity of some plant extracts against bacterial strains causing food poisoning diseases. Saudi J Biol Sci [Internet] 25(2):361 [cited 2024 Feb 11]. (Articles: PMC5815983)
5. Bumbudsanpharoke N, Ko S (2019) Nanoclays in food and beverage packaging. J Nanomater 2019
6. Wagner A, Rakesh Gupta, Dinu CZ (2017) Nanoclays. Nanotechnology commercialization [Internet] :369–93 [cited 2024 Feb 11]. https://doi.org/10.1002/9781119371762.ch11
7. Brandelli A (2018) Toxicity and safety evaluation of nanoclays. Nanomater Ecotoxicity Saf Public Percept [Internet] :57–76 [cited 2024 Feb 11]. https://doi.org/10.1007/978-3-030-05144-0_4
8. Abulyazied DE, Ene A (2021) An investigative study on the progress of nanoclay-reinforced polymers: preparation, properties, and applications: a review. Polymers (Basel) [Internet] 13(24) [cited 2024 Feb 11]. (Articles: PMC8704753)
9. Nanoclays—an overview|ScienceDirect topics [Internet] [cited 2024 Feb 11]. https://www.sciencedirect.com/topics/agricultural-and-biological-sciences/nanoclays
10. Kumari N, Mohan C (2021) Basics of clay minerals and their characteristic properties. Clay Clay Miner
11. Al Ani T, Sarapää O (2008) Clay and clay mineralogy. Geochem Surv Finl
12. Kumar P, Gupta P (2023) Halloysite-based magnetic nanostructures for diverse applications: a review. ACS Appl Nano Mater [Internet] 6(15):13824–13868 [cited 2024 Feb 11]. https://doi.org/10.1021/acsanm.3c02573
13. Gładysz-Płaska A, Majdan M, Tarasiuk B, Sternik D, Grabias E (2018) The use of halloysite functionalized with isothiouronium salts as an organic/inorganic hybrid adsorbent for uranium(VI) ions removal. J Hazard Mater 15(354):133–144
14. Wan Ikhsan SN, Yusof N, Aziz F, Ismail AF, Jaafar J, Wan Salleh WN et al (2021) Superwetting materials for hydrophilic-oleophobic membrane in oily wastewater treatment. J Environ Manage 15(290):112565
15. Hamza A, Hussein IA, Mahmoud M (2023) Introduction to reservoir fluids and rock properties. Dev Pet Sci 1(78):1–19
16. Ismadji S, Soetaredjo FE, Ayucitra A (2015) Natural clay minerals as environmental cleaning agents:5–37 [cited 2024 Feb 11]. https://doi.org/10.1007/978-3-319-16712-1_2

17. Kotal M, Bhowmick AK (2015) Polymer nanocomposites from modified clays: recent advances and challenges. Prog Polym Sci 1(51):127–187
18. Bandyopadhyay J, Ray SS (2019) Are nanoclay-containing polymer composites safe for food packaging applications?—an overview. J Appl Polym Sci [Internet] 136(12):47214 [cited 2024 Feb 11]. https://doi.org/10.1002/app.47214
19. Ahari H, Anvar AA, Ataee M, Naeimabadi M (2021) Employing nanosilver, nanocopper, and nanoclays in food packaging production: a systematic review. Coatings [Internet] 11(5):509 [cited 2024 Feb 11]. https://www.mdpi.com/2079-6412/11/5/509/htm
20. Liao L, Li X, Wang Y, Fu H, Li Y (2016) Effects of surface structure and morphology of nanoclays on the properties of jatropha curcas oil-based waterborne polyurethane/clay nanocomposites. Ind Eng Chem Res [Internet] 55(45):11689–11699 [cited 2024 Feb 11]. https://doi.org/10.1021/acs.iecr.6b02527
21. Katti KS, Jasuja H, Jaswandkar SV, Mohanty S, Katti DR (2022) Nanoclays in medicine: a new frontier of an ancient medical practice. Mater Adv [Internet] 3(20):7484 [cited 2024 Feb 11]. (articles: PMC9577303)
22. Lau K, Gu C, Hui D (2006) A critical review on nanotube and nanotube/nanoclay related polymer composite materials. Compos B Eng 37(6):425–436
23. Elmi C (2023) Physical-chemical properties of nano-sized phyllosilicates: recent environmental and industrial advancements. Encyclopedia [Internet] 3(4):1439–1460 [cited 2024 Feb 11]. https://www.mdpi.com/2673-8392/3/4/103/htm
24. Jayrajsinh S, Shankar G, Agrawal YK, Bakre L (2017) Montmorillonite nanoclay as a multifaceted drug-delivery carrier: a review. J Drug Deliv Sci Technol
25. Ogunlaja SB, Pal R (2020) Effects of bentonite nanoclay and cetyltrimethyl ammonium bromide modified bentonite nanoclay on phase inversion of water-in-oil emulsions. Colloids Interfaces 4(1)
26. Ben Dhieb F, Dil EJ, Tabatabaei SH, Mighri F, Ajji A (2019) Effect of nanoclay orientation on oxygen barrier properties of LbL nanocomposite coated films. RSC Adv 9(3)
27. Paara T, Lange S, Saal K, Lõhmus R, Krumme A, Mändar H (2022) Improving the oxygen barrier of polyamide food packaging by using nanoclay. Medziagotyra 28(2)
28. Merah N, Ashraf F, Shaukat MM (2022) Mechanical and moisture barrier properties of epoxy–nanoclay and hybrid epoxy–nanoclay glass fibre composites: a review. Polymers 14
29. Kang DJ, Park GU, Park HY, Park JU, Im HG (2018) A high-performance transparent moisture barrier using surface-modified nanoclay composite for OLED encapsulation. Prog Org Coat 118
30. Saucedo-Zuñiga JN, Sánchez-Valdes S, Ramírez-Vargas E, Guillen L, Ramos-deValle LF, Graciano-Verdugo A et al (2021) Controlled release of essential oils using laminar nanoclay and porous halloysite/essential oil composites in a multilayer film reservoir. Microporous Mesoporous Mater:316
31. Saucedo-Zuñiga JN, Sánchez-Valdes S, Ramírez-Vargas E, Guillen L, Ramos-DeValle LF, Rodriguez JA (2021) Controlled release of essential oils using laminar nanoclay and halloysite/essential oil composites in a multilayer film. In: Annual technical conference—ANTEC, conference proceedings
32. Elsacker E, De Laet L, Peeters E (2022) Functional grading of mycelium materials with inorganic particles: the effect of nanoclay on the biological, chemical and mechanical properties. Biomimetics 7(2)
33. Rajabifar N, Rostami A (2023) Investigation of the effect of hybrid nanofiller on the mechanical performance and surface properties of bio-based polylactic acid/polyolefin elastomer (PLA/POE) blend. Polymers (Basel) 15(12)
34. Osman AF, Tuty TF, Rakibuddin M, Hashim F, Tuan Johari SAT, Ananthakrishnan R et al (2017) Pre-dispersed organo-montmorillonite (organo-MMT) nanofiller: morphology, cytocompatibility and impact on flexibility, toughness and biostability of biomedical ethyl vinyl acetate (EVA) copolymer. Mater Sci Eng C:74
35. Bahari-Sambran F, Meuchelboeck J, Kazemi-Khasragh E, Eslami-Farsani R, Arbab Chirani S (2019) The effect of surface modified nanoclay on the interfacial and mechanical properties of basalt fiber metal laminates. Thin-Walled Struct:144

36. Mohammadzadeh-Vazifeh M, Hosseini SM, Mohammadi A, Jahanfar M, Maleki H (2020) Investigation of the antimicrobial properties of nanoclay and chitosan based nanocomposite on the microbial characteristics of Gouda cheese. Iran J Microbiol 12(2)
37. Shin H, Thanakkasaranee S, Kambiz S, Seo J (2023) Preparation and applicability of polylactic acid/polyethylene glycol/nanoclay composite films for smart steam release in microwave packaging. Food Packag Shelf Life:40
38. Abu-Thabit N, Hakeem AS, Mezghani K, Ratemi E, Elzagheid M, Umar Y et al (2020) Preparation of ph-indicative and flame-retardant nanocomposite films for smart packaging applications. Sensors (Switzerland) 20(19)
39. Kim HJ, Kim SW (2016) Enhancement of physical and optical performances of polycarbonate-based diffusers for direct-lit LED backlight unit by incorporation of nanoclay platelets. J Appl Polym Sci 133(6)
40. Joseph A, Joseph S, Mathew N, Koshy A, Jayalakshmi N, Mathew V (2019) Effect of incorporation of nanoclay on the properties of heat cure denture base material: an in vitro study. Contemp Clin Dent 10(4)
41. Alipal J, Saidin S, Abdullah HZ, Idris MI, Lee TC (2022) Physicochemical and cytotoxicity studies of a novel hydrogel nanoclay EPD coating on titanium made of chitosan/gelatin/halloysite for biomedical applications. Mater Chem Phys:290
42. Hakimi F, Jafari H, Hashemikia S, Shabani S, Ramazani A (2023) Chitosan-polyethylene oxide/clay-alginate nanofiber hydrogel scaffold for bone tissue engineering: preparation, physical characterization, and biomimetic mineralization. Int J Biol Macromol 233
43. Palak H, Aktürk B, Kayaoğlu BK, Göcek I (2022) Fabrication of montmorillonite nanoclay-loaded electrospun nanofibrous mats for UV protection. J Ind Text 51(3_suppl)
44. Bernardo V, Mugica M, Perez-Tamarit S, Notario B, Jimenez C, Rodriguez-Perez MA (2018) Nanoclay intercalation during foaming of polymeric nanocomposites studied in-situ by synchrotron X-ray diffraction. Materials 11(12)
45. Aly KI, Younis O, Mahross MH, Tsutsumi O, Mohamed MG, Sayed MM (2019) Novel conducting polymeric nanocomposites embedded with nanoclay: synthesis, photoluminescence, and corrosion protection performance. Polym J 51(1)
46. Samariha A, Hemmasi AH, Ghasemi I, Bazyar B, Nemati M (2015) Effect of nanoclay contents on properties, of bagasse flour/reprocessed high density polyethylene/nanoclay composites. Maderas: Ciencia y Tecnologia 17(3)
47. Yao H, You Z, Li L, Goh SW, Lee CH, Yap YK et al (2013) Rheological properties and chemical analysis of nanoclay and carbon microfiber modified asphalt with Fourier transform infrared spectroscopy. Constr Build Mater 38
48. Mahand SN, Yazdanbakhsh A, Tayouri MI, Zarei A, Nouranian S, Ruckdäschel H et al (2023) Theoretical and experimental investigation of selective gas permeability in polystyrene/polyolefin elastomer/nanoclay nanocomposite films. Polym Test 120
49. Wagner A, Eldawud R, White A, Agarwal S, Stueckle TA, Sierros KA et al (2017) Toxicity evaluations of nanoclays and thermally degraded byproducts through spectroscopical and microscopical approaches. Biochim Biophys Acta [Internet] 1861(1 Pt A):3406 [cited 2024 Feb 11] (articles: PMC5148634)
50. Radoor S, Karayil J, Shivanna JM, Jayakumar A, Varghese SA, Krishnankutty RE et al (2021) Environmental and toxicological aspects of nanostructures in food packaging. Nanotechnol Enhanc Food Packaging [Internet]:361–378 [cited 2024 Feb 11]. https://doi.org/10.1002/9783527827718.ch15
51. Broughton WR, Koukoulas T, Woolliams P, Williams J, Rahatekar SS (2013) Assessment of nanoparticle loading and dispersion in polymeric materials using optical coherence tomography. Polym Test 32(7)
52. Saba N, Jawaid M, Asim M (2016) Recent advances in nanoclay/natural fibers hybrid composites. In: Engineering materials
53. LeCorre D, Hohenthal C, Dufresne A, Bras J (2013) Comparative sustainability assessment of starch nanocrystals. J Polym Environ 21(1)

54. Jin Y, Xiong R, Antonelli PJ, Long CJ, McAleer CW, Hickman JJ et al (2021) Nanoclay suspension-enabled extrusion bioprinting of three-dimensional soft structures. J Manuf Sci Eng Trans ASME 143(12)
55. Diaz AJ, Noh H, Meier T, Solares SD (2017) High-stress study of bioinspired multifunctional PEDOT: PSS/nanoclay nanocomposites using AFM, SEM and numerical simulation. Beilstein J Nanotechnol 8(1)
56. Li W, Qamar SA, Qamar M, Basharat A, Bilal M, Iqbal HMN (2021) Carrageenan-based nano-hybrid materials for the mitigation of hazardous environmental pollutants. Int J Biol Macromol 190:700–712
57. Cheikh D, Majdoub H, Darder M (2022) An overview of clay-polymer nanocomposites containing bioactive compounds for food packaging applications. Appl Clay Sci 1(216):106335

Quantum Dots in Food Packaging

Suzan Uzun

Abstract Quantum dots are zero-dimensional semiconductor nanoparticles with ultrafine sizes in a range of 2–10 nm. Such small sizes give quantum dots (QDs) unique optical, physical, and chemical properties. QDs are made of heavy metals in the core and covered with a shell. Due to an increase in concerns about the cytotoxicity of QDs, there has been an increased focus on the synthesis of more biocompatible QDs, such as carbon quantum dots (CQDs). Carbon quantum dots (CQDs) are carbon-based fluorescent nanoparticles. CQDs mainly consist of carbon; therefore, the synthesis procedure requires carbon sources that are non-toxic and abundant in nature. Furthermore, chemical stability, biocompatibility, low toxicity, ecological safety, high solubility, excellent photoluminescence, and optical characteristics are some of the remarkable properties of CQDs. These features make CQDs excellent tools for various research fields including bioimaging, biosensors, drug delivery systems, solar cells, and food packaging. Recently, CQDs have been used as nanofillers, photoluminescent, antimicrobial, antioxidant, and UV-light-blocking agents in food packaging materials. This chapter overviews the synthesis methods along with functionalization and CQDs applications for food preservation. Moreover, its potential for the development of active and intelligent food packaging and contributions to film properties are presented.

Keywords Carbon quantum dots (CQDs) · Active packaging · Intelligent packaging · Food safety · Food quality

S. Uzun (✉)
Department of Food Engineering, Faculty of Agriculture, Tekirdag Namik Kemal University, Tekirdag, Turkey
e-mail: suzanuzun@nku.edu.tr

M. Alle et al. (eds.), *Functional Nanomaterials and Nanocomposites for Biodegradable Food Packaging*, Smart Nanomaterials Technology,
https://doi.org/10.1007/978-981-97-9084-5_10

1 Introduction

Food waste is one of the major global issues with social, economic, and environmental consequences. According to the reports, approximately one-third of the food produced worldwide is wasted each year. Food packaging plays a crucial role in preventing food waste by ensuring the safety, quality, and integrity of food products during transportation, storage, distribution, and marketing. Packaging materials serve as a barrier, preventing the migration of gases, moisture, and light, protecting against damage or contamination, extending the shelf life of products, and providing information to consumers about food products and their condition. Therefore, the properties of packaging materials have become crucial in preventing or delaying food deterioration.

Consumer demand for packaged food products continues to rise around the world. This leads to an increase in the accumulation of packaging material waste and causes environmental pollution as well as other related issues. Today, plastic is the most used material in food packaging. While plastics offer numerous advantages in packaging, such as transparency, low weight, durability, low cost, and barrier properties, their decomposition takes hundreds of years resulting in plastic pollution in ecosystems. Increasing concerns over the environmental impacts of plastics have led to investigations for sustainable alternatives such as biodegradable materials. Unlike plastics, biodegradable materials naturally decompose over time. Biopolymers have been widely used as biodegradable materials for packaging films, offering numerous advantages as a new type of packaging material [1–3]. However, using biopolymers for packaging presents some challenges when compared to plastics, primarily due to limitations in barrier, mechanical, and thermal properties. To address these weaknesses in biopolymer films, various approaches [4–9] have been employed including preparation of blends with other biopolymers, crosslinking, surface modifications, multilayer application, and incorporation of nanoparticles or additives.

Carbon dots, also known as carbon quantum dots (CQDs), are very small carbon-based nanoparticles with sizes typically ranging from 1 to 10 nm. CQDs resemble quantum dots, exhibiting unique optical, electrical, and chemical properties that distinguish them from other types of nanoparticles [10]. Unlike conventional quantum dots comprised of heavy metals, CQDs are metal-free, more environmentally friendly, and have low toxicity. Therefore, CQDs have a wide range of potential applications in optoelectronic devices, sensing, bioimaging, energy storage, drug delivery, and food packaging [11, 12].

The use of carbon dots in food packaging is still in its infancy, despite this, the research on the utilization of carbon dots in food packaging and safety applications has increased in the last decade. Due to changes in the fluorescence intensity of CQDs during the detection process, they have been intensely used as fluorescent probes to determine the presence and quality of various substances in food products [13–17]. In substance detection, fluorescence intensity increases, or decreases as a result of analyte interactions with CQDs, making them effective turn-on/turn-off probes. The fluorescence quenching mechanism of CQDs can be adopted to food packaging films which can be used as safety monitoring probes and maintain freshness [18]. Food spoilage leads to alterations in pH levels, primarily driven by the generation of spoilage metabolites, with volatiles being the predominant contributors. Smart films are engineered to identify spoilage by integrating pH-sensitive elements such as CQDs, as CQDs can react to pH shifts by changing color in natural light, reducing fluorescence intensity, or causing fluorescence quenching [19–21].

CQDs are introduced in food packaging materials to improve packaging films in terms of mechanical, thermal, and barrier properties [10, 22–24]. Biopolymers are generally brittle, less thermostable, have low resistance to prolonged process operations, exhibit no UV protection and almost most of them have no antimicrobial properties. To tailor the properties of biopolymer films, additional substances, or fillers such as CQDs, are incorporated into the film. CQDs show strong photoluminescence, excellent photostability, high water solubility, UV light absorption, good thermal stability, antimicrobial activity, and antioxidant properties. These characteristics make them promising and innovative additives for the development of active food packaging materials. CQDs are biocompatible with many biopolymers due to functional groups on the surface, leading to interactions with biopolymer molecules and resulting in improved additional functional properties of the packaging films [10, 12].

This chapter provides an overview of carbon dots, including their synthesis methods, physicochemical characteristics, and current applications in food packaging technology. The utilization of CQDs in active and intelligent food packaging systems is also summarized, with a focus on how CQDs incorporation impart functional properties to the films. Furthermore, the advantages and challenges of integrating CQDs into packaging films have been discussed through a comprehensive analysis of the most recent research and developments in this field.

2 Synthesis of CQDs

CQDs can exist in various food products such as coffee, hamburgers, beer, honey, pizza, barbecue, and caramels but in low amounts [25–31]. Many approaches have been formulated to produce CQDs at a substantial scale. In general, two main approaches have been used for preparing CQDs: top-down and bottom-up approaches. These methods use chemical, electrochemical, or physical techniques to generate CQDs.

2.1 Top-Down Synthesis

The top-down approach involves breaking down larger carbon materials such as graphite oxide, graphene, and activated carbon into carbon nanoparticles smaller than 10 nm. The decomposition of carbon sources can take place through physical, chemical, and electrochemical processes. Top-down methods include arc discharge, laser ablation, and electrochemical oxidation methods. These methods are suitable for precise control of CQD's size and shape with the desired geometry and allow mass production of CQDs. Although the top-down strategy is an effective approach for manufacturing CQDs, substances composed of carbon atoms are exposed to harsh conditions, which can make it challenging to control reactions and impurities during CQD synthesis, resulting in structural defects [32]. Nevertheless, top-down approaches are still important strategies in CQD fabrication.

2.1.1 Arc Discharge

The first reported CQDs have been accidentally synthesized using the arc discharge method during the purification of single-walled carbon nanotubes. However, they were not named 'carbon dots' or 'carbon quantum dots' until two years later [33]. Hence, the arc discharge was the first method employed for CQD fabrication. In this method, a constant current is applied between two electrodes immersed in an inert gas at high pressure. The discharge of the arc leads to the vaporization of carbon precursor on the anode, initiated by the high current and resulting in a high-energy plasma. The gaseous carbon is aggregated and collected from the cathode. Controlling the size of CQDs and enabling uniform size distribution are difficult in the arc discharge process. Regardless of low yield and high cost, arc discharge offers the synthesis of carbon nanomaterials with low-density structural defects desirable for high-performance nanoelectronics.

2.1.2 Laser Ablation

In laser ablation, a solid carbon precursor is irradiated with high-energy laser beam. When the laser beam hits the surface, the target material heats up and evaporates into a plasma state followed by condensing as CQDs [32]. In 2006, Sun et al. [34] achieved fabricating highly photoluminescent carbon dots from the mixture of graphite and cement by employing laser ablation in the presence of argon gas carrying water vapor, followed by acid oxidation and surface passivation to improve luminescence. Thereafter, many studies have been conducted to exploit CQDs in various applications. Laser irradiation is a facile process to produce nanoparticles from bulk carbon material by avoiding contamination and incorporation of impurities [35]. Fabricating CQDs in different solvents can affect the photoluminescence of CQDs. Furthermore, the size of CQDs can be controlled by tuning laser energy, laser pulse width, and ablation time [36]. However, low quantum yield, heterogeneity in size and shape limit their further applications.

2.2 Bottom-Up Synthesis

The bottom-up synthesis approach consists of the polymerization of organic precursors into carbon core and eventually formation of larger units, carbon quantum dots. The conversion of carbon sources into CQDs can be employed through different bottom-up strategies such as hydrothermal, solvothermal, microwave-assisted, and pyrolysis [35].

2.2.1 Hydrothermal/Solvothermal Method

Hydrothermal or solvothermal method is the most common technique used in CQD synthesis because the procedure is simple, inexpensive, and environmentally friendly. In this technique, the carbon precursor solution is prepared in aqueous or organic solvents and then sealed in a container, mostly an autoclave, followed by a thermal reaction at high temperature and pressure. When water is used as a reaction medium, this method is called hydrothermal but if organic solvents are used in precursor solution, then the method is called solvothermal. Fabrication of CQDs via hydrothermal or solvothermal method includes subsequent stages including polymerization, carbonization, and passivation. The characteristics of carbon dots can be tuned by process temperature, pressure, the ratio between solvent/carbon source, and surface functionalization with heteroatom doping [10]. Potato skin was one of the food wastes used as raw material for conversion to value-added functional material, CQDs [37]. In CQDs synthesis, the hydrothermal reaction was carried out at 200 °C for 6 h. The obtained CQDs were uniform spherical particles with 9.27–9.41 nm particle size and exhibited brown color under day light whereas they emitted green fluorescence when excited with UV light. They also showed antioxidant and antibacterial activity allowing the CQDs to be used in the development of multifunctional packaging materials. Synthesis of CQDs from sugarcane bagasse pulp with multiple surface functional groups increased solubility in water and production with a high quantum yield (QY) of 17.98% [38]. Travlou et al. [39] fabricated nitrogen and sulfur-doped carbon dots (N-CQDs and S-CQDs) via hydrothermal technique. The bactericidal activity of the functionalized CQDs relied on specific surface chemistry, and their sizes in the range of nanometers. N-CQDs displayed better antibacterial activity because of amides and amines in the structure.

Hydrothermal or solvothermal synthesis is the most effective process for the conversion of biomaterials into CQDs. Recently, many research groups have conducted hydrothermal or solvothermal methods to produce CQDs from biological sources such as orange juice, garlic, tea leaves, grapes, coffee, lemon, and starch [40–46]. Highly photoluminescent carbon dots were prepared from orange juice with hydrothermal treatment at low temperatures and a short time [40]. The CQDs were monodispersed with an average size of ~2.5 nm and highly photoluminescent with a quantum yield of 26%. The photoluminescence (PL) of CQDs was affected by the pH shifts where PL intensity noticeably decreased at high pH and remained stable in a range of pH 4.8–8.7 which was suitable for bioimaging. Tyagi et al. [47] prepared water-soluble CQDs from lemon peel through hydrothermal treatment at 200 °C for 12 h. Synthesized CQDs had spherical morphology and narrow size distribution (1–3 nm) with a quantum yield of 14%. The CQDs were selective and sensitive fluorescent probes to Cr^{6+} due to the oxygen-rich surface.

2.2.2 Microwave Method

Microwave-assisted synthesis has attracted attention in the fabrication of CQDs as it holds many advantages over other methods. In microwave irradiation, polar molecules in the reaction medium absorb energy and begin to oscillate, resulting in intermolecular friction that generates heat. Unlike other heating methods, microwave irradiation uniformly heats the reaction mixture, enabling precise control of CQD synthesis. Microwave-assisted synthesis is a rapid way to produce CQDs. While hydrothermal processes typically take hours, CQDs can be prepared in just a few minutes with microwave-assisted methods. In a study by Ramezani et al. [48] CQDs were synthesized from quince fruit using a one-pot microwave treatment and a hydrothermal method. One-pot microwave treatment was conducted at 220 °C for 30 min while the hydrothermal treatment required 4 h at 200 °C. The results showed that 30 min was insufficient for CQD synthesis via the hydrothermal method, whereas fluorescent CQDs were successfully produced using the microwave-assisted method within the same timeframe. Furthermore, microwave-assisted synthesis is a green method, as it consumes less energy and is suitable for industrial-scale production. Ding et al. [49] reported a rapid, microwave-assisted synthesis of nitrogen and sulfur co-doped carbon dots (N,S-CQDs) using 1,6-hexanediamine dihydrochloride and dimethyl sulfoxide as precursors. These N,S-CQDs displayed low cytotoxicity, biocompatibility, and outstanding photoluminescent properties. In addition, they exhibited excellent ion detection capabilities for MnO_4^- and $Cr_2O_7^{2-}$ with detection limits of 0.34 μM and 0.23 μM, respectively.

2.2.3 Pyrolysis Method

Pyrolysis is a thermal process in which carbon-containing materials are decomposed into carbon residues at high temperatures. In this process, carbon precursors undergo irreversible physical and chemical changes during dehydration, polymerization, and carbonization. Pyrolysis is considered a simple, inexpensive, and environmentally friendly method that requires no solvent for the decomposition of carbon material. Liu et al. [50] used citric acid and glycine to prepare CQDs and studied the effect of pyrolysis process conditions on CQDs properties. The maximum fluorescence was observed for CQDs produced at 220 °C and 3 h pyrolysis reaction where high product yield was obtained. The molar ratio of citric acid to glycine was a determining factor of fluorescence color and quantum yield of CQDs (84%). Ahlawat et al. [51] investigated the impact of carbon sources (urea and citric acid) on the characteristics of synthesized CQDs via the pyrolysis method. The fluorescent spectra showed that maximum emission was observed at different wavelengths for citric acid derived and urea derived CQDs. CQDs prepared from both sources were monodispersed and had similar morphology in size and shape.

The simplicity of the method enables pyrolysis applicable to a wide range of carbonaceous materials including food items and food wastes. Peanut shell, a waste of the peanut butter process, was utilized for CQDs synthesis using pyrolysis treatments [52]. The optimum pyrolysis parameters for CQDs synthesis were determined as 400 °C, 4 h for 70 g of peanut shell. At optimum conditions, synthesized CQDs exhibited excitation-dependent emission behavior, good water dispersibility, high photostability, and possessed a potential for Cu^{2+} detection.

3 Properties of CQDs

Carbon dots have first become popular due to their unique optical properties like quantum dots but different than other nanomaterials. CQDs are carbon particles at nanoscales and are classified as zero-dimensional nanomaterials. CQDs are quasi-spherical nanoparticles with core–shell arrangement consisting of crystalline and amorphous matrix [25]. The core consists of a sp^2 hybridized network, like graphene, which forms the basic framework of the carbon dots. This core structure provides the CQDs with excellent electronic and optical properties. Surrounding the core, carbon dots usually have a surface passivation layer or shell. This shell is formed by the functional groups present on the surface of the CQDs, which can be derived from various sources. These functional groups may include oxygen-containing groups such as hydroxyl (OH), carboxyl (–COOH), amine (NH_3), sulfur (S), or carbonyl (–C=O) groups [10]. It is known that surface functional groups are covalently bonded to the carbon core, and this may change the chemical and physical properties of CQDs.

The presence of these functional groups on the surface of CQDs makes them hydrophilic, which is advantageous for their applications in aqueous environments. Moreover, functional groups can tune the band gaps of CQD which results in photoluminescence shifting [26]. As these nanoparticles are smaller than 10 nm, they exhibit fluorescence with different colors where the quantum confinement effect is responsible [11]. Tuning photoluminescence wavelength is possible with alteration in the size of CQDs.

In addition to their excellent optical features, CQDs have various physical and chemical properties. Other properties of CQDs include good biocompatibility, high quantum yield, light absorption, good water solubility, non-toxicity, excellent antimicrobial, and antioxidant, easy functionalization, and abundant carbon source. CQDs exhibit low cytotoxicity and are well-tolerated by living cells, making them suitable for bioimaging, drug delivery, and other medical applications [38]. Their high quantum yield, or the efficiency with which they emit light upon excitation, makes them excellent fluorescent probes [16]. Their strong light absorption capabilities, especially in the visible and ultraviolet range, enable efficient excitation and subsequently intense fluorescence emission. Furthermore, carbon dots are water-soluble, facilitating their use in aqueous environments and biological systems. Their non-toxic nature further enhances their safety for biological and environmental applications [27]. In addition to these attributes, carbon dots have shown promise as antimicrobial agents, combating bacteria and other pathogens [41]. They also exhibit antioxidant properties, effectively scavenging free radicals and reducing oxidative stress [28]. These multifaceted properties position carbon dots as versatile and valuable nanomaterials in the fields of nanotechnology, biomedicine, optoelectronics, and beyond.

4 Carbon Quantum Dots for Packaging Applications

Carbon Quantum Dots (CQD) have recently gained interest in food packaging applications. They are incorporated in packaging films to enhance the film properties. CQDs have outstanding properties including antimicrobial, antioxidant, anti-ultraviolet, non-toxicity, strong photoluminescence, and high biocompatibility, therefore, they are used for developing functionally active and intelligent packaging films. The addition of CQDs has impacts on the physical, chemical, and mechanical properties of packaging films. Therefore, this new kind of packaging film enables to increase shelf life of the foods, maintains the freshness and quality of the foods, or senses food spoilage. Table 1 summarizes the applications of CQDs in active, and intelligent food packaging films.

4.1 Effect of Carbon Dots on Properties of Packaging Films

4.1.1 Mechanical Properties

Foods can undergo various physical damages during processing, preservation, transportation, and distribution. Packaging plays a crucial role in safeguarding foods against such damage, making the mechanical properties of food packaging materials vital. These properties include tensile strength, elongation at break, and elastic modulus values. The use of thinner packaging films with increased mechanical strength can reduce material costs. Mechanical properties are improved by incorporating fillers such as carbon quantum dots in the polymer matrix.

CQDs have been introduced to polymeric matrices to modify the mechanical properties of films (Table 2). In polymeric matrices, the intermolecular interactions between CQDs and polymers, as well as the interactions among polymer molecules, play a crucial role in determining the tensile strength of the composite films. For instance, the inclusion of CQDs derived from tangerine peel in cellulose nanofiber significantly increased the tensile strength of the film. The tensile strength increased

Table 1 Summary of CQDs applications in the active and intelligent food packaging films

Biopolymer matrix	CQDs source	Food sample	Application	References
Chitosan	Lemon peel	Blueberries, pork	Active packaging	[29]
Filter paper	Green tea leaves	Shrimp	Intelligent packaging	[30]
Chitosan	Turmeric	Pork	Active and intelligent packaging	[31]
Cellulose	Tangerine peel	Shrimp	Active and intelligent packaging	[53]
Cellulose nanofiber	Tangerine peel and resazurin	Shrimp	Intelligent packaging	[53]
Chitosan and gelatin	Green tea leaves	Pork	Active packaging	[54]
Bacterial nanocellulose	Cell-free supernatant of *Lactobacillus acidophilus*	–	Active packaging	[55]
Carboxymethyl cellulose	Ascorbic acid and chitosan	Lemon	Active packaging	[56]
Cellulose and polylactic acid	Polyvinyl alcohol, citric acid, and polyethylene imine	–	Active packaging	[57]
Cellulose nanofibers	*Brassica oleracea* anthocyanin	Fish, pork, shrimp	Active and intelligent packaging	[58]
Chitosan	Citric acid monohydrate	Pork	Intelligent packaging	[59]
Filter paper	Sodium citrate and ammonium chloride	–	Intelligent packaging	[15]
Cellulose	Sodium citrate and urea	Milk	Intelligent packaging	[60]
Gelatin and Persian gum	Grape leaves	Trout fillets	Intelligent packaging	[61]
Chitosan and gelatin	Glucose	Avocado	Active packaging	[62]
Carrageenan and sodium alginate	Garlic	Beef meat	Active packaging	[63]
Starch	Soy protein isolate	Pork	Active and intelligent packaging	[64]

by approximately 30% with the addition of 3% resazurin functionalized-CQDs. This improvement can be attributed to the uniform dispersion of CQDs in the polymer matrix and the formation of a more compact structure [53]. The mechanical properties of the films can be influenced differently by factors such as compatibility, dispersion states, and CQD concentration. The concentration of CQDs influences the structural development and, consequently, the mechanical properties of the films. Chitosan films reinforced with CQDs exhibited increased tensile strength with higher CQD concentrations [54]. On the other hand, the addition of CQDs to silk sericin/chitosan films led to a decrease in tensile strength but an increase in elongation at break. This change may result from the interaction between CQDs, sericin, and glycerol, which reduced protein–protein interactions and enhanced film mobility [22]. In the case of gelatin films, the tensile strength and elongation at break were not significantly affected by the addition of 2% CQDs, but the films exhibited reduced stiffness [37]. A significant reduction in the tensile strength, elongation at break, and elastic modulus was evident in the gelatin films containing 4% CQDs. The starch films, when incorporated with CQDs, showed improved mechanical properties in comparison to neat starch films, primarily due to the uniform distribution of CQDs within the films [64]. However, the addition of more than 5% of CQDs resulted in a significant reduction in tensile strength.

Although the presence of CQDs plays an essential role in the enhancement of the mechanical properties of films, the distribution of CQDs within the polymer matrix also exerts an influence on these properties. In gelatin films, it has been reported that the interaction between the hydroxyl groups of CQDs and the carboxylic and amino groups of Persian gum and gelatin can increase the tensile strength of the composite film [61]. However, the aggregation of CQDs caused a significant decrease in the tensile strength of Persian gum/gelatin films suggesting that there is an optimal concentration of CQDs that can be incorporated into the films. The decrease in tensile strength of carboxylated acrylonitrile butadiene latex has been attributed to CQD agglomeration above a certain concentration [24]. When high a concentration of the CQDs was added to the polymer matrix, nanoparticles aggregated within the film, compromising the film's integrity, and reducing its strength. In other words, excessive integration of CQDs into the films could lead to a decline in mechanical properties. Similarly, the tensile parameters of starch films were relatively low when more than 17% graphene carbon dots were added [69].

The functionalization of carbon dot surfaces influences the interaction of molecules in the film matrix resulting in changes in the mechanical properties of the films. Ezati and coworkers [70] prepared pectin/gelatin films and added CQDs and sulfur-functionalized (S-CQDs) CQDs to the matrix to enhance the mechanical properties of the films. The interaction of CQDs with the biopolymer matrix led to an increase in the tensile strength, while a negative impact was observed in the case of S-CQDs. The films with incorporated S-CQDs had significantly low tensile strength due to the weak bonding between S-CQDs and polymers. Another distinct difference was observed in the values of elongation at break (EB) and elastic modulus (EM) of the films. In CQDs-added films, EB reduced, and EM increased, while the opposite

Table 2 Mechanical and other functional properties of quantum dots incorporated packaging films

Biopolymer	Precursor	Tensile strength		Elongation (%)		Active properties	References
		Control film	Active film	Control film	Active film		
Gelatin carrageenan	Enoki mushrooms	52.8 MPa	81.2 MPa	3.9	6.4	Antioxidant, UV-blocking	[65]
Chitosan and starch	Green tea	–	–	–	–	Antibacterial, antioxidant, UV-blocking	[66]
Starch and κ-carrageenan	Cloves	~32 MPa	40 MPa	28	26	Antioxidant UV-blocking	[28]
Gelatin	Potato skin	60.7 MPa	50.3 MPa	12.2	11.3	Antibacterial, antioxidant, UV-blocking	[37]
Cellulose	Glucose	73.3 MPa	75.2 MPa	3.8	3.9	Antimicrobial, antioxidant, UV-blocking	[67]
Chitosan	Sublimed sulfur powder	60.3 MPa	78.2 MPa	6.2	7.0	Antibacterial, antioxidant, UV-blocking	[68]
Starch	Soy protein isolate	~7 MPa	~10 MPa	~9	~11	pH sensitivity, antioxidant	[64]
Sericin	Cocoon silk	1.23 MPa	3.41 MPa	18.53	54.70	Antibacterial, antioxidant	[22]
Chitosan and gelatin	Glucose	79.5 MPa	82.3 MPa	7.3	8.2	Antimicrobial, antioxidant, UV-blocking	[62]
Chitosan	Citric acid monohydrate	17.4 MPa	29.3 MPa	–	–	Antibacterial, antioxidant, UV-blocking, pH sensitivity	[59]
Carrageenan and alginate	Garlic	45.7 MPa	58.6 MPa	4.5	3.8	Antibacterial, antioxidant, UV-blocking	[63]

trend was observed in S-CQDs-containing films, which was attributed to the formation of crystalline structures. This suggests that surface functionalization was a key factor in determining the mechanical characteristics of the films. In another study, the soy protein isolate films were reinforced with carboxyl functionalized carbon dots (C-CQDs) [71]. It was reported that the tensile strength increased with the addition of C-CQDs, reaching a maximum of 0.15% C-CQDs. However, adding more C-CQDs led to a significant reduction in tensile strength, likely due to the self-aggregation of C-CQDs.

4.1.2 Barrier Properties

Food packaging is the proper way to protect food from chemical, biological, and physical degradation, prolong shelf life, and maintain food safety and quality. As every food requires special protection due to its physicochemical properties, the barrier properties of packaging material should be appropriate which ensures high protection from degradation. UV radiation is one of the most responsible factors that lead to the onset of degradation processes in packaged foods. Food is constantly exposed to light in the food production-consumption chain including handling, production, storage, transportation, and grocery retail. Ultraviolet (UV) radiation causes the formation of reactive chemical species, such as free radicals. Free radicals thereafter initiate photolysis and photooxidation reactions which break down food components. As a result, exposure to UV radiation leads to the formation of off-flavors and unpleasant odors, loss of desirable color, texture, and nutritional value.

Food products are protected from light damage with UV-light-blocking packaging materials such as paper, amber glass bottles, aluminum foil, and paperboard. However, packaging materials with UV-light blocking materials are mostly translucent or opaque which does not allow to see the visual appearance of the product. Consumers generally prefer the see-through properties of the packages over their UV-blocking properties. Therefore, there is a demand for transparent packaging materials that prevent the UV-mediated deterioration of food.

Polymers, both synthetic and biopolymers are popular food packaging materials due to their low cost, abundance, recyclability, excellent gas barrier properties, softness, tensile strength, lightness, and transparency [72]. Transparency enables visual inspection to assess the appearance and freshness of food. For example, polyethylene is commonly used to produce transparent food packaging materials like bags, bottles, and stretch films, but it has poor UV resistance [73]. Most polymers used in food packaging allow UV light to pass through. To reduce UV-light transmittance, various biomaterials and nanoparticles are employed to absorb or scatter light within the packaging material.

Biomaterials from natural sources possess UV protection properties, which are attributed to functional groups like hydroxyl, carboxyl, amino, alcohol, ketone, and aldehyde. These properties are primarily due to the presence of π–π^* transitions. Biomaterials such as lignin, chitosan, and pectin have been reported to naturally have UV-blocking properties. Chitosan has been extensively used as a film-forming material due to its biodegradability, nontoxicity, good antimicrobial activity, and UV-blocking properties. For instance, Roy and Rhim prepared chitosan films blended with alizarin using the solution casting method and evaluated the films' functional properties [74]. Alizarin, a natural colorant, was added to the film as an acid–base indicator, creating a smart film that changed color from yellow to purple in response to pH changes caused by food spoilage. This addition significantly improved the thermal, mechanical, and barrier properties of the chitosan films. Although chitosan had the potential of UV blocking, the UV protection significantly increased in the presence of alizarin. This enhanced UV-blocking property of alizarin can be attributed to its phenolic compounds. Additionally, alizarin decreased the light transmittance at

660 nm but the reduction at 280 nm was more pronounced than at 660 nm. This indicates that the chitosan films not only provide good UV protection but also maintain transparency. Another natural UV-blocker polymer is lignin. Lignin, a byproduct of the paper industry, possesses an aromatic structure that makes it an effective UV absorber. The UV protective properties of cellulose films were enhanced with an increased lignin content [74]. When the lignin content exceeded 20%, cellulose films effectively shielded all UV radiation in the UVA and UVB ranges. The UV protection offered by lignin is associated with its phenolic structure and chromophores. There are limited reports on the UV-blocking properties of films made from pectin. For example, Ezati and Rhim developed a fluorescent pectin film reinforced with CQDs [75]. While the addition of CQDs enhanced the UV resistance of the films, it's worth noting that neat pectin films also proved effective at significantly blocking UV radiation compared to other biomaterials.

Among other nanoparticles, CQDs are excellent UV-light absorbents that can absorb UV light and emit visible light under UV excitation. CQDs exhibit an increased absorption of light at short-wavelength than long-wavelength in the composite films indicating high protection of products from UV-light damages. Recent studies have focused on developing packaging films based on the high UV barrier properties of CQDs [22, 37, 76].

Chitosan is a potent UV-blocking polysaccharide capable of effectively absorbing UV light across all ranges, including UVA, UVB, and UVC. A biodegradable packaging film was produced from chitosan, with the addition of nitrogen and phosphorus-doped CQDs to the matrix [77]. The films exhibited high UV-light blocking and were remarkably transparent. The UV-light blocking behavior increased with the increasing concentration of CQDs in the chitosan film. The UV-A light was shielded by 90.2 and 96.5% of UV-B light was absorbed with the addition of 4% of CQDs. The transparency decreased with adding more CQDs, but the films were still apparently transparent. Ardini and coworkers [57] decorated polyvinyl alcohol (PVA) and cellulose films with CQDs though spray or spin coating. The CQDs-deposited films were capable of blocking UV light in the range of 300–400 nm wavelengths. The light transmission of neat cellulose and PVA films was 90% in the 350–700 nm range which dropped to 0% at 360 nm in spin-coated and spray-coated cellulose films. In contrast, spray-coated and spin-coated PLA films had 10% and 30% light transmittance, respectively. This indicated that CQDs-deposited cellulose films had better UV barrier properties compared to the PVA films. UV-photo reduction activity of CQD was studied in bacterial nanocellulose film [55]. In bacterial cellulose films loaded with CQDs, no green fluorescence was observed under UV light at lower CQDs concentrations (6.44 ± 0.94 mg cm^{-2} and below), while BNC films emitted green fluorescence at higher CQDs contents. The effect of CQDs concentration on the light transmittance of BCN films was also assessed for the range of 200–800 nm. The UV transmittance decreased with an increase in the amount of CQDs in BCN films. The BCN films were capable of blocking >99% of UV-light at higher CQDs concentrations. Similarly, Lin et al. [59] reported that the light transmission of chitosan films was dependent on concentration of N-doped CQDs. In the UVB region, the UV blocking property of the films increased up to 66% with increasing CQDs concentration. CQDs acted as fillers, promoting light scattering in the chitosan matrix, and higher transmission values (91%) at 600 nm. Green tea was used to synthesize CQDs with hydrothermal reactions. These green tea derived carbon dots (G-CQDs) were embedded in chitosan/gelatin films to prepare multifunctional active packaging films [54]. The UV blocking property of G-CQDs was confirmed by transmission spectra of CQDs in an aqueous solution. The UV-light blocking was 74% for UVA and 90.6% for UVB region at 0.01% of G-CQDs while the concentration of G-CQDs was increased to 0.03%, UV-light protection was increased against UVA and UVB radiations by 99.1% and 93.8% respectively. A similar trend was observed in G-CQDs incorporated chitosan/gelatin films. These films were effective in the range of 200–400 nm, blocking 90.2% of UVA and 99.7% of UVB when 3% of G-CQDs were added. On the other hand, the transparency of the films was not affected by the increasing concentration of G-CQDs. Another set of UV-light blocking films was produced from gelatin/carrageenan with the addition of CQDs [65]. CQDs did not significantly affect the light transmittance for visible light but exponentially decreased the transmittance at 280 nm. The films shielded more UV light with increasing CQD concentrations, yet the transparency of the films remained unchanged with the addition of more CQDs. This suggests that CQDs prevented the transmission of short-wavelength UV light while allowing the transmission of long-wavelength visible light.

Functionalization of CQDs is a useful strategy for tuning their surface structure. CQDs can be synthesized from natural sources containing nitrogen, sulfur, and phosphorus, allowing the preparation of N-, S-, and P-functional groups on their surface. Ezati and colleagues [70] recently developed pectin/gelatin composite films by introducing sulfur-doped CQDs (S-CQDs) and CQDs to evaluate the physicochemical, antioxidant, and antimicrobial properties of the films. The UV transmittance of the neat pectin/gelatin films was notably low in the UVB range (280–320 nm), possibly due to the UV-blocking properties of gelatin. The light transmittance of the films at 660 nm remained stable in the presence of CQDs, but S-CQDs notably reduced the transmittance. At 280 nm, the UV-light absorption of the pectin/gelatin films became stronger with the addition of both CQDs and S-CQDs. The films incorporating CQDs were effective in significantly blocking both UVA (320–400 nm) and UVB radiation.

4.1.3 Water Vapor Permeability

Gas permeability of packaging material is an important factor in preventing food spoilage. It regulates the exchange of gases such as water vapor, oxygen, carbon dioxide, and volatile aromas between the interior and exterior of the package, thereby influencing the food quality and shelf life.

Water vapor permeability measures the diffused water vapor from a higher vapor pressure to a lower vapor pressure area. It is necessary to understand water vapor permeability characteristics to prolong the shelf life and quality of food products. For example, dry food products, such as flour, nuts, beans, dried fruits, or vegetables, should avoid water absorption, while fresh food products like meats, fruits, and vegetables, require protection from water loss to maintain their quality. The physical and chemical deteriorations as well as microbial spoilages, are directly related to the water activity of food products, which can be regulated by the choice of packaging material. However, while biodegradable materials can be an alternative to replace petroleum-based packaging materials, they often exhibit poor water vapor permeability properties due to their high hydrophilic nature. Most biopolymers display high hydrophilicity, causing internal structural changes in the films when water vapor passes through as polymer chains contact with water molecules. The water vapor transmission rate in hydrophilic films depends on the humidity and temperature of the environment [78]. Additionally, the arrangement of polymers in film structure determines the water vapor diffusion pathway, thus influencing the rate of water vapor diffusion. The inclusion of nanoparticles in the polymer matrix represents a novel approach to modifying the water vapor permeability properties of biopolymer films [79–81]. CQDs serve as excellent nanofillers for enhancing the barrier properties of packaging materials [79]. The involvement of CQDs in the film matrix induces the formation of a dense inner structure, effectively blocking the passage of water vapor and thus improving the films' WVP [59].

Khan et al. [66] used nitrogen and phosphorus-doped carbon dots (NP-CQDs) to improve the barrier properties of chitosan/starch films. Incorporating NP-CQDs into the chitosan/starch films provided functional properties such as UV-light blocking, antibacterial, and antioxidant activity but had no influence on thermal stability and water vapor permeability. While a slight increase in water vapor permeability (WVP) was observed with the addition of up to 3% NP-CQDs into the film matrix, this increase was not significant. The increase in WVP was related to the hydrophilic nature of the NP-CQDs. Both starch and chitosan are rich in hydroxyl groups, and the content of free –OH group increased with the inclusion of NP-CQDs. Min et al. [37] observed a 28% increase in WVP of gelatin films with the addition of CQDs produced from potato skin as a filler. This was ascribed to hydrogen bonding between CQDs and gelatin, resulting in increased hydrophilicity (as evidenced by a 13% reduction in water contact angle) and an increase in WVP of the resulting film. In another study, Persian gum/gelatin blend films were prepared with varying amounts of CQDs, and WVP was reduced after adding CQDs up to 5% [61]. On the other hand, incorporating more CQDs resulted in an increase in the WVP of the films. This result was attributed to the abundance of hydroxyl groups in CQDs, while the crosslinking of Persian gum with gelatin was more effective, and CQDs contributed to the formation of a compact film structure at low CQDs concentrations. Carrageenan/alginate-based composite films were functionalized with CQDs derived from garlic cloves [63]. CQDs were successful in enhancing the mechanical properties of the films and UV-light blocking but demonstrated no effect on the WVP.

Koshy et al. [82] previously prepared soy protein films reinforced with chitin nanowhiskers and silver nanoparticle (AgNPs) anchored CQDs. AgNPs stabilized with CQDs through interactions of OH and COOH groups on CQDs surface and Ag^+. Incorporating a specific amount of chitin nanowhiskers decreased the WVP and adding more chitin nanowhiskers resulted in an increase in WVP. However, the combination of chitin nanowhiskers and AgNPs, had a synergistic effect on WVP, leading to a decrease in the water vapor transmission of the films. The homogenously distributed nanoparticles generated a torturous pathway for water vapor, increasing the distance of the diffusion pathway.

4.1.4 Antibacterial Activity

Developing antimicrobial food packaging materials has received attention in extending the shelf life of food. This is usually achieved by the incorporation of antimicrobial agents in the packaging material. Although various natural substances have antimicrobial activity, attempts have been made to find new antimicrobial agents. Carbon quantum dots are carbon nanomaterials widely used in bioimaging due to their fluorescent properties. Recent studies focus on their other properties such as UV-light blocking, antioxidants, nanofillers, and antimicrobial. All these properties make CQDs a new tool for developing active packages with improved properties.

Carbon dots (CQDs) are a new class of photo-activated antimicrobial agents, gaining increasing attention for their potential in food protection. Recent studies highlight the antimicrobial activity of CQDs when exposed to natural or visible light. This activity is closely linked to the generation of reactive oxygen species (ROS), and CQDs have been reported to promote ROS production (Fig. 1). ROS are a group of reactive molecules including superoxide anion ($O_2^{\bullet-}$), hydrogen peroxide (H_2O_2), and hydroxyl radicals ($^{\bullet}OH$). In the inactivation mechanism, CQDs interact with the phospholipid bilayer, eventually leading to the disruption of the cell membrane. This interaction involves a high rate of electron transfer between CQDs and cell membranes upon attachment. The excess free electrons generated by CQDs result in the production of ROS. However, this leads to overproduction of ROS which exceeds cellular tolerance and induces oxidative stress in the cell. Oxidative stress causes damage to lipids, DNA, and proteins, negatively impacting various biological processes, destroying the cell membrane, causing cytoplasm leakage, and consequently cell death [83].

CQDs may exhibit varying antimicrobial activity against a broad spectrum of foodborne pathogens and spoilage microorganisms. This activity can be influenced by the sources from which they are derived. For example, CQDs produced from glucose were found to effectively inhibit *Aspergillus flavus* and *Colletotrichum orbiculare*, while those derived from white mulberry exhibited inhibition against *Listeria monocytogenes*. When *p*-phenylenediamine was used for the synthesis of CQDs, they displayed strong bactericidal activity against both *Staphylococcus aureus* and *Escherichia coli* [62, 84, 85].

The bacterial nanocellulose films (BCN) showed antimicrobial activity against *L. monocytogenes* and *E. coli* when the CQDs loading in the films was maintained at 71.74 ± 4.13 mg cm^{-2} [55]. Below this concentration or in the absence of CQDs, BCN films displayed no antimicrobial properties, indicating the presence of a threshold limit for antimicrobial activity. Furthermore, the average zone of inhibition against *L. monocytogenes* was larger compared to *E. coli*. This was attributed to

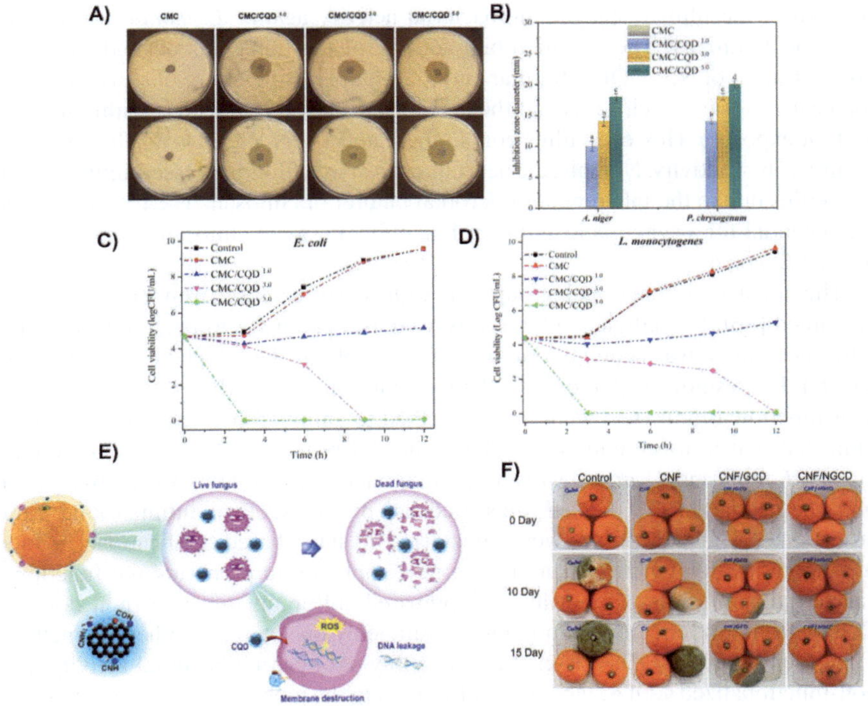

Fig. 1 **a** Antifungal activity of the carboxymethyl cellulose-based films (CMC) loaded with varying concentrations of CQDs against *Aspergillus niger* and *Penicillium chrysogenum* by agar well diffusion method (Reprinted from Ref. [56] with permission from Elsevier, Copyright 2022); **b** inhibition zone of the CMC-based films, antibacterial activity of CMC-based films against; **c** *Escherichia coli* (Reprinted from Ref. [56] with permission from Elsevier, Copyright 2022) and **d** *Listeria monocytogenes* (Reprinted from Ref. [56] with permission from Elsevier, Copyright 2022). **e** The antifungal activity mechanism of CQDs on coated tangerines (Reprinted from Ref. [11] with permission from Elsevier, Copyright 2022), and **f** appearance change of tangerines coated with the cellulose nanofiber-based films (CNF) containing CQDs and N-CQDs during storage at 25 °C (Reprinted from Ref. [11] with permission from Elsevier, Copyright 2022)

the production of reactive oxygen species (ROS) induced by CQDs right after attachment to the surface of bacteria. ROS were more harmful to gram-positive bacteria due to their adverse effects on cell wall integrity. Khan et al. [54] synthesized carbon dots from green tea (G-CQDs) via hydrothermal reaction. The antimicrobial activity of G-CQDs was investigated against selected microorganisms, both gram-positive (*L. monocytogenes*) and gram-negative (*E. coli*) bacteria. The minimum inhibitory concentration/minimum bactericidal concentration results of G-CQDs were lower for *L. monocytogenes* than for *E coli* and the inhibition zone tests showed that G-CQDs inhibited *L. monocytogenes* growth more than *E. coli.* Chitosan/starch films were functionalized with nitrogen and phosphorus-doped carbon dots derived from green tea (N,P-CQDs) [66]. In addition to their enhanced mechanical and UV-blocking properties, the films displayed antibacterial activity against *L. monocytogenes, E. coli*, and *S. aureus*. The antimicrobial activity of the films was dependent on the concentration of N,P-CQDs. Similar antimicrobial activities were observed for both the neat chitosan/starch films and those loaded with 1% N,P-CQDs within the first 12 h of exposure. However, films containing 2 and 3% N,P-CQDs displayed higher antimicrobial activity. Notably, the neat chitosan/starch films possessed antimicrobial properties due to the inherent antimicrobial natural of chitosan. The results suggest the potential for a synergistic antimicrobial effect between chitosan and N,P-CQDs against *E. coli*.

The impact of functional groups on CQDs' surfaces on film properties was examined [86]. Polyethylene glycol was used as a precursor to prepare CQDs, and ethylenediamine was added during the synthesis of nitrogen-doped carbon dots (N-CQDs). The resulting CQDs or N-CQDs were introduced to carboxymethyl cellulose/agar functional films, and were then subjected to antimicrobial testing. The results indicated that films containing N-CQDs showed pronounced bactericidal activity against *E. coli* and *L. monocytogenes*, while bactericidal effects were observed in both the neat and CQDs-added films. Because of the excellent antimicrobial properties, N-CQDs-added films were considered to have broad applications in food packaging. Sulfur functionalization is a new approach to preparing the functional carbon quantum dots. Roy et al. [87] prepared sulfur functional carbon dots (S-CQDs) from turmeric and ammonium persulfate by one-step hydrothermal method. They found that S-CQDs possessed remarkably stronger antibacterial activity than non-functionalized CQDs, which was attributed to the presence of functional groups such as sulfonates and sulfonic acids. Therefore, S-CQDs were able to bind the enzymes of microorganisms, leading to the production of ROS, which resulted in protein denaturation and cell death. CQDs are not only effective on bacteria, but they can also have the potential to prevent the growth of fungus. Ezati et al. [67] used glucose as a carbon source to produce carbon dots (CQDs), and urea was included in the synthesis of nitrogen-doped carbon dots (N-CQDs). CQDs-added cellulose nanofiber composite films demonstrated antibacterial activity by reducing the growth of microorganisms in contrast to neat cellulose nanofiber films. The N-CQDs embedded films had stronger antibacterial activity compared to CQDs-added films. N-CQDs were not only effective on bacteria, but the results also showed that the cellulose nanofiber films incorporated with N-CQDs inhibited the growth of *Aspergillus flavus* suggesting high antifungal activity of the films.

In various studies, the antimicrobial properties of CQDs have been reported both in aqueous solutions and in films. However, some studies revealed that CQDs had no antimicrobial efficacy either when used alone or when embedded into polymeric films. For example, when carboxyl functionalized carbon dots (C-CQDs) were introduced into soy protein isolate films, no inhibition zone against *E. coli* and *L. monocytogenes* was observed, indicating a lack of antibacterial properties [71]. Moreover, another study reported that CQDs produced from *Lactobacillus plantarum* biomass showed no antibacterial activity in inhibiting the growth of *E. coli*. Interestingly, they were effective in preventing biofilm formation by *E. coli* [88].

In conclusion, all these studies suggest that CQDs have a broad spectrum of antimicrobial activity which is attributed to factors such as particle size, surface functional groups, the type of carbon source, and synthesis process. CQDs are considered an emerging approach to developing new types of food packaging materials with high antimicrobial activity.

4.1.5 Antioxidant Activity

Oxidation is one of the major quality deterioration reactions in food products. Foods containing polysaccharides, proteins, fats, and phenolic compounds are more prone to oxidation, particularly during processing, storage, and distribution. The oxidation reaction causes a loss of nutrients and aroma compounds, alters the color, and increases the presence of rancid and off-flavor substances in food products. All these changes in food quality are unacceptable to consumers and have the potential to contribute to food waste. To prevent the oxidation reactions in food products, antioxidants are introduced into packaging films to enhance the shelf life of the packed food. The selection of antioxidants is important because there may be migration from packaging film to food, therefore, antioxidants must be non-toxic and safe for health. Recent studies highlight the antioxidant properties of CQDs. CQDs are not only promising particles due to their contribution to the barrier and mechanical properties of the packaging films but also their antioxidant properties (Fig. 2). Therefore, integration of CQDs into the film matrix offers many advantages for developing a new class of active food packaging films.

Ezati et al. [62] produced CQDs with superior antioxidant activity and then embedded CQDs in chitosan/gelatin coating film which was used for the improvement of avocado shelf life after harvesting. CQDs showed excellent antioxidant activity (~95%) against DPPH• and ABTS• within minutes. The superior antioxidant activity of CQDs was explained by their surface hydroxyl groups acting as hydrogen donors for scavenging free radicals. The antioxidant activity of the neat chitosan/gelatin film increased from 16.1 to 98.1% for ABTS assay and DPPH scavenging ability increased from 2.8 to 74.7% when 2% CQDs were incorporated into the film. CQDs had higher antioxidant activity in the ABTS assay. It was attributed to the higher release rate of CQDs in aqueous solution than methanol because the lower release rate led to the interference of CQDs interaction with DPPH radicals. Recent findings confirmed that the antioxidant activity of CQDs relied on their concentration in biopolymer films [89]. Purkayastha et al. [89] utilized oil-and-protein spent rapeseed waste to synthesize CQDs and prepared biodegradable films from rapeseed protein which was functionalized with different amounts of CQDs (20 and 30 mg/100 mL film forming solution). These CQDs-introduced films emitted green fluorescence under UV light. For real food samples, rapeseed oil was stored within containers made from CQDs containing protein films, and the effect of the films on oil oxidation was monitored for 28 days. During the first 14 days, the free fatty acids content of rapeseed oil remained unchanged when stored in containers with higher CQD concentrations. In contrast, rapeseed oils stored in neat film and containers with

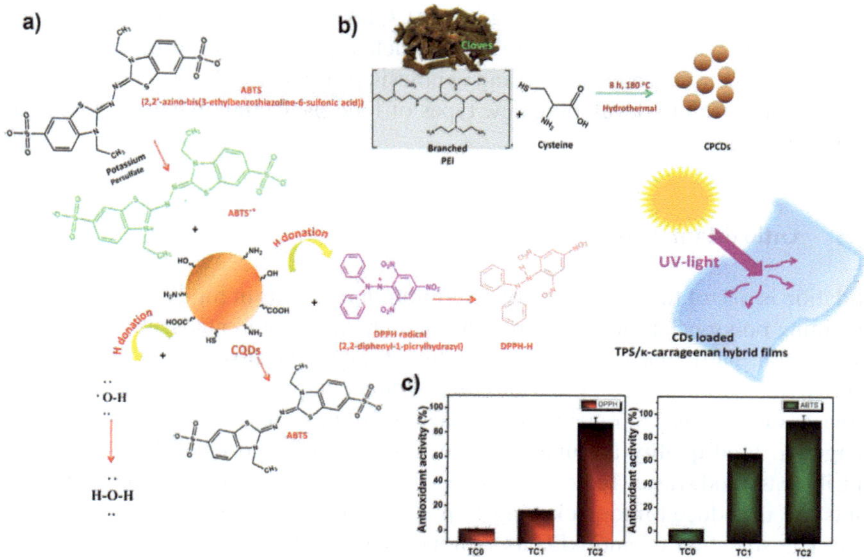

Fig. 2 **a** Schematic diagram of the antioxidant mechanism of CQDs derived from cloves, **b** CQDs synthesis procedure, and **c** antioxidant activity of CQDs unloaded and loaded films, TC0: unloaded, TC1: 4 mg of CQDs, TC2: 20 mg of CQDs (Reprinted from Ref. [28] with permission from the American Chemical Society, Copyright 2022)

lower CQD levels showed higher free fatty acid content from the beginning. By the end of the storage period, the lowest free fatty acid content was observed in rapeseed oils stored in containers with higher CQD levels Li et al. [90] fabricated antioxidant sunlight conversion film with red emissive carbon dots that absorbed UV light, converting it to red light. This unique property induced photosynthesis in plants. It was noted that the films' fluorescence diminished over time due to the oxidation of CQDs on their surface. However, this oxidation and fluorescence quenching of the CQDs were prevented by the addition of ascorbic acid, making the films antioxidant.

Antioxidant activity can also be influenced by the carbon source used in CQD synthesis. For example, CQDs were synthesized using apple pomace and rosemary powder (resulting in A-CQDs and R-CQDs, respectively), and their morphological, optical, and antioxidant properties were characterized [91]. DPPH assay results indicated that R-CQDs exhibited significantly higher antioxidant activity (twice as much) compared to A-CQDs at all concentrations. Both types of CQDs showed a similar trend in antioxidant activity, reaching their maximum with increasing CQD concentration and decreasing at higher concentrations. Additionally, these CQDs were incorporated into sodium caseinate and pectin biopolymer films. The pure biopolymer films demonstrated radical-inhibiting capacity, which increased with the addition of CQDs. As anticipated, films containing R-CQDs exhibited higher antioxidant activity compared to those with A-CQDs. Notably, CQDs derived from coriander leaves displayed concentration-dependent scavenging activity, which could reach up to 94% [92]. Murru et al. [93] utilized vegetable wastes, including tea and grape pomace, to produce CQDs. Although both tea waste and grape pomace derived CQDs had similar total phenolic contents, CQDs obtained from tea waste exhibited significantly higher antioxidant activity compared to those from grape pomace.

Carbon dots can be synthesized not only from natural biological sources but also during food processing. The Maillard reaction involves a series of subsequent reactions that occur with the condensation of reducing sugars and substances containing free amino groups, resulting in condensation products. These products are generally undesirable in food products due to their negative effects on color, flavor, and taste. However, recent studies have revealed the fluorescence properties of the Maillard reaction products [94–97]. CQDs, Maillard reaction products, were fabricated through the reaction of glucose and lysine under high temperature (180 °C) [94]. Maillard reaction derived CQDs were effective on scavenging hydroxy radicals and DPPH even at low concentrations.

4.1.6 Cytotoxicity of CQDs

Food packaging acts as a barrier between food and the environment, safeguarding food from external factors, including chemical, physical, and microbiological risks. This protection helps maintain food safety, quality, and freshness. In the meantime, the food packaging material must be safe for food and should not pose a risk of contaminating food with harmful substances intended for human consumption because it comes into direct contact with the food. However, safety concerns have arisen in recent decades regarding the use of nanomaterials in packaging due to the increased utilization of various types of nanomaterials in packaging development. As CQDs have recently used as nanofillers in packaging materials due to their multifunctional properties, determining the toxicity of CQDs is crucial for its applications.

Cytotoxicity assays are essential for assessing the health risks associated with various substances. Researchers have recently conducted numerous studies to investigate the cytotoxicity of CQDs to determine their safety when used in packaging materials. For instance, Alas et al. [98] developed active food packaging materials using polyvinyl alcohol and multicolored fluorescent CQDs through a solvent-casting method. These composite films, known for flexibility, transparency, and UV-light blocking abilities, underwent in vitro cytotoxicity assays. The in vitro cytotoxicity assays of the neat polyvinyl alcohol (PVA) and CQD-added PVA composite films were conducted against a human lung fibroblast healthy cell line (MRC-5) exposing varying concentrations of composite films. According to the results, the neat PVA films and CQDs containing yellow films exhibited reduced cell viability at the highest concentration (500 mg/L) after 24 h, while the 0.05CQD/PVA samples showed no cytotoxicity even at higher concentrations. However, at 48 h, both PVA and 0.05CQD/PVA films at the highest concentration showed a significant decrease in cell viability, while other 0.05CQD/PVA films (red, blue, and green) had no impact on cell viability at any concentration tested. In another study, spent coffee grounds were used to fabricate CQDs, and these CQDs were used to functionalize gelatin/poly (vinyl alcohol) (Gel/PVA) films along with grapefruit seed extract (GSE) [99]. The multifunctional films showed low cytotoxicity at various concentrations of CQDs ranging from 0.1 to 1 mg/mL when tested against the mouse fibroblast line. *Actinidia deliciosa* (*A. deliciosa*) fruit extract was used as a carbon precursor to produce CQDs through the hydrothermal method [27]. The biocompatibility tests showed that the cell viability was over 90% for the mouse fibroblast cell line, and approximately 80% of the MCF-7 human breast adenocarcinoma cell line remained viable after incubation at different concentrations of CQDs for 24 h. Lin et al. [59] developed active and intelligent chitosan films by incorporating pH-sensitive fluorescent nitrogen-doped carbon dots (N-CQDs) to track food freshness. The chitosan film exhibited antioxidant, antimicrobial, and good UV-blocking properties. Furthermore, the cell viability was high (>93% on the 9th day) for 9 days suggesting good biocompatibility. However, it decreased remarkably after extended storage.

Numerous research studies have reported that CQDs are non-toxic or show low cytotoxicity. The safety of CQDs is mostly related to concentration, as up to a certain concentration, CQDs are generally considered non-toxic to both human and animal cells. For example, CQD extracted from kvass, a fermented beverage, displayed concentration-dependent cytotoxicity as determined by MTT assay using a Chinese hamster ovary cell line [100]. The cytotoxicity of kvass CQDs was negligible at concentrations up to 20 mg/mL, but cell death increased at higher CQDs concentrations. Another study confirmed that the CQDs exhibited negligible cytotoxicity for the mouse fibroblast L929 cell line at low concentrations, with only low cytotoxicity observed when CQDs concentrations reached up to 0.5 mg/mL [37].

5 Intelligent Packaging Applications

Carbon dots are considered emerging nanomaterials, distinguished by their distinct features, particularly their fluorescence. CQDs can emit various fluorescent colors such as blue, green, yellow, or red when excited with UV light. This property makes CQDs an excellent choice for fluorescent imaging and sensing applications. Numerous studies have highlighted their use in in vivo cellular imaging, because of their small particle size, which enables them to penetrate cellular membranes. The surfaces of CQDs can be functionalized with various groups such as carbonyl, ether, epoxy, amino, carboxylic acid, and a hydroxyl group. These modifications enable them to interact with a variety of substances, often in a highly selective and specific manner [12]. The fluorescence of CQDs changes with the interaction of the substances making CQDs ideal for sensing. The fluorescence of CQDs changes when they interact with these substances, making CQDs ideal for sensing applications. As a result, CQDs can be tailored for the detection of analytes. To date, various strategies have been developed using CQDs for food safety, allowing for the detection of the presence and quantity of various substances in food products such as vitamins, toxins, allergens, and pesticides.

Researchers have recently focused on presenting real-time safety and quality monitoring systems using CQDs to maintain food quality and safety. Intelligent food packaging is an innovative packaging system that enables the monitoring of food freshness and spoilage, while also extending the shelf life of the product. Intelligent packaging is considered a future food packaging system, but it is also important to develop intelligent systems from eco-friendly materials. As combating environmental pollution caused by plastic packaging materials has become a significant global issue, future packaging is expected to be produced from biodegradable materials to reduce pollution.

Today, most of the packaging materials are produced from petroleum-based plastic materials. Plastics are dangerous pollutants to the environment, particularly when they break down into small pieces called microplastics. Unfortunately, they don't decompose easily, and recycling rates are very low. Many efforts have been made to develop alternatives to replace plastics and reduce environmental pollution. Therefore, the focus on the production of safe packaging materials has shifted to the development of biodegradable materials from natural sources. Natural sources are abundant, sustainable, and non-toxic making biomaterials from these sources easily degradable by microorganisms. Similarly, the interest in biodegradable materials has been increased in food packaging.

With an increase in demand for using biodegradable materials, biodegradable packaging films have been prepared not only to improve the quality or shelf life of food but also to monitor the safety of food. To produce intelligent biodegradable films, intelligent components are incorporated into biodegradable packaging films. These components, such as CQDs, have the capability to detect alterations in factors such as pH, temperature, humidity, or oxygen both within and outside of the packaging. They provide valuable information about these changes, which can be directly related to the status of food quality or safety [101]. A humidity-sensing nanocomposite film produced from guar gum-sodium alginate was reported. CQDs were added into the film as a fluorescent sensing agent capable of responding to humidity changes by altering their on–off fluorescence status. At high relative humidity, CQDs were quenched, and the fluorescent intensity decreased while the opposite behavior was observed at a low relative humidity rate in which the fluorescent intensity increased [102].

Recently, a color-changing mechanism has been widely used in the development of freshness indicators for intelligent packaging systems. The color-changing indicators provide qualitative and quantitative information and show the food's freshness and quality status. The chromatic change feature of the indicators is addressed to the substances that respond to the alterations towards the surrounding changes such as pH (Fig. 3). Interestingly, pH is a reliable indicator of food spoilage since various acidic and alkaline substances are generated during spoilage, and pH-responsive materials interact with these metabolites, leading to a change in color. This straightforward operational principle allows for the utilization of pH-responsive materials in developing intelligent food packaging materials. Natural colorants have received the attention of researchers in intelligent packaging due to their capacity for pH-dependent color-changes. These pH-responsive natural colorants exhibit a reddish hue in acidic pH, a pink shade in neutral pH, and blue or green in basic conditions. However, the stability of natural colorants is influenced by several factors, which currently restrict their application in intelligent packaging materials. More stable indicators are essential for generating pH-sensitive materials. CQDs are relatively stable nanoparticles against various physical and chemical conditions. Recent studies have unveiled that CQDs are fluorescent, non-toxic, and biocompatible particles responding to alterations in pH by changing the intensity of fluorescence. This fluorescence property has made CQDs valuable in analyte detection, employing techniques like fluorescence quenching or fluorescence enhancement (Fig. 4) [103]. Intelligent food packaging systems are a new research area for the use of the pH-dependent fluorescent CQDs. There have been a limited number of reports concerning the application of CQDs as fluorescent probes in intelligent packaging materials [15, 59, 104–109].

Fu et al. [18] evaluated the potential of green-synthesized CQDs to detect pH changes and subsequently integrated CQDs into gelatin/chitosan film, resulting in the development of an active-intelligent packaging film. The results suggested that the developed nanocomposite film had excellent active packaging properties such as

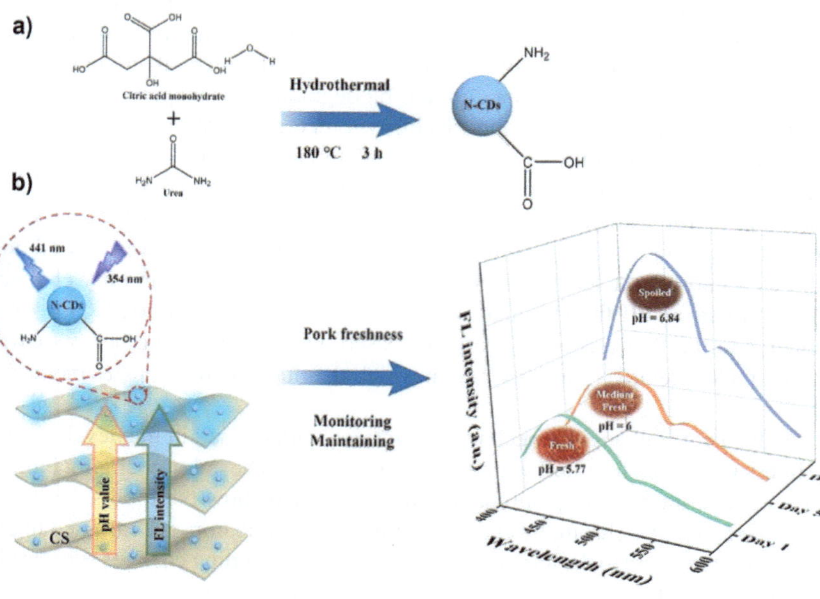

Fig. 3 Schematic illustration of chitosan/N-doped carbon dots (CS/N-CQDs) packaging films for pork freshness monitoring and maintaining **a** synthesis process of N-CQDs, **b** fluorescence emission spectra of CS/N-CQDs films with different pH treatment (Reprinted from Ref. [59] with permission from Elsevier, Copyright 2022)

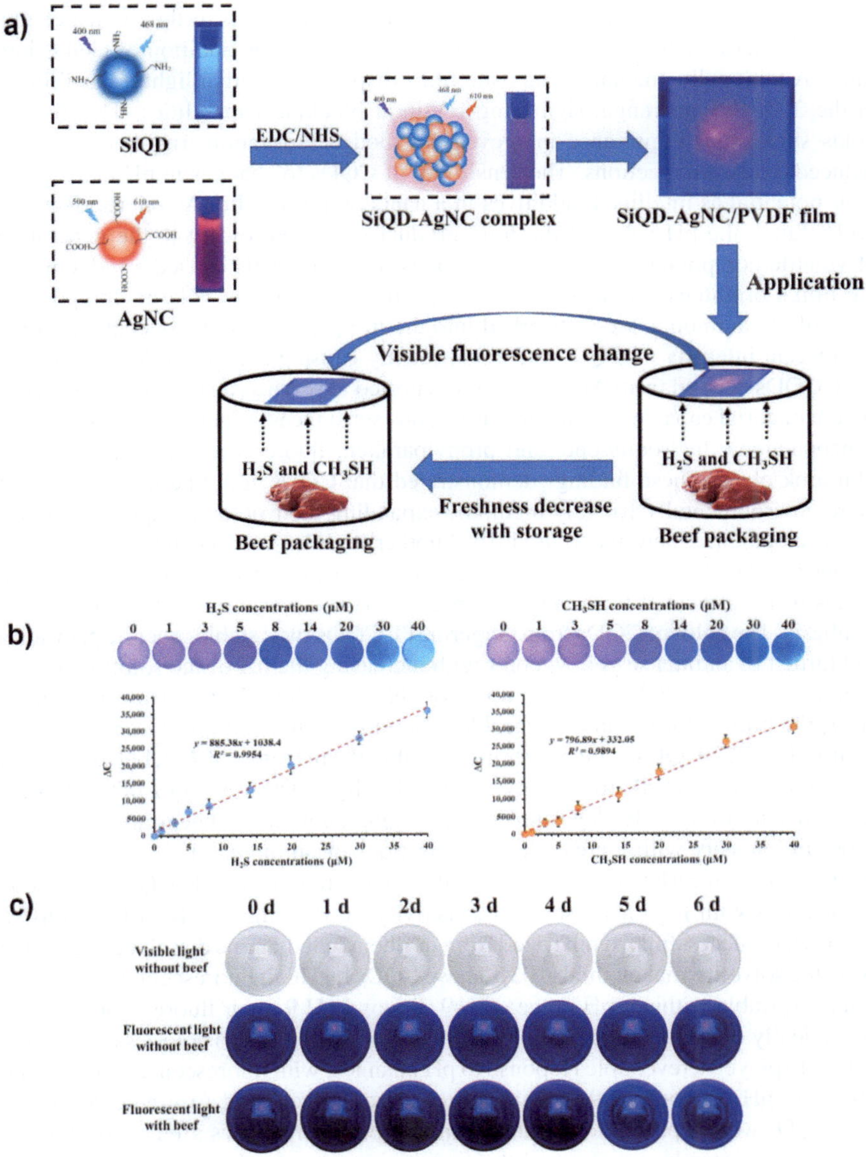

Fig. 4 **a** The synthesis and food spoilage monitoring mechanism of a silicon quantum dots-silver nanoclusters film (SiQD-AgNC/PVDF), **b** the fluorescence changes of the SiQD-AgNC/PVDF film in response to H_2S and CH_3SH. The relation between total color change (ΔC) of SiQD-AgNC/PVDF film and the concentrations of H_2S and CH_3SH, **c** the visible light photos and fluorescent light photos of a polypropylene packaging box integrated with a SiQD-AgNC/PVDF film, during storage at 4 °C. The fluorescent light photos of the packaging system with beef sample, during storage at 4 °C (Reprinted from Ref. [109] with permission from MDPI, Copyright 2023)

UV-blocking, antimicrobial, and antioxidant. The in vitro tests further indicated that the antibacterial activity of chitosan was increased with the addition of CQDs into films. Additionally, the nanocomposite films displayed minimal light transmittance in the 200–400 nm range, signifying UV-light blocking capabilities. This feature holds substantial importance in preventing food deterioration triggered by light-induced oxidation reactions. The sensitivity of CQDs to changes in pH highlighted their potential as intelligent additives in a nanocomposite film. As microorganisms proliferated, the pH value of the food product increased, owing to the generation of volatile compounds during spoilage. This increase in pH caused a reduction in the film's brightness under UV light (365 nm). Notably, the volatile compounds, particularly ammonia, were absorbed into the film, leading to the quenching of the fluorescent intensity of CQDs and consequently causing the film to lose its brightness. CQDs derived from Aloe vera were reported to serve as an on–off fluorescence switch material capable of detecting hazardous substances. When exposed to PABA, benzophenone, hydroquinone, and propylparaben, fluorescence quenching of the film took place. These findings demonstrated that CQDs could be produced using more environmentally friendly methods, expanding their potential applications for the development of new materials [110]. Another study has utilized the pH-responsive properties of CQDs to design paper-based indicators to detect biogenic amine [105]. CQDs were fabricated from tangerine peel (Tan) and resazurin with hydrothermal synthesis. The color of CQDs from tangerine (T-CQDs) was stable at acidic pH values and turned to an intense yellow color with increasing the pH of the solution. CQDs from resazurin (R-CQDs) showed no color change within pH 4–6. However, above pH 7, the pink color became noticeable. The most significant color transformation occurred for T,R-CQDs, which were green within the pH range of 4–6, but exhibited a vibrant orange hue at alkaline pH levels under UV light. The evaluation of the fluorescent intensity of the T,R-CQDs towards changing ammonia concentrations showed a loss in intensity with increasing concentrations of ammonia (1–100 μM) due to the quenching of carbon dots confirming the potential of using T,R-CQDs in various applications with high accuracy and sensitivity. In the other study, pH-responsive CQDs were successfully synthesized from m-phenylenediamine dissolved in acetone using the solvothermal method [103]. These CQDs had high fluorescent intensity and remained stable within a pH range of 1–9. Beyond pH 9, their fluorescent intensity dramatically decreased, and the color shifted from pale green to colorless. However, CQDs displayed a reversible response to pH changes, with fluorescence re-emerging when the pH dropped to 9 or below, and the color returned to green. Additionally, CQDs were reported to maintain similar optical properties after adsorption on the surface of starch particles and their addition to the starch films, suggesting the potential of CQDs for use as a pH-sensitive on–off fluorescent probe.

6 Conclusion

This chapter has presented recent research on the utilization of carbon quantum dots in food packaging. Carbon dots have been intensively used in bioimaging and sensing applications, but over the last decade, research has shifted towards incorporating CQDs into packaging. Since CQDs are biocompatible particles, the surface functional groups interact with polymer molecules, thereby altering the film properties. Introducing CQDs in film matrix contributes to the development of enhanced packaging film properties including barrier, mechanical, thermal, antioxidant, and antimicrobial properties. This impact is important for the development of active food packaging films, which have the potential to enhance the quality and extend the shelf life of food products. CQDs are fluorescent nanoparticles that are highly sensitive to environmental factors, making them ideal for intelligent packaging. Recent reports have pointed out the potential of developing intelligent food packaging films using CQDs as indicator substances. By incorporating intelligent packaging with CQDs, consumers will be aware of the food quality and safety status, prevent or reduce food waste, and improve consumer confidence. The advantages and possible challenges of using CQDs in intelligent materials have also been discussed. The utilization of CQD in intelligent packaging materials is still in its infancy, and further research is required to understand the detection concept. Integrating these findings into smartphones will open a path for developing novel intelligent films.

References

1. Mohamed SAA, El-Sakhawy M, El-Sakhawy MAM (2020) Polysaccharides, protein and lipid-based natural edible films in food packaging: a review. Carbohydr Polym 238:116178
2. Hadidi M, Jafarzadeh S, Forough M, Garavand F, Alizadeh S, Salehabadi A, Khaneghah AM, Jafari SM (2022) Plant protein-based food packaging films; recent advances in fabrication, characterization, and applications. Trends Food Sci Technol 120:154–173
3. Mir SA, Dar BN, Wani AA, Shah MA (2018) Effect of plant extracts on the techno-functional properties of biodegradable packaging films. Trends Food Sci Technol 80:141–154
4. Zhang J, Han Y, Ben Z, Han T, Yin P (2023) Effect of branched polyethyleneimine and citric acid on the structural, physical and antibacterial properties of corn starch/chitosan films. Int J Biol Macromol 231:123186
5. Zheng H, Zhao M, Dong Q, Fan M, Wang L, Li L (2022) Extruded transglutaminase-modified gelatin–beeswax composite packaging film. Food Hydrocoll 132:107849
6. Xia Y, Meng F, Wang S, Li P, Geng C, Zhang X, Zhou Z, Kong F (2022) Tough, antibacterial fish scale gelatin/chitosan film with excellent water vapor and UV-blocking performance comprising liquefied chitin and silica sol. Int J Biol Macromol 222:3250–3260
7. Ordoñez R, Atarés L, Chiralt A (2023) Multilayer antimicrobial films based on starch and PLA with superficially incorporated ferulic or cinnamic acids for active food packaging purposes. Food Chem Adv 2:100250
8. Westlake JR, Tran MW, Jiang Y, Zhang X, Burrows AD, Xie M (2022) Biodegradable active packaging with controlled release: principles, progress, and prospects. ACS Food Sci Technol 2:1166–1183
9. Garcia CV, Shin GH, Kim JT (2018) Metal oxide-based nanocomposites in food packaging: applications, migration, and regulations. Trends Food Sci Technol 82:21–31

10. Deepika KL, Gaikwad KK (2023) Carbon dots for food packaging applications. Sustain Food Technol 1:185–199
11. Ezati P, Priyadarshi R, Rhim JW (2022) Prospects of sustainable and renewable source-based carbon quantum dots for food packaging applications. Sustain Mater Technol 33:e00494
12. Moradi M, Molaei R, Kousheh SA, T. Guimarães J, McClements DJ (2021) Carbon dots synthesized from microorganisms and food by-products: active and smart food packaging applications. 63:1943–1959. https://doi.org/101080/1040839820212015283
13. Liu JM, Lin LP, Wang XX, Lin SQ, Cai WL, Zhang LH, Zheng ZY (2012) Highly selective and sensitive detection of Cu^{2+} with lysine enhancing bovine serum albumin modified-carbon dots fluorescent probe. Analyst 137:2637–2642
14. Singh H, Singh S, Bhardwaj SK, Kaur G, Khatri M, Deep A, Bhardwaj N (2022) Development of carbon quantum dot-based lateral flow immunoassay for sensitive detection of aflatoxin M1 in milk. Food Chem 393:133374
15. Huang X, Sun W, Li Z, Shi J, Zhang N, Zhang Y, Zhai X, Hu X, Zou X (2022) Hydrogen sulfide gas sensing toward on-site monitoring of chilled meat spoilage based on ratio-type fluorescent probe. Food Chem 396:133654
16. Liu K, Guo Y, Yu H, Cheng Y, Xie Y, Yao W (2023) Sulfhydryl-functionalized carbon dots as effective probes for fluorescence enhancement detection of patulin. Food Chem 420:136037
17. Zhang YZ, Zhang B, Chen QQ, Qin L, Hu YQ, Bai CH, Le LY, Qin B (2023) N-doped carbon dots coupled with molecularly imprinted polymers as a fluorescent sensor for ultrasensitive detection of genistein in soya products. Food Chem 424:136133
18. Fu B, Liu Q, Liu M, Chen X, Lin H, Zheng Z, Zhu J, Dai C, Dong X, Yang DP (2022) Carbon dots enhanced gelatin/chitosan bio-nanocomposite packaging film for perishable foods. Chin Chem Lett 33:4577–4582
19. Yang M, Li B, Zhong K, Lu Y (2018) Photoluminescence properties of N-doped carbon dots prepared in different solvents and applications in pH sensing. J Mater Sci 53:2424–2433
20. Zhu X, Jin H, Gao C, Gui R, Wang Z (2017) Ratiometric, visual, dual-signal fluorescent sensing and imaging of pH/copper ions in real samples based on carbon dots-fluorescein isothiocyanate composites. Talanta 162:65–71
21. Kilic B, Dogan V, Kilic V, Kahyaoglu LN (2022) Colorimetric food spoilage monitoring with carbon dot and UV light reinforced fish gelatin films using a smartphone application. Int J Biol Macromol 209:1562–1572
22. Mei S, Fu B, Su X, Chen H, Lin H, Zheng Z, Dai C, Yang DP (2022) Developing silk sericin-based and carbon dots reinforced bio-nanocomposite films and potential application to litchi fruit. LWT 164:113630
23. Li X, Liu X, Su Y, Jiang T, Li D, Ma X (2022) Green synthesis of carbon quantum dots from wasted enzymatic hydrolysis lignin catalyzed by organic acids for UV shielding and antioxidant fluorescent flexible film. Ind Crops Prod 188:115568
24. Sreenath PR, Singh S, Satyanarayana MS, Das P, Dinesh Kumar K (2017) Carbon dot—unique reinforcing filler for polymer with special reference to physico-mechanical properties. Polymer (Guildf) 112:189–200
25. Jorns M, Pappas D, Tagmatarchis N, Kelarakis A (2021) A review of fluorescent carbon dots, their synthesis, physical and chemical characteristics, and applications. Nanomaterials 11:1448
26. Tian R, Hu S, Wu L, Chang Q, Yang J, Liu J (2014) Tailoring surface groups of carbon quantum dots to improve photoluminescence behaviors. Appl Surf Sci 301:156–160
27. Arul V, Sethuraman MG (2018) Facile green synthesis of fluorescent N-doped carbon dots from Actinidia deliciosa and their catalytic activity and cytotoxicity applications. Opt Mater (Amst) 78:181–190
28. Das P, Ganguly S, Ahmed SR, Sherazee M, Margel S, Gedanken A, Srinivasan S, Rajabzadeh AR (2022) Carbon dot biopolymer-based flexible functional films for antioxidant and food monitoring applications. ACS Appl Polym Mater 4:9323–9340
29. Chen S, Zeng Q, Tan X et al (2023) Photodynamic antibacterial chitosan/nitrogen-doped carbon dots composite packaging film for food preservation applications. Carbohydr Polym 314:120938

30. Khan A, Ezati P, Rhim J-W, Kim JT, Molaei R (2023) pH-sensitive green tea-derived carbon quantum dots for real-time monitoring of shrimp freshness. Colloids Surf A Physicochem Eng Asp 666:131242
31. Wen F, Li P, Yan H, Su W (2023) Turmeric carbon quantum dots enhanced chitosan nanocomposite films based on photodynamic inactivation technology for antibacterial food packaging. Carbohydr Polym 311:120784
32. Al-Douri Y (2022) Graphene, nanotubes and quantum dots-based nanotechnology: fundamentals and applications
33. Xu X, Ray R, Gu Y, Ploehn HJ, Gearheart L, Raker K, Scrivens WA (2004) Electrophoretic analysis and purification of fluorescent single-walled carbon nanotube fragments. J Am Chem Soc 126:12736–12737
34. Sun YP, Zhou B, Lin Y et al (2006) Quantum-sized carbon dots for bright and colorful photoluminescence. J Am Chem Soc 128:7756–7757
35. Khan R, Murali S, Gogoi S (2022) Carbon dots in agricultural systems strategies to enhance plant productivity
36. Cui L, Ren X, Wang J, Sun M (2020) Synthesis of homogeneous carbon quantum dots by ultrafast dual-beam pulsed laser ablation for bioimaging. Mater Today Nano 12:100091
37. Min S, Ezati P, Rhim JW (2022) Gelatin-based packaging material incorporated with potato skins carbon dots as functional filler. Ind Crops Prod 181:114820
38. Pandiyan S, Arumugam L, Sakthy, Srirengan P, Pitchan R, Sevugan P, Kannan K, Pitchan G, Hegde TA, Gandhirajan V (2020) Biocompatible carbon quantum dots derived from sugarcane industrial wastes for effective nonlinear optical behavior and antimicrobial activity applications. ACS Omega 5:30372
39. Travlou NA, Giannakoudakis DA, Algarra M, Labella AM, Rodríguez-Castellón E, Bandosz TJ (2018) S- and N-doped carbon quantum dots: surface chemistry dependent antibacterial activity. Carbon N Y 135:104–111
40. Sahu S, Behera B, Maiti TK, Mohapatra S (2012) Simple one-step synthesis of highly luminescent carbon dots from orange juice: application as excellent bio-imaging agents. Chem Commun 48:8835–8837
41. Wang Z, Sheng L, Yang X et al (2023) Natural biomass-derived carbon dots as potent antimicrobial agents against multidrug-resistant bacteria and their biofilms. Sustain Mater Technol 36:e00584
42. Ge G, Li L, Chen M, Wu X, Yang Y, Wang D, Zuo S, Zeng Z, Xiong W, Guo C (2022) Green synthesis of nitrogen-doped carbon dots from fresh tea leaves for selective Fe^{3+} ions detection and cellular imaging. Nanomaterials 12:986
43. Tang X, Wang H, Yu H, Bui B, Zhang W, Wang S, Chen M, Yuan L, Hu Z, Chen W (2022) Exploration of nitrogen-doped grape peels carbon dots for baicalin detection. Mater Today Phys 22:100576
44. Hong WT, Yang HK (2021) Anti-counterfeiting application of fluorescent carbon dots derived from wasted coffee grounds. Optik (Stuttg) 241:166449
45. Ding H, Ji Y, Wei JS, Gao QY, Zhou ZY, Xiong HM (2017) Facile synthesis of red-emitting carbon dots from pulp-free lemon juice for bioimaging. J Mater Chem B 5:5272–5277
46. Dong G, Lang K, Ouyang H, Zhang W, Bai L, Chen S, Zhang Z, Gao Y, Mu Z, Zhao X (2020) Facile synthesis of N, P-doped carbon dots from maize starch via a solvothermal approach for the highly sensitive detection of Fe^{3+}. RSC Adv 10:33483–33489
47. Tyagi A, Tripathi KM, Singh N, Choudhary S, Gupta RK (2016) Green synthesis of carbon quantum dots from lemon peel waste: applications in sensing and photocatalysis. RSC Adv 6:72423–72432
48. Ramezani Z, Qorbanpour M, Rahbar N (2018) Green synthesis of carbon quantum dots using quince fruit (Cydonia oblonga) powder as carbon precursor: application in cell imaging and As^{3+} determination. Colloids Surf A Physicochem Eng Asp 549:58–66
49. Ding C, Deng Z, Chen J, Jin Y (2020) One-step microwave synthesis of N, S co-doped carbon dots from 1,6-hexanediamine dihydrochloride for cell imaging and ion detection. Colloids Surf B Biointerfaces 189:110838

50. Liu X, Jin X, Deng H, Sha Z, Zhou X (2021) Formation of nitrogen-doped blue- and green-emitting fluorescent carbon dots via a one-step solid-phase pyrolysis. J Nanopart Res 23:1–17
51. Ahlawat A, Dhiman TK, Solanki PR, Rana PS (2023) Facile synthesis of carbon dots via pyrolysis and their application in photocatalytic degradation of rhodamine B (RhB). Environ Sci Pollut Res 1:1–8
52. Ma X, Dong Y, Sun H, Chen N (2017) Highly fluorescent carbon dots from peanut shells as potential probes for copper ion: the optimization and analysis of the synthetic process. Mater Today Chem 5:1–10
53. Ezati P, Khan A, Rhim JW (2023) Cellulose nanofiber-based pH indicator integrated with resazurin-modified carbon dots for real-time monitoring of food freshness. Food Biosci 53:102679
54. Khan A, Ezati P, Rhim JW (2023) Chitosan/gelatin-based multifunctional film integrated with green tea carbon dots to extend the shelf life of pork. Food Packag Shelf Life 37:101075
55. Kousheh SA, Moradi M, Tajik H, Molaei R (2020) Preparation of antimicrobial/ultraviolet protective bacterial nanocellulose film with carbon dots synthesized from lactic acid bacteria. Int J Biol Macromol 155:216–225
56. Riahi Z, Rhim JW, Bagheri R, Pircheraghi G, Lotfali E (2022) Carboxymethyl cellulose-based functional film integrated with chitosan-based carbon quantum dots for active food packaging applications. Prog Org Coat 166:106794
57. Ardini B, Manzoni C, Squeo B, Villafiorita-Monteleone F, Grassi P, Pasini M, Bollani M, Virgili T (2023) Spectral imaging of UV-blocking carbon dot-based coatings for food packaging applications. Coatings 13:785
58. Wagh RV, Khan A, Priyadarshi R, Ezati P, Rhim JW (2023) Cellulose nanofiber-based multi-functional films integrated with carbon dots and anthocyanins from Brassica oleracea for active and intelligent food packaging applications. Int J Biol Macromol 233:123567
59. Lin W, Huang G, Yang W, Zeng S, Luo X, Huang J, Li Z (2022) A dual-function chitosan packaging film for simultaneously monitoring and maintaining pork freshness. Food Chem 392:133242
60. Kong Y, Lei T, He Y, Song G (2022) Background-free room temperature phosphorescence and digital image colorimetry detection of melamine by carbon nitride quantum dots in cellulose matrix with smartphone-based portable device. Food Chem 390:133135
61. Khoshkalampour A, Ghorbani M, Ghasempour Z (2023) Cross-linked gelatin film enriched with green carbon quantum dots for bioactive food packaging. Food Chem 404:134742
62. Ezati P, Rhim JW, Molaei R, Rezaei Z (2022) Carbon quantum dots-based antifungal coating film for active packaging application of avocado. Food Packag Shelf Life 33:100878
63. Khan A, Priyadarshi R, Bhattacharya T, Rhim JW (2023) Carrageenan/alginate-based functional films incorporated with allium sativum carbon dots for UV-barrier food packaging. Food Bioproc Tech 1:1–15
64. Koshy RR, Koshy JT, Mary SK, Sadanandan S, Jisha S, Pothan LA (2021) Preparation of pH sensitive film based on starch/carbon nano dots incorporating anthocyanin for monitoring spoilage of pork. Food Control 126:108039
65. Roy S, Ezati P, Rhim JW (2021) Gelatin/carrageenan-based functional films with carbon dots from enoki mushroom for active food packaging applications. ACS Appl Polym Mater 3:6437–6445
66. Khan A, Ezati P, Rhim JW (2023) Chitosan/starch-based active packaging film with N, P-doped carbon dots for meat packaging. ACS Appl Bio Mater. https://doi.org/10.1021/ACSABM.3C00039/SUPPL_FILE/MT3C00039_SI_001.PDF
67. Ezati P, Rhim JW, Molaei R, Priyadarshi R, Han S (2022) Cellulose nanofiber-based coating film integrated with nitrogen-functionalized carbon dots for active packaging applications of fresh fruit. Postharvest Biol Technol 186:111845
68. Min S, Priyadarshi R, Ezati P, Rhim JW, Kim JT (2023) Chitosan-based multifunctional coating combined with sulfur quantum dots to prevent Listeria contamination of enoki mushrooms. Food Packag Shelf Life 35:101014

69. Javanbakht S, Namazi H (2017) Solid state photoluminescence thermoplastic starch film containing graphene quantum dots. Carbohydr Polym 176:220–226
70. Ezati P, Roy S, Rhim JW (2022) Pectin/gelatin-based bioactive composite films reinforced with sulfur functionalized carbon dots. Colloids Surf A Physicochem Eng Asp 636:128123
71. Rani S, Kumar KD, Mandal S, Kumar R (2020) Functionalized carbon dot nanoparticles reinforced soy protein isolate biopolymeric film. J Polym Res 27:1–10
72. Tajeddin B, Arabkhedri M (2020) Polymers and food packaging. Polym Sci Innov Appl: Mater Tech Future Dev 525–543
73. Wang M, Zhang M, Pang L, Yang C, Zhang Y, Hu J, Wu G (2019) Fabrication of highly durable polysiloxane-zinc oxide (ZnO) coated polyethylene terephthalate (PET) fabric with improved ultraviolet resistance, hydrophobicity, and thermal resistance. J Colloid Interface Sci 537:91–100
74. Ezati P, Rhim JW (2020) PH-responsive chitosan-based film incorporated with alizarin for intelligent packaging applications. Food Hydrocoll 102:105629
75. Ezati P, Rhim JW (2022) Pectin/carbon quantum dots fluorescent film with ultraviolet blocking property through light conversion. Colloids Surf B Biointerfaces 219:112804
76. Zhang X, Wang H, Niu N, Chen Z, Li S, Liu SX, Li J (2020) Fluorescent poly(vinyl alcohol) films containing chlorogenic acid carbon nanodots for food monitoring. ACS Appl Nano Mater 3:7611–7620
77. Wang L, Liu X, Qi P, Sun J, Jiang S, Li H, Gu X, Zhang S (2022) Enhancing the thermostability, UV shielding and antimicrobial activity of transparent chitosan film by carbon quantum dots containing N/P. Carbohydr Polym 278:118957
78. Spatafora Salazar AS, Sáenz Cavazos PA, Mújica Paz H, Valdez Fragoso A (2019) External factors and nanoparticles effect on water vapor permeability of pectin-based films. J Food Eng 245:73–79
79. Chavan P, Sinhmar A, Sharma S, Dufresne A, Thory R, Kaur M, Sandhu KS, Nehra M, Nain V (2022) Nanocomposite starch films: a new approach for biodegradable packaging materials. Starch Stärke 74:2100302
80. Roy S, Priyadarshi R, Rhim JW (2022) Gelatin/agar-based multifunctional film integrated with copper-doped zinc oxide nanoparticles and clove essential oil Pickering emulsion for enhancing the shelf life of pork meat. Food Res Int 160:111690
81. Wu H, Wu H, Qing Y, Wu C, Pang J (2022) KGM/chitosan bio-nanocomposite films reinforced with ZNPs: colloidal, physical, mechanical and structural attributes. Food Packag Shelf Life 33:100870
82. Koshy RR, Reghunadhan A, Mary SK, Sadanandan S, Jose S, Thomas S, Pothen LA (2022) AgNP anchored carbon dots and chitin nanowhisker embedded soy protein isolate films with freshness preservation for active packaging. Food Packag Shelf Life 33:100876
83. Lam PL, Wong RSM, Lam KH et al (2020) The role of reactive oxygen species in the biological activity of antimicrobial agents: an updated mini review. Chem Biol Interact 320:109023
84. Ye Z, Li G, Lei J, Liu M, Jin Y, Li B (2020) One-step and one-precursor hydrothermal synthesis of carbon dots with superior antibacterial activity. ACS Appl Bio Mater 3:7095–7102
85. Salimi F, Moradi M, Tajik H, Molaei R (2021) Optimization and characterization of eco-friendly antimicrobial nanocellulose sheet prepared using carbon dots of white mulberry (Morus alba L.). J Sci Food Agric 101:3439–3447
86. Tammina SK, Rhim JW (2023) Carboxymethylcellulose/agar-based functional film incorporated with nitrogen-doped polyethylene glycol-derived carbon dots for active packaging applications. Chemosphere 313:137627
87. Roy S, Ezati P, Rhim JW, Molaei R (2022) Preparation of turmeric-derived sulfur-functionalized carbon dots: antibacterial and antioxidant activity. J Mater Sci 57:2941–2952
88. Lin F, Li C, Chen Z (2018) Bacteria-derived carbon dots inhibit biofilm formation of Escherichia coli without affecting cell growth. Front Microbiol 9:259
89. Das Purkayastha M, Manhar AK, Das VK, Borah A, Mandal M, Thakur AJ, Mahanta CL (2014) Antioxidative, hemocompatible, fluorescent carbon nanodots from an "end-of-pipe" agricultural waste: exploring its new horizon in the food-packaging domain. J Agric Food Chem 62:4509–4520

90. Li D, Kou E, Li W et al (2021) Oxidation-induced quenching mechanism of ultrabright red carbon dots and application in antioxidant RCDs/PVA film. Chem Eng J 425:131653
91. Rodríguez-Varillas S, Murru C, Díaz-García ME, Badía-Laíño R (2022) Green carbon dots as additives of biopolymer films for preserving from oxidation of oil-based products. Antioxidants 11:2193
92. Sachdev A, Gopinath P (2015) Green synthesis of multifunctional carbon dots from coriander leaves and their potential application as antioxidants, sensors and bioimaging agents. Analyst 140:4260–4269
93. Murru C, Badía-Laíño R, Díaz-García ME (2020) Synthesis and characterization of green carbon dots for scavenging radical oxygen species in aqueous and oil samples. Antioxidants 9:1147
94. Li D, Na X, Wang H, Xie Y, Cong S, Song Y, Xu X, Zhu BW, Tan M (2018) Fluorescent carbon dots derived from Maillard reaction products: their properties, biodistribution, cytotoxicity, and antioxidant activity. J Agric Food Chem 66:1569–1575
95. Wei W, Xu C, Wu L, Wang J, Ren J, Qu X. Non-enzymatic-browning-reaction: a versatile route for production of nitrogen-doped carbon dots with tunable multicolor luminescent display. https://doi.org/10.1038/srep03564
96. Li D, Na X, Wang H, Wang C, Yuan Z, Zhu BW, Tan M (2020) The effects of carbon dots produced by the Maillard reaction on the HepG2 cell substance and energy metabolism. Food Funct 11:6487–6495
97. Ma Z, Ma Y, Gu M, Huo X, Ma S, Lu Y, Ning Y, Zhang X, Tian B, Feng Z (2020) Carbon dots derived from the maillard reaction for pH sensors and Cr (VI) detection. Nanomaterials 10:1924
98. Alaş MÖ, Doğan G, Yalcin MS, Ozdemir S, Genç R (2022) Multicolor emitting carbon dot-reinforced PVA composites as edible food packaging films and coatings with antimicrobial and UV-blocking properties. ACS Omega 7:29967–29983
99. Min S, Ezati P, Yoon KS, Rhim JW (2023) Gelatin/poly(vinyl alcohol)-based functional films integrated with spent coffee ground-derived carbon dots and grapefruit seed extract for active packaging application. Int J Biol Macromol 231:123493
100. Liao H, Jiang C, Liu W, Vera JM, Seni OD, Demera K, Yu C, Tan M (2015) Fluorescent nanoparticles from several commercial beverages: their properties and potential application for bioimaging. J Agric Food Chem 63:8527–8533
101. Cheng H, Xu H, Julian McClements D, Chen L, Jiao A, Tian Y, Miao M, Jin Z (2022) Recent advances in intelligent food packaging materials: principles, preparation and applications. Food Chem 375:131738
102. Rahman S, Chowdhury D (2022) Guar gum-sodium alginate nanocomposite film as a smart fluorescence-based humidity sensor: a smart packaging material. Int J Biol Macromol 216:571–582
103. Chao D, Chen J, Dong Q, Wu W, Qi D, Dong S (2020) Ultrastable and ultrasensitive pH-switchable carbon dots with high quantum yield for water quality identification, glucose detection, and two starch-based solid-state fluorescence materials. Nano Res 13:3012–3018
104. Si L, Shi ZK, Hou JY, Miao CX, Hou Q, Xu Z, Ai S (2022) Lignin-derived carbon dot/cellulose nanofiber films for real-time food freshness monitoring. ACS Appl Nano Mater 5:16620–16632
105. Ezati P, Khan A, Rhim JW, Kim JT, Molaei R (2023) pH-responsive strips integrated with resazurin and carbon dots for monitoring shrimp freshness. Colloids Surf B Biointerfaces 221:113013
106. Yang S, Lou J, Jing L, Ding Q, Li X, Jiang Y, Liu Z, Han W (2023) Blue/red dual emission based ratiometric fluorescent intelligent labels for real-time food freshness monitoring. Food Control 150:109775
107. Lu Z, Li M, Chen M et al (2023) Deep learning-assisted smartphone-based portable and visual ratiometric fluorescence device integrated intelligent gel label for agro-food freshness detection. Food Chem 413:135640

108. Yan J, Fu Q, Zhang S, Liu Y, Shi X, Hou J, Duan J, Ai S (2022) A sensitive ratiometric fluorescent sensor based on carbon dots and CdTe quantum dots for visual detection of biogenic amines in food samples. Spectrochim Acta A Mol Biomol Spectrosc 282:121706
109. Sun Y, Zhai X, Zou X, Shi J, Huang X, Li Z (2023) A ratiometric fluorescent sensor based on silicon quantum dots and silver nanoclusters for beef freshness monitoring. Foods 12:1464
110. Deb A, Saikia R, Chowdhury D (2019) Nano-bioconjugate film from aloe vera to detect hazardous chemicals used in cosmetics. ACS Omega 4:20394–20401

Functional Food Packaging—Active, Intelligent, and Smart Bio-nanocomposites

Harish Sharma, Aaditya Pandey, Sharad Raghuvanshi, Ashish A. Kadam, Dharm Dutt, and Anurag Kulshreshtha

Abstract The food packaging business is rapidly expanding, necessitating careful attention to the problem of sustainability when substituting plastic products with biodegradable packaging. The functional features of biobased packaging, such as mechanical and barrier qualities, are significant challenges. The same package with varying functionality may be created using different nanomaterials as fillers in packaging composites. This chapter focuses on applying bio-nanocomposites in food packaging for active, intelligent, and smart functioning. These Bio-nanocomposites can be made from natural or synthetic polymers with biodegradable characteristics. The development of bio-nanocomposites in food packaging is covered, along with a detailed analysis of several forms of nanomaterials used in food packaging. Bio-nanocomposites are presently employed in food packaging research to build packages with various functions such as oxygen scavenging, time temperature indication, controlled release of nutraceuticals and bioactive substances, antibacterial properties, and sensing capabilities. Because the toxicological impact of using nanoparticles has minimal scientific validity, these biobased nanocomposite packages have limited interaction with food products. As a result, additional experimental work on this intriguing subject is required.

Keywords Active packaging · Smart and intelligent packaging · Bio-nanocomposites · Functional films

H. Sharma · S. Raghuvanshi · D. Dutt · A. Kulshreshtha (✉)
Department of Paper Technology, Indian Institute of Technology Roorkee, Saharanpur Campus, Saharanpur 247001, India
e-mail: anuragk77@gmail.com

A. Pandey
Department of Polymer and Process Engineering, Indian Institute of Technology Roorkee, Saharanpur Campus, Saharanpur 247001, India

A. A. Kadam
Department of Food Technology, Koneru Lakshmaiah Education Foundation, Vaddeswaram, Guntur District 522302, Andhra Pradesh, India

M. Alle et al. (eds.), *Functional Nanomaterials and Nanocomposites for Biodegradable Food Packaging*, Smart Nanomaterials Technology,
https://doi.org/10.1007/978-981-97-9084-5_11

1 Introduction

The way people connect with brands and products and shop for them is changing, thanks to functional packaging. In the next 3–5 years, the food industry will switch to flexible packaging for 190 billion retail units, predicts Euromonitor International. According to Euromonitor, pressure-sensitive reclosures will also experience faster growth, with a compound annual growth rate (CAGR) of 6%.

Food packaging is a part of the global packaging sector, which is expanding fast. Food packaging must perform diverse tasks for the food it contains, which is why it has become both a necessary and useful tool for various functions like storage, marketing, transportation, preservation, and containment. While necessary packaging contains vital components for food safety and quality, functional food packaging adds value to packaged food items. One should not only pick food packaging based on its appearance on the shelf. Additionally, other factors must be considered, such as its ability to contain and preserve goods over long and short periods, convenience and simplicity of the package for the consumer, and its ability to successfully convey the product's function and advantages to consumers. Innovation in functional food packaging is essential to a product's success once it hits the market. "Functional Packaging" offers consumers value-added benefits, including increased usability or convenience. Flexible packaging, pressure-sensitive films, reclosure techniques, and intelligent label technologies might all be incorporated inventively. Figure 1 illustrates the various functions of a food packaging material [1].

Trends in food packaging are observed in primary research and development in the field of smart packaging, including active packaging and intelligent packaging.

Active packaging refers to a packaging system in which the container/package, used to protect the food product, interacts with the environment and food to improve/

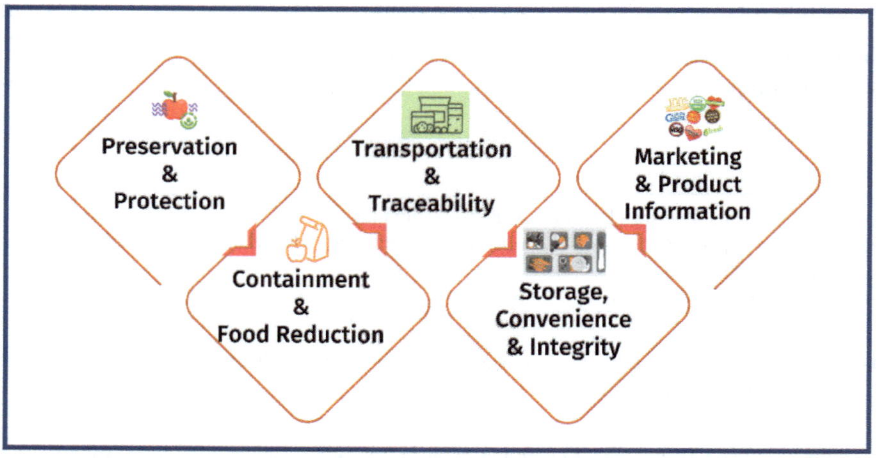

Fig. 1 Functions of food packaging

maintain the shelf life or quality of the food product. In other words, the package possesses features other than merely passively containing and protecting the product [2] and adapts to changes in the environment both within and outside the box.

Physiology (e.g., respiration, ripening, and transpiration in fresh produce), physical processes (e.g., powders caking and bread staling), chemistry (e.g., oxidation of oils and fats) of the food product, and its microbiological spoilage due to bacteria, fungi, and yeast, are the factors that are being influenced by the package to improve the shelf life while maintaining quality [3]. Scavengers, emitters, etc., are typical examples of active packaging systems. Critical active packaging technologies include flavor-releasing/absorbing systems, moisture absorbers, ethylene absorbers, oxygen scavengers, and carbon dioxide emitters/absorbers [4].

The intelligent packaging system is similar to active packaging, in the manner with which it interacts with the environment, but it is different in its response to this interaction. It can be understood by an example of a pH sensor incorporated in packaging. When it is clubbed in the packaging system, it interacts with food products to sense their pH, but it doesn't influence physiology, chemistry, etc.; it is made just to indicate the pH of food by some indicators or color-changing labels. Thus, it can be concluded that these packaging systems don't influence the shelf life or quality of food; they are just indicators of food's current state. Typical intelligent packaging systems can be classified as sensors, indicators, and RFIDs. Biosensors, gas sensors, chemical sensors, time–temperature indicators, freshness indicators, integrity indicators, and RFID are the commonly used systems in functional food packaging [5].

Smart packaging can be seen as a combination of the previously discussed two forms of packaging. It creates a packaging system that observes changes in a food product or its environment (intelligent), and on the other hand, it responds to these changes (active) if needed. Chemical or biological sensors are used in smart packaging to track the quality and safety of food from manufacturers to consumers [6].

Developing smart, active, and intelligent packaging technologies relies on several factors, such as materials and new sensor technologies. Out of these factors, materials play a crucial role. Development in packaging materials dramatically influences the growth of smart, active, and intelligent packaging for food products. Customers are increasingly paying greater attention to sustainable packaging, which has led to the popularity of biobased plastics and biopolymers. Functionality requirements for packages play a big part in the food packaging industry. Since biobased polymers lack the mechanical and barrier qualities needed for packaging, employing functional nanomaterials to enhance the capabilities of biobased polymers is now the popular trend in the packaging sector. Nanocomposites based on natural biopolymers (such as cellulose, starch, and lactic acid) are being investigated for various uses, with food packaging being one such growing field.

Also, some nanomaterials can be modified to impart different functionality in packaging materials. This creates an opportunity for packaging technologists to use the same biopolymer for additional varying functionalities in food packaging. For example, cellulose-based packaging material possesses antibacterial properties when hydroxypropyl methylcellulose (HPMC) matrix added with silver nanoparticles (AgNPs) as fillers were used [7]. It exhibits better mechanical properties in the form of α-hydroxy sulfonic acid cellulose (HSAC) films [8]. This wide range of applications is possible due to the functionality spectrum provided by various nanomaterials. Despite the improvement in material characteristics, consumers' concerns about the consequences of using nanoparticles in food packaging materials have increased. The nanoparticles could possibly leach out from the packaging material into the food, and consequently could be ingested by the consumers. Therefore, it is necessary to understand how these nanoparticles (NPs) behave within the human body, as well as how they are metabolized and eliminated. In addition, describing the difficulties in their monitoring, regulating their potential migration into food matrices, and their toxicity, is paramount. Academic research is heavily focused on enhancing the mechanical and barrier qualities of packaging material, even if practical uses of these bio nanocomposites are still under investigation for commercial feasibility studies.

In this chapter, applications of bio-nanocomposites in the food packaging segment are discussed in detail for various types of packages such as active, intelligent, and smart packaging. This discussion summarises the diverse research and commercial trends in bio-nanocomposite development so that readers will have a better understanding. In the end, possibilities for future research scope and commercial feasibility are also discussed.

2 Bio-nanocomposites: Alternative to Traditional Food Packaging Polymers

The non-biodegradable polymers are heavily used in the food packaging industry, and the waste produced by packaging materials is a significant issue in terms of sustainability. Environmental concerns have put polyethylene terephthalate (PET), polypropylene (PP), high-density polyethylene (HDPE), low-density polyethylene (LDPE), polystyrene (PS), and all other non-biodegradable packaging materials under scrutiny; thus, it is imperative to find biodegradable alternatives for these polymers. Governments worldwide have tried to regulate the usage of these non-biodegradable polymers by imposing several restrictions. In such attempts, several countries have banned plastic carry bags less than 75 μm, and policies are updated from time to time for single-use plastics and disposal of used plastics. Extended Producer Responsibility (EPR) for plastic packaging is another program that enhanced accountability, traceability, and transparency and made it easier for producers, importers, and brand owners to report compliance with EPR responsibilities. However, all the steps mentioned above are therapeutic. To reduce the environmental effect of packaging materials, protective measures are required to deal with the trash that amounts to more than 5 billion tonnes annually. This may be achieved by discovering substitutes in biodegradable packaging materials, which is why biopolymers and bio-nanocomposites have gained such significance in today's food packaging industry.

Regarding their structure, bio-nanocomposites are created by mixing two or more materials with different properties, much like other conventional composites. The concentration of these ingredients must determine the impact on the overall effectiveness of bio-nanocomposites. The fusion of organic and inorganic materials is increasing in demand nowadays. Synthetic materials are typically replaced with more environmentally friendly, biodegradable substitutes. The homogeneous dispersion of nanofillers during the preparation stages significantly impacts the better performance of bio-nanocomposites. Bio-nanocomposites are generally prepared by solvent casting, in situ polymerization, and melt intercalation methods [9–11].

Classification of bio-nanocomposites can be done in two ways. The first is based on biopolymers used, and the second is based on nanomaterials used.

Classification based on biopolymers used:

(a) Polysaccharides-based bio-nanocomposites
(b) Polypeptide-based bio-nanocomposites
(c) Aliphatic polyester-based bio-nanocomposites

This classification can also be done based on the categorization of base biopolymers used. Biopolymers are classified based on the source (plant-based, animal-based, or microorganism-based) and production (natural or synthetic).

Polysaccharides-based bio-nanocomposites consist of composites made up using polysaccharides and their derivatives. Their derivative modification makes them suitable to be used for packaging applications. Among all the cellulose-derivative sources, hydroxypropyl cellulose (HPC) is a strong contender that has been researched for its edible film-forming ability as a food protection agent. Similarly, chitosan (cationic polysaccharide), alginate (anionic polysaccharide), pectin (plant-derived polysaccharide), and guar gum (galactomannan polysaccharide) have been used as base polymers for composite matrix and that's how their classification can be done. All the above-mentioned biopolymers are obtained from plants, and that's why they have biodegradable properties. Incorporating nanomaterials in these biopolymers results in better properties of food packaging materials. Most nanofillers utilized to enhance the characteristics of the matrix include clays, carbon nanotubes (CNTs), graphene, layered double hydroxides (LDHs), polyhedral oligomeric silsesquioxane (POSSs), and silica. Due to their affordability, accessibility, adaptability, and environmental friendliness, clays like Montmorillonite (MMT) are now the subject of most of the studies.

Polypeptide-based bio-nanocomposites are prepared by using plant or animal-derived proteins. Proteins are very complex amphoteric polymers with many diverse functions, where the amino acid monomers are linked together by amide bonds to form the protein's polymeric structure. They have decreased water vapor permeability and excellent gas barrier qualities. Currently, due to localized charge and multifunctionality, it is possible to create nanocomposites based on milk protein, soybean protein, gelatine, maize zein, and wheat gluten.

In the last type, aliphatic polyester-based bio-nanocomposites, polyesters derived from renewable resources or petroleum are used as the base polymer in nanocomposites. In this category, polylactides (PLAs), polyhydroxyalkanoates (PHAs), polyesteramide (PEA), poly (butylene succinate) (PBS), polybutylene adipate terephthalate (PBAT), etc., are the polyesters that are used as base polymers for making nanocomposites. These polymers have biodegradable characteristics, and that's why they are used in making bio-nanocomposites with improved properties.

Classification based on types of nanomaterials used is performed as:

(a) Inorganic nanomaterial-based bio nanocomposites
(b) Organic nanomaterial-based bio-nanocomposites
(c) Nanohybrid bio-nanocomposites

Inorganic nanomaterials-based nanocomposites include the utilization of metallic nanoparticles and metal oxide nanoparticles to prepare the nanocomposites. Silver nanoparticles (AgNPs), copper nanoparticles (CuNPs), gold nanoparticles (AuNPs), and sulfur nanoparticles (SNPs) are commonly used NPs in bio nanocomposites. These NPs have various functional properties such as antimicrobial and antibacterial. Metal oxide-based nanoparticles such as silicon dioxide NPs (SiO_2NPs), ZnO, and CuO NPs are also used in biopolymer matrices to form bio-nanocomposites. Other than the above-mentioned, inorganic NPs also include nano clays. Sepiolite (also known as palygorskite and sepiolite), illite (also known as glauconite), smectite (also known as montmorillonite and beidellite), chlorite, and kaolinite are the several types of pure clays that makeup nano clays (kaolinite, halloysite). Nano clays are used in many different fields, including medicine, pharmaceuticals, cosmetics, catalysis, food packaging, and textiles.

Organic nanoparticles for bio-nanocomposites involve the utilization of cellulose NPs, starch NPs, chitosan NPs, etc. Cellulose nanocrystals (CNC), cellulose nanofibers (CNF), and bacterial cellulose are the three primary subtypes of nanocellulose due to their availability, abundance, high aspect ratio, mechanical characteristics, renewability, biocompatibility, and lack of toxicity. The starch NPs are non-toxic, biodegradable, cost-effective, and biocompatible. Additionally, they have a high biological penetration rate, reactive surface, decent solubility, and absorption capacity. Nano-cellulose materials have attracted increasing attention from businesses and academia.

As discussed earlier in this chapter, smart packaging systems can be developed with superior bio-nanocomposites. It is proven that bio-nanocomposites can offer superior convenience, containment, protection, preservation, marketing, and communication properties than traditional food packaging systems. It can offer better mechanical, thermal, gas, water, and antimicrobial barriers than biopolymer-based packaging systems. This is the reason for ongoing extensive research in this field. Nowadays, most researchers are utilizing these bio-nanocomposites in active and intelligent packaging systems due to the properties offered by these materials. The next section of the chapter discusses developments in the field of bio-nanocomposites for smart packaging systems. Challenges and regulations with these bio-nanocomposites are also discussed in later sections. This will help future researchers to understand the opportunities and challenges offered by these materials in functional packaging systems.

3 Bio-nanocomposites for Active Food Packaging Systems

Food packaging applications require good barrier and preservation qualities in the material used in direct contact with food. These properties range from antimicrobial properties and antioxidant properties to barrier properties, which protect the food from external environmental factors and microorganisms. Nowadays, these properties are imparted in packaging films/materials with the utilization of bio-based materials. Biopolymers such as starch and chitosan are blended with various additives for better functional properties. Food is subjected to multiple changes throughout its shelf life, and packaging should be able to withstand these changes. That's where active packaging comes into the picture. Bio-based polymers are nowadays being extensively utilized for this purpose. This segment of the chapter deals with these innovations, and the use of bio-nanocomposites in active packaging is discussed in detail.

Active packaging comprises materials that may provide a solid barrier to external environmental conditions and guard against microorganisms that cause food to deteriorate. [11, 12]. Various antibacterial substances [7], antioxidants [13], food preservation agents, and gas absorbers [14] have been recommended by several researchers. Antimicrobial compounds can be added to food to prevent bacteria development on the surface. The integration of nanoparticles during the creation of bio nanocomposites can have a significant influence on the antibacterial surface. The majority of the inserted nanoparticles can interact directly with microbial pathogens and can stop their development. Metallic nanoparticles can stop bacterial pathogens from growing or killing them by interacting with their cell wall components, breaking down their cell walls and other proteins, deactivating their enzymes, and causing oxidative stress by releasing superoxide radicals [15]. This will ultimately lead to altered gene expression and further prevent bacterial reproduction.

Many researchers have used various composite matrices to develop active food packaging solutions using these bio-nanocomposites. Because all foods have various spoiling processes, each matrix has unique features that lead to various food applications. Such types of different matrices are summarized in Table 1, showing their food packaging applications. This table shows that cellulose-based matrices are widely used in active food packaging systems with different nanotechnological versions, such as cellulose nanocrystals and cellulose nanofibrils. Also, chitosan-based bio-nanocomposites are used in various antioxidant food packaging systems. Similarly, several researchers are exploring starch-based composites due to their biodegradable nature. Graphene oxide and its derivatives have also been used widely in active food packaging systems. Reduced graphene oxide (rGO) is one such derivative. The ability of rGO to scavenge free radicals can give bio-nanocomposites antioxidant properties. The effectiveness of rGO can be seen with the fact that in their study,

Barra et al. concluded that after 8 h of incubation, chitosan-based films with 20–33 wt% of rGO showed an increase in inhibition ranging from 54 to 82% and this outcome is favorable for usage in food packaging that requires oxidation resistance [16]. The ability of graphene-based nanostructures to cause cell membrane rupture and oxidative stress, which impair bacterial growth and sporulation, is the basis for their antibacterial and antifungal activities.

Nanomaterials are the second most crucial ingredient in bio-nanocomposite. Several nanomaterials have been investigated to be used as nanofillers in the fabrication of bio-nanocomposite materials. These nanomaterials are added to bio nanocomposite to impart some important functional properties. Nowadays, controlled release mechanisms are getting popular in packages. The most valuable features in these releasing packaging methods relate to antibacterial substances and oxygen scavengers in bio-nanocomposites. The bio-nano composite's stability, bioavailability, and controlled release qualities are all aided by encapsulation, which also helps to hide the compound's unfavorable taste and odor. Maftoonazad and Ramaswamy beautifully reviewed bio-nanocomposites in active food packaging systems and discussed the trends in nanoencapsulation application in functional food packaging systems [35]. Active chemical encapsulation is a rapidly expanding field of study with important implications for transporting and preserving bioactive substances. Entrapping an active substance, such as a food bioactive chemical or live cells, inside a wall material is what encapsulation is all about (e.g., bio-nanocomposite). Macrocapsules, microcapsules and nanocapsules are the three different types of capsules that have been postulated according to their particle size. Particles produced by encapsulation procedures should have a high degree of encapsulation, a favorable polydispersity index (PDI), and a long shelf life. An encapsulating process must also adhere to other requirements and be compatible with the food components. Additionally, it should possess particular qualities to endure environmental impacts [36]. Quercetin, a model molecule for encapsulation investigations, has been encapsulated using nanoparticulate systems, as described by several publications. The quercetin encapsulation effectiveness into chitosan/lecithin nanoparticles showed values higher than 95% for all studied doses. Encapsulated quercetin also enhanced antioxidant effects [37].

Priyadarshi et al. critically reviewed nanomaterials in active food packaging systems emphasizing antimicrobial nanomaterials [38]. They concluded that all nanofillers interact with microbial cells but lack antibacterial capabilities due to their diverse chemical makeup. Inorganic nanostructures made of metal or metal oxide often have the highest antibacterial potential, making them the most popular antimicrobial fillers for food packaging. Metals have been recognized to have antibacterial properties against both bacteria and fungi since the dawn of time. For instance, food and water have been stored and preserved in copper and silver pots. Today, antimicrobial metal compounds are still frequently employed in business, agriculture, and medicine. Silver nanoparticles have been used in a variety of food packaging techniques that have been developed. When added to biopolymer films made of cellulose, chitosan, starch, and alginate, silver nanoparticles consistently display antibacterial action against both gram-positive and gram-negative bacteria [39].

Table 1 Biocomposite materials for active food packaging application

S. no.	Composite matrix	Application	References
1	Chitosan—reduced graphene oxide	Food packaging applications with antioxidant and electrically conductive packaging requirements	[16]
2	Wheat gluten/nanocellulose/titanium dioxide	Active (antimicrobial properties) food packaging material	[17]
3	Zinc oxide–neem oil–chitosan bio-nanocomposite	Active packaging of carrot pieces	[18]
4	Carbon dots enhanced gelatin/chitosan bio-nanocomposite	Active packaging of perishable foods	[19]
5	Chitosan film impregnated with turmeric essential oil (TEO)-loaded magnetic-silica nanocomposites	Active packaging of surimi preservation	[20]
6	Chitosan/negatively charged graphitic carbon nitride self-activation bio-nanocomposite films	Active preservation of tangerine fruits	[21]
7	Cellulose-based bactericidal nanocomposites containing silver nanoparticles	Active food packaging	[7]
8	Carboxymethyl cellulose-based multifunctional film combined with zinc oxide nanoparticles and grape seed extract	Preservation of high-fat meat products	[22]
9	Bio-nanocomposite films based on potato starch/halloysite nanoclay	Active food packaging applications	[23]
10	Silver nanoparticles immobilized on cellulose nanofibrils for starch-based nanocomposites	High antibacterial, biocompatible, and mechanical properties	[24]
11	Soy protein-based films incorporated with cellulose nanocrystals and pine needle extract	Active packaging applications	[25]
12	Agar/TiO_2/radish anthocyanin/neem essential oil bio-nanocomposite	Active packaging of banana	[26]
13	Antioxidant and antibacterial lignin nanoparticles in polyvinyl alcohol/ chitosan films	Active packaging applications	[27]
14	Nano reinforced gelatinized starch	Biobased edible films for active packaging applications	[28]
15	ZnO nanoparticles in a chitosan/gum Arabic edible coating	Active packaging of banana preservation	[29]
16	Starch-based bio-nanocomposites films reinforced with cellulosic nanocrystals	Active food packaging applications	[30]

(continued)

Table 1 (continued)

S. no.	Composite matrix	Application	References
17	Antioxidant film based on chitosan and Tulsi essential oil	Active packaging of fried potato fingers	[31]
18	Retention of cinnamaldehyde in poly(vinyl alcohol) films	Preservation of faba beans through vapor-phase antimicrobial effect	[32]
19	Pine needle and halloysite nanotube-based ethylene gas scavenging paper	Quality preservation in banana fruits	[33]
20	Chitosan-based antimicrobial micro-capsules	Active packaging applications	[34]

Antimicrobial packaging materials are also prepared with bio-nanocomposites from numerous antimicrobial agents such as modified clay minerals, metallic nanoparticles, isothiocyanates, bacteriocins, antimicrobial enzymes, etc. [40]. The literature shows that halloysite (HS), montmorillonite (MMT), and bentonite (BT) are the three forms of nano clays that are employed in food packaging [41]. The existence of its intrinsic lumen structure accounts for halloysite nanotubes' exceptional support as bioactive carriers. The lumen is a valuable structure for encapsulating medications and bioactive substances and aids in sustained release [42]. Salicylic acid may be adsorbed to composite alginate-pectin films using halloysite nano clay, and the release patterns of these films were observed. Better release profiles and antibacterial activity were demonstrated by alginate films [43]. Additionally, pullulan-chitosan-based active packaging materials may be made to prevent food product quality degradation and storage deterioration by combining rutin with chitosan-modified halloysite nanotubes [44]. Different quantities of halloysite (0.5–2.0 mg) were added to cellulose nanofiber (CNF) composite films, and these films reinforced the mechanical, barrier, and thermal characteristics. With the increase in HS concentration, mechanical properties up to a specific concentration start decreasing due to aggregate formation [45]. Similarly, when HS is added to fish gelatin films, the qualities of the film are improved due to reinforcement. As the amount of glycerol (a common plasticizer) was raised (0–40%), the gelatin films' tensile strength (TS) decreased, and their elongation at break (EAB) rose. However, the inclusion of HS enhanced these qualities, which was credited to the development of a robust 3D network, including the chains of gelatin, HS, and glycerol [46]. MMT, another kind of nano clay, is frequently employed in active food packaging applications. Comparing bio-nanocomposite films made only of starch and glycerol contents devoid of nano clay filler materials to bio-nanocomposite films generated simply with starch and glycerol contents resulted in lower water absorption rates (hydrophobic nature). The increased hydrophobicity and water repellence of the films are caused by intercalation generated between starch and sodium MMT [47]. There are other sides to this sort of variance in water absorption rates. The functional characteristics of the starch-based bio-nanocomposite films vary depending on the glycerol content and the type of clay (hydrophilic or hydrophobic). Both hydrophilic and hydrophobic clays showed increased tensile strength when the glycerol content was between 20 and 30%. The hydrophobic nanoclay materials used to support the plasticized starch reduced the permeability of the water vapor and created an antimicrobial effect [48]. The incorporation of MMT in a bio-nanocomposite matrix imparts hydrophobic and hydrophilic properties, but excessive nanoclay inclusion in the matrix affects its strength [49]. Similarly, adding bentonite (BT) to cellulose nanofibrils at various loads caused the films to acquire polar functions, which increased their hydrophilicity. This bio-nanocomposite film creates a barrier between the molecules of O_2 and H_2O. The clay fillers significantly reduced the oxygen transfer rate in the films at low and moderate relative humidity levels [50]. A study by Zheng et al. demonstrated an interesting fact about bio-nanocomposites loaded with MMT and halloysite nanotubes (HSNT) [50]. While water vapor permeability (WVP) was not enhanced by the inclusion of nano clays in either case, increasing amounts of nano clay lowered the moisture content and oxygen permeability of the nanocomposite films.

The above discussion shows that nanomaterials in biopolymer matrix have ample scope in active food packaging for commercialization. Their limitations and future scope are discussed in the later segments of this chapter.

4 Bio-nanocomposites for Intelligent Food Packaging Systems

The main difference between active and intelligent packaging is that intelligent packaging monitors, communicates, and provides information on packaged food and its environment (pH, temperature, humidity, etc.) inside the package [51]. However, active packaging takes action to protect the food from spoilage and extend its shelf life. Figure 2 shows the model of packaging functions. Intelligent packaging systems are capable of performing intelligent functions such as tracing, detecting, recording, communicating, and applying scientific logic to aid decision-making in order to extend shelf life, safety, and quality, as well as inform us about potential problems in food packaging [52, 53]. Nowadays, biocomposites are gaining much attention in food packaging as they are good and safe for the environment and human health. However, due to their poor mechanical and barrier properties, they are not efficient enough for food packaging. Nanofillers have been applied as nano-enforcements to improve the properties of biocomposites and have proven to be very efficient [54]. Therefore, a new type of composite material, known as bio-nanocomposites, is introduced in packaging.

Bio-nanocomposites are a promising packaging material for active packaging but can also function as intelligent materials. Bio-nanocomposites offer good mechanical, thermal, and barrier properties to active and intelligent packaging systems, and due to their biodegradability and eco-friendly nature, they are also referred to as green

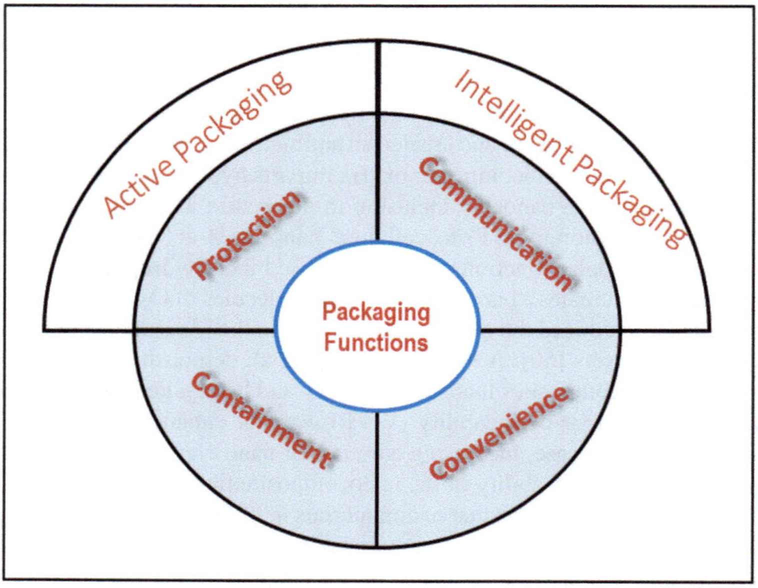

Fig. 2 Model of packaging functions and the role of active and intelligent packaging

sustainable materials [55, 56]. Nanoparticles can be employed in bio-nanocomposite to retrieve information about the condition of the package by detecting the gases, volatile compounds, release of chemicals, etc., inside the package environment [51].

Basically, four types of biopolymers are being used in making bio-nanocomposites: (1) polysaccharides like cellulose and its derivative, starch, chitin, pectin, and gums (xanthan, alginate, carrageenan, etc.); (2) proteins like gelatin, collagen, meat myofibril, glutenin, etc.; (3) lipids like essential oils, vegetable and animal fats, waxes, glycerides, etc.; (4) polyesters like polylactic acid. The biodegradability of these biopolymers is high, but mechanical strength is poor. Nano-enforcements are required as fillers to make them suitable for packaging [54].

Nanofillers include organic (chitin, chitosan, starch nanocrystals, cellulose nanofibers, etc.) and inorganic (Ag, TiO_2, $CaCO_3$, CuO, ZnO nanoparticles) materials [11]. The incorporation of nanofillers enhances the barrier properties (impermeability) of the packaging films as well as improves the mechanical and thermal properties (melting point, glass transition temperature, etc.). Nanofillers impregnated on biopolymer matrix along with intelligent materials have proven their potential for intelligent packaging.

Intelligent materials or indicators in bio-nanocomposites include natural pigments or color markers, which show variation in color during changes in pH. Indicators provide information on food through color change by reacting with volatile compounds or with the chemicals released during microbial spoilage of food. Natural dyes like anthocyanin, alizarin, beta-carotene, betaine, curcumin, etc., have been increasingly used as indicator dyes instead of synthetic dyes to make sustainable bio-nanocomposites films [57]. For example, Pereira et al. and his co-workers developed a chitosan film incorporated with anthocyanin pigment to determine the temperature deviations through monitoring of pH change in packaged food during storage [58].

In another study, Sun et al. prepared konjac glucomannan/chitosan-based bio-nanocomposites films impregnated with ZnO nanoparticles (~30 nm) and mulberry anthocyanin extract (MAE) for multifunctional response as active and intelligent packaging of chicken fillets. The MAE provided pH-sensitivity (red at pH 2–4; blue at pH 10–12) to the films, making them act as freshness indicators, while ZnO provided protection against microbial spoilage (Fig. 3A) [59]. Shavisi and Shahbazi have incorporated *Rosa damascena* anthocyanins extract into the chitosan/gum Arabic nanofiber to make an active and intelligent bio-nanocomposite film. The bio-based film showed great potential for monitoring the freshness of raw chicken fillets by visual color indication due to pH change (acidic-pale red; neutral-khaki color) (Fig. 3B–D) [60].

Recently, Koshy et al. have developed a bio-nanocomposite film with potato starch and carbon nanodots (~20 nm) incorporated with anthocyanin from *Clitoria ternatea* flower [61]. The film possesses good thermal, mechanical, antioxidant, and multiple pH-responsive properties (reddish pink at pH 2–3; purple at pH 4–7; bluish green at pH 8–11; yellow at pH 12) (Fig. 4B) that can be used to monitor the spoilage of pork (Fig. 4A). Moreover, carbon dots (~10 nm) were added to gelatin/chitosan biopolymer to develop the bio-nanocomposites film with exceptional mechanical, barrier, antibacterial, and antioxidant properties as well as anticounterfeiting and pH-responsive features due to fluorescence emission by carbon dots (Fig. 4C) [19]. Other indicators like gas indicators (O_2 and CO_2) can also be used to monitor the food quality in the package. Most indicators are based on pH change resulting from the enzymatic and chemical reactions due to the chemical reactions taking place between the indicator and the volatile compounds or gases produced by the packaged food product over time. Traditional indicators use synthetic dyes like methylene blue, which changes its color upon redox reactions [62]. However, the use of synthetic dyes is restricted due to increasing awareness about the issues of using synthetic products in food contact applications. Natural dyes are safe and very sensitive to environmental variations and undergo color change in the package with alterations in humidity, temperature, and chemical composition. Metal nanoparticles (silver and titanium oxide) are also capable of forming color-based indicators, but they raise concerns about their migration into the food [63]. Therefore, presently, only natural dyes are being used to fabricate bio-nanocomposites because of their eco-friendly nature. However, metal and metal oxide nanoparticles are still being used in many active and intelligent films as nanofillers.

Over the years, many sensors have been developed for intelligent packaging systems, such as sensors (biosensors, chemical, gas sensors, etc.), freshness indicators, and temperature and humidity indicators [51, 64]. Biofunctional magnetic nanoparticles-based sensors are quite well-known nowadays in the detection of microbial spoilage and toxins in food [65, 66]. For the readers' convenience,

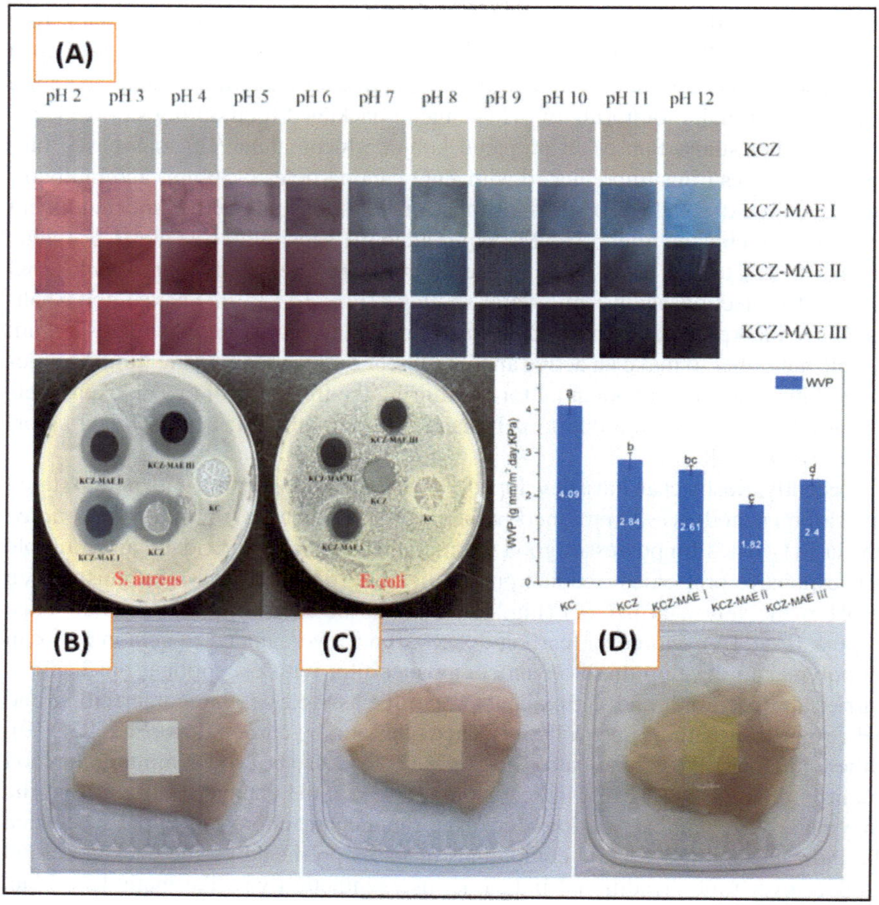

Fig. 3 **A** Konjac glucomannan/chitosan/ZnO/nanoparticles/mulberry anthocyanin extract (KCZ-MAE) based nanocomposite film for active (antibacterial and antioxidant) and intelligent packaging (pH-based freshness indicator); adapted from [59]. **B–D** Chitosan/gum Arabic nanofiber based bio-nanocomposites film impregnated *with Rosa damascena* anthocyanins extract as freshness indicator for chicken fillets during storage after **B** 0 days, **C** 4 days (light khaki color), and **D** 8 days (dark khaki color); adapted from [60]

some recently developed bio-nanocomposites films for intelligent packaging are summarized in Table 2.

Due to the early stage of development, bio-nanocomposites are facing challenges due to the use of nanoparticles and their impact on the environment and human health. Nanoparticle accumulation in the body can lead to lung damage, cancer induction, asthma, or even death [85–87]. Therefore, life cycle assessment (LCA) and risk assessment of bio-nanocomposites are always recommended before their commercial use. Also, there is a lack of proper methodologies and guidelines for LCA analysis and

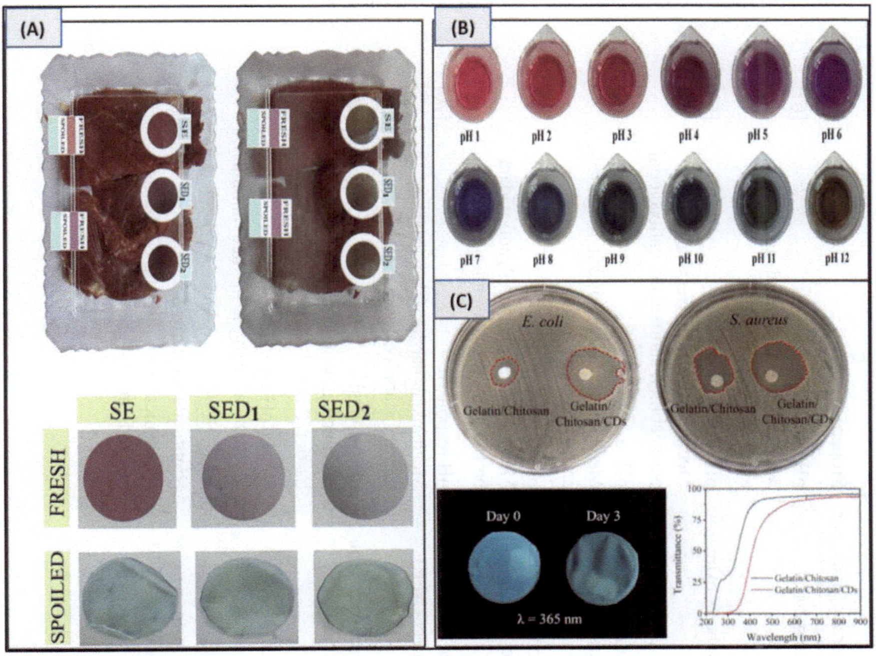

Fig. 4 Starch/carbon dots/*Clitoria ternatea* flower extract (SED) based bio-nanocomposites film. **A** Freshness monitoring of pork sample with pH-sensitive SED film showing a green color due to ammonia release from pork during storage. **A–B** Color variation of SED_1 (carbon dots 9% w/v) and SED_2 (carbon dots 17% w/v) film at different pH (1–12); adapted from [61]. **C** Antibacterial activity of carbon dots with gelatin/chitosan/CDs film and monitoring of fish meat and color change of film during storage (0–3 days); adapted from [19]

risk assessment for bio-nanocomposites. International Organization for Standardization (ISO) has just provided general guidelines on the use of bio-nanocomposites (ISO 14040 and ISO 14044). One study on life cycle assessment and impact assessment has shown that polylactic acid-based bio-nanocomposites are 100% compostable while conventional polyethylene terephthalate (PET) based composites are 60% incinerated and remaining landfilled. In PLA-based bio-nanocomposites, approximately 20% of PLA is composted, 40% is incinerated, and 40% is landfilled (biodegradable). Additionally, the study has proven that bio-nanocomposite packaging materials increase the shelf life of fruits during storage and are much more efficient than PET-based composites. PLA-based bio-nanocomposites are also eco-friendly and can be used in packaging [88].

Most of the research work on bio-nanocomposites is going on towards active packaging rather than towards intelligent packaging. Due to the early phase development of bio-nanocomposites in intelligent packaging, there is huge scope for research in this area, and its life cycle assessment and impact assessment.

Table 2 Recently developed bio-nanocomposites films for intelligent packaging

S. no.	Biopolymer	Nanofillers	Indicator used	Source of indicator	Properties	References
1	Chitosan	Mesoporous silica nanoparticles	Curcumin	Commercial	pH responsive Mesoporous, mechanical, and antibacterial properties	[67]
2	Konjac glucomannan/ Chitosan	ZnO nanoparticles	Anthocyanin	Mulberry extract	pH responsive Mechanical, UV barrier, water–vapor permeability, antioxidant, antibacterial	[59]
3	Chitosan	Surface deacylated Chitin nanofibers	Anthocyanin	Purple potato extract	pH responsive Mechanical, water resistant, roughness, antioxidant	[68]
4	Chitosan	Nano-TiO_2	Anthocyanin	Rich black plum peel extract	pH responsive Mechanical, antibacterial, antioxidant, barrier, ethylene scavenging	[69]
5	Chitosan	TiO_2 nanoparticles	Anthocyanin	Red apple pomace extract (APE)	pH responsive Mechanical, barrier, UV barrier, thermal stability, antimicrobial	[70]
6	Chitosan	Gum Arabic nanofiber	Anthocyanin	*Rosa damascene* extract	pH responsive Mechanical, water vapor permeability, barrier, thermal	[60]
7	Alginate and guar gum	Carbon dots (CD)	Carbon dots (CD)	Glucose and glycerol	Highly pH responsive Mechanical strength, thermal, humidity, fluorescence	[71]
8	Potato starch	Carbon nanodots	Anthocyanin	*Clitoria ternatea* flower	Highly pH response Mechanical, moisture and UV barrier, thermal, antioxidant	[61]

(continued)

Table 2 (continued)

S. no.	Biopolymer	Nanofillers	Indicator used	Source of indicator	Properties	References
9	Furcellaran/glycerol	Selenium and ZnO nanoparticles	Chlorogenic acid	Green tea and yerba mate extract	pH-responsive Mechanical, thermal, UV barrier, antioxidant, antibacterial	[72]
10	κ-carrageenan/gelatin	TiO_2 nanoparticles	Anthocyanin	Saffron petals and barberry fruit	Highly pH response Mechanical, moisture, and UV barrier, antibacterial, antioxidant	[73]
11	Gelatin	Chitin nanocrystals	Anthocyanin	Black rice bran	pH-responsive UV barrier, antioxidant	[74]
12	Gelatin/chitosan	Carbon nanoparticles	Carbon dots (CD)	Chitosan precursor	pH-responsive Mechanical, structural, UV shielding, antioxidant, antibacterial	[19]
13	Zein protein	Mesoporous silica nanoparticles	Anthocyanin with tea tree essential oil	Anthocyanin extracted from blueberries	Highly pH-responsive Mechanical, antibacterial, antioxidant	[75]
14	Soy protein isolate	Copper oxide nanorods and cellulose nanofibrils	Anthocyanin	*Clitoria ternatea* flower	pH-responsive Mechanical, barrier, antibacterial, antioxidant	[76]
15	Nano-liposome/gelatin/chitosan/glycerol	Chitosan nanofibers and ZnO nanoparticles	Betanin	Commercial	Highly pH-responsive Mechanical, water barrier, antibacterial, an antioxidant	[77]
16	Bacterial cellulose	Cellulose nanofibers	Anthocyanin	Black carrot extract	pH responsive	[78]
17	Bacterial cellulose (BC)	Cellulose nanofibrils	Anthocyanin	*Echium amoenum* extract	pH responsive	[79]
18	Methyl cellulose (MC)	Chitosan nanofiber (ChNF)	Anthocyanin	Barberry extract	Highly pH-responsive Mechanical, water barrier, UV barrier	[80]

(continued)

Table 2 (continued)

S. no.	Biopolymer	Nanofillers	Indicator used	Source of indicator	Properties	References
19	Starch/PVA	TiO_2 nanoparticles	Anthocyanin	*Elderberry* extract	pH-responsive Mechanical, functional, water barrier, antimicrobial	[55]
20	Pullulan	Chitin nanofibers	Hybrid (anthocyanin and curcumin)	Commercial	Strong pH response Mechanical, thermal, antibacterial, antioxidant	[81]
21	Gellan gum	Silver nanoparticles	Silver nanoparticles	Gellan gum as a reducing agent	Highly pH-responsive for H_2S detection	[82]
22	Cellulose/chitosan	ZnO nanoparticles	Anthocyanin	Black carrot extract	High pH response Swelling and solubility	[83]
23	Polylactic acid (PLA)	TiO_2 nanoparticles	Lycopene	Tomato extract	Color change on lycopene oxidation Antimicrobial, antioxidant, and water barrier	[84]

5 Challenges with Bio-nanocomposites for Food Packaging Applications

From the discussion above, it is clear that nanotechnology and biopolymers are assisting the food packaging industry in maintaining sustainability while providing various functionalities in food packages. As bio-nanocomposites gain significant interest in academic and commercial applications, several difficulties exist. The effects on human health cannot be disregarded in this field of study since it involves compounds that directly interact with food. Thus, it is crucial to identify safety issues and environmental consequences thoroughly. The leaching of nanoparticles from packaging into food raises serious concerns for human health. The nanoscale size of the particles causes them to have physicochemical characteristics distinct from those of their macro-sized counterparts. Because toxicity concerns change with size, it is challenging to estimate toxicity profiles based on macroscopic particles because of these variances in characteristics [89, 90], and this is the reason why several bio-nanocomposites can't be certified as GRAS (generally recognized as safe).

The migration of toxic materials from nanocomposites to food involves two stages. Those nanoparticles, which are enclosed inside the surface layers, must be the cause of the first migration. The nanomaterial trapped in the specimen's interior must travel via voids and other spaces between the polymer molecules, and release following the surface nanomaterials. The release of nanomaterials depends on the polymer's density, crystallinity, degree of crosslinking, and branching. In some circumstances, the well-encapsulated nanomaterials in the film must oxidize and migrate out via the polymer matrix. These nanomaterials mainly cause the release at a later period. It suggests that the base biopolymers and nanofillers have a significant impact on the matrix of bio-nanocomposites; a slight change in concentration may change the functionality and properties required for the particular application, particularly in the case of foods where the main target of these bio-nanocomposites in active and intelligent food packaging applications is to reduce the spoilage. The anticipated functionalities of nanocomposites may occasionally vary due to specific alterations. Nanoparticle aggregation in the matrix of bio-nanocomposites may be caused by inappropriate mixing and incompatibility of the components employed. Additionally, the functional additives may lose functionality throughout the nanocomposites' preparation steps [11].

The most significant difficulty faced by food researchers and the food business is the risk evaluation of nanocomposite packaging materials [89]. Although it has been claimed that some nanoparticles may be damaging to human health, there is insufficient evidence to determine if this is related to any uses for nanomaterials, particularly in the food packaging sector. Priyadarshi et al. discussed various issues that needed to be addressed before using bio nanocomposites for a complete safety assessment [38]. These are illustrated in Fig. 5.

Several researchers have studied the advantages and disadvantages of both sides of bio-nanocomposites. Numerous studies have shown that nanoparticles are hazardous

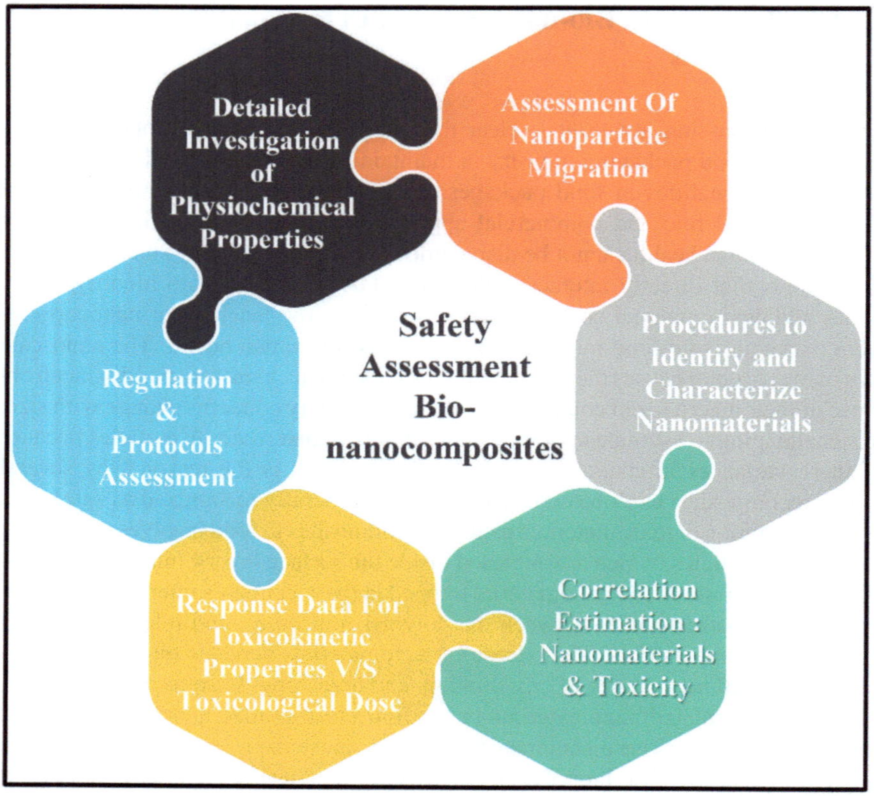

Fig. 5 Important issues regarding safety assessment of bio-nanocomposites [38]

to the environment and living things due to toxic effects. These become highly reactive because of the high surface area-to-volume ratio. These nanomaterials can also interact with the elements of cells. Nanomaterials become highly reactive because of the high surface area-to-volume ratio. In this way, human health may be impacted by the inclusion of nanoparticles in food packaging systems. Inhalation of these nanomaterials causes lung infection and sometimes heart-related issues. The size of the toxic particles plays a crucial role when we talk about their impact on living organisms. Compared to silver nanoparticles of 100 nm size, silver nanoparticles of 20 nm diameter exhibit more significant toxicity to lung tissue [91]. Several attempts have been made to overcome this issue, and encapsulation is one such attempt. The migration of nanoparticles into food can be slowed by completely encapsulating them in a polymer matrix [92]. Thus, strict regulations must be imposed by government regulation authorities of countries across the globe before permitting the production and usage of bio-nanocomposites for functional food packaging applications, especially toxic nanoparticles, which must not be allowed for food packaging applications.

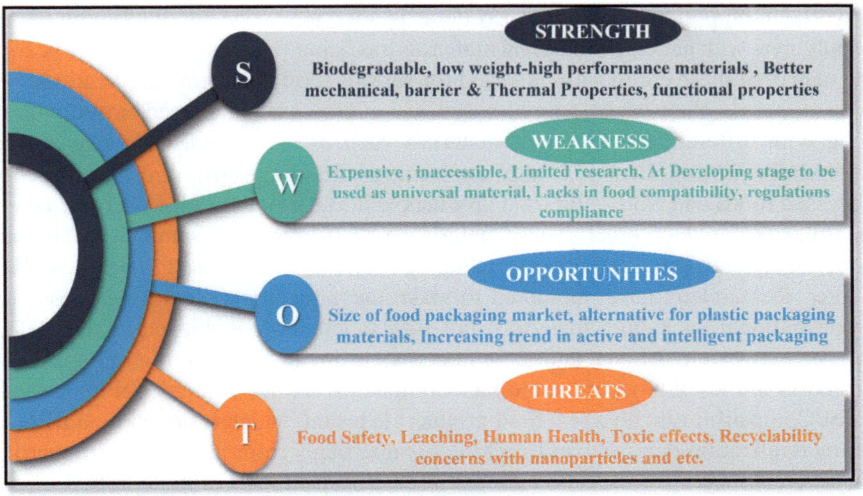

Fig. 6 SWOT analysis of bio-nanocomposites in food packaging [93]

The success of any research or trend can be analyzed by SWOT (strengths, weaknesses, opportunities, and threats) analysis to guide future research direction. The functional food packaging applications of bio-nanocomposites and the related challenges described above show that this food packaging segment is under development. Many collective efforts are needed to guide future research in this field. In such actions, Perera et al. [93] critically reviewed the use of nanomaterials in bio-nanocomposites for food packaging applications. They concluded that bio-nanocomposites commercialization needs serious attention in future research as the migration issue is still unresolved in the case of most of the nanoparticles. Also, regulations and legislation must be improved for bio-nanocomposites, as GRAS was employed when nanocomposites were not discovered. In their SWOT analysis, they highlighted these concerns and future opportunities in this study field. This analysis is illustrated in Fig. 6. for a better understanding of readers.

6 Conclusion and Future Aspects

The functional food packaging market is growing rapidly across the globe. Still, the environmental concerns raised due to the non-biodegradable nature of conventional food packaging materials such as PP, LDPE, and HDPE, are forcing packaging industries to look forward to replacing these non-ecofriendly packaging materials with materials with similar characteristics but minimal environmental impacts. This push resulted in the evolution of several new technologies in functional food packaging, and bio-nanocomposite is one of those attempts. The utilization of natural or synthetic biopolymers to pack food items is not a new concept; combining these biopolymers with nanotechnological materials has gained attention in previous years. The matrix of these composites is tested for various food packaging applications by using numerous plant-based, animal-based, and micro-organism-based bio-polymers along with different types of nanomaterials such as nanotubes, nano clays, and nanohybrids. The application area of these composites ranges from fresh produce to processed foods and from solid food products to liquid products. However, the latter one is not explored too much. These materials are used for various unique functional properties in active and intelligent packaging systems, as discussed in detail in this chapter. Researchers have attempted to make these materials as close as possible to conventional packaging materials, but still, much work is needed in this field. The gap between traditional and these materials is due to cost and migration issues. Migration studies of most of the bio-nanocomposites are not available for most food items. Compatibility with a few food products is tested, which is not enough from a commercialization perspective. Leaching nanoparticles from the composite matrix is a severe concern for metallic nanoparticles. Encapsulation of nanoparticles into biopolymers is adopted, and with the increase in encapsulation efficiency, composites have resulted in better mechanical, thermal, barrier, and functional properties. However, still, the results available are limited. Future research should study the migration of bio-nanocomposites with various food applications. Also, developed bio-nanocomposites should be investigated in detail for physiochemical properties; correlations between their characteristics and toxicity and enough accumulated data for their toxicokinetic properties and dose–response relation should be established. The researchers should test their products per standard regulations for packaging materials. As no resources are available to categorize these bio-nanocomposites as GRAS materials, different regulation authorities should update their definitions per the best available data from research in this field. The packaging industry has observed that only materials with universal qualities for a wide range of food items can thrive in the market. Therefore, for bio-nanocomposites to be an excellent replacement for traditional non-biodegradable packaging materials, future research must be able to impart universal properties in these materials for food packaging applications.

References

1. Pascall MA, DeAngelo K, Richards J, Arensberg MB (2022) Role and importance of functional food packaging in specialized products for vulnerable populations: implications for innovation and policy development for sustainability. Foods 11
2. Lopez-Rubio A, Almenar E, Hernandez-Muñoz P et al (2004) Overview of active polymer-based packaging technologies for food applications. Food Rev Int 20:357–387
3. Scott T (2022) Active packaging: what it is and why it's important—felix instruments. In: Felix
4. Ozdemir M, Floros JD (2010) Active food packaging technologies. 44:185–193. https://doi.org/10.1080/10408690490441578
5. Negi YS, Priyadarshi R, Kulshreshta DA (2019) Intelligent packaging systems for food applications. Packaging 360

6. Schaefer D, Cheung WM (2018) Smart packaging: opportunities and challenges. Procedia CIRP 72:1022–1027. https://doi.org/10.1016/J.PROCIR.2018.03.240
7. De Moura MR, Mattoso LHC, Zucolotto V (2012) Development of cellulose-based bactericidal nanocomposites containing silver nanoparticles and their use as active food packaging. J Food Eng 109:520–524. https://doi.org/10.1016/J.JFOODENG.2011.10.030
8. Sirviö JA, Liimatainen H, Niinimäki J, Hormi O (2013) Sustainable packaging materials based on wood cellulose. RSC Adv 3:16590. https://doi.org/10.1039/c3ra43264e
9. Fernandes EM, Pires RA, Mano JF, Reis RL (2013) Bionanocomposites from lignocellulosic resources: properties, applications and future trends for their use in the biomedical field. Prog Polym Sci 38:1415–1441. https://doi.org/10.1016/J.PROGPOLYMSCI.2013.05.013
10. Puiggalí J, Katsarava R (2017) Bionanocomposites. Clay-Polym Nanocompos 239–272. https://doi.org/10.1016/B978-0-323-46153-5.00007-0
11. Jayakumar A, Radoor S, Kim JT et al (2022) Recent innovations in bionanocomposites-based food packaging films—a comprehensive review. Food Packag Shelf Life 33:100877. https://doi.org/10.1016/J.FPSL.2022.100877
12. Sharma H, Sharma B, Raghuvanshi S et al (2023) Development of acrylic based antimicrobial water-based ink using curcumin for active packaging applications. Polym Eng Sci. https://doi.org/10.1002/PEN.26552
13. Janani N, Zare EN, Salimi F, Makvandi P (2020) Antibacterial tragacanth gum-based nanocomposite films carrying ascorbic acid antioxidant for bioactive food packaging. Carbohydr Polym 247:116678. https://doi.org/10.1016/J.CARBPOL.2020.116678
14. Pirsa S (2021) Nanocomposite base on carboxymethylcellulose hydrogel: simultaneous absorbent of ethylene and humidity to increase the shelf life of banana fruit. Int J Biol Macromol 193:300–310. https://doi.org/10.1016/J.IJBIOMAC.2021.10.075
15. Shaikh S, Nazam N, Mohd S et al (2019) Molecular sciences mechanistic insights into the antimicrobial actions of metallic nanoparticles and their implications for multidrug resistance. https://doi.org/10.3390/ijms20102468
16. Barra A, Ferreira NM, Martins MA et al (2019) Eco-friendly preparation of electrically conductive chitosan—reduced graphene oxide flexible bionanocomposites for food packaging and biological applications. Compos Sci Technol 173:53–60. https://doi.org/10.1016/J.COMPSCITECH.2019.01.027
17. El-Wakil NA, Hassan EA, Abou-Zeid RE, Dufresne A (2015) Development of wheat gluten/nanocellulose/titanium dioxide nanocomposites for active food packaging. Carbohydr Polym 124:337–346. https://doi.org/10.1016/J.CARBPOL.2015.01.076
18. Sanuja S, Agalya A, Umapathy MJ (2015) Synthesis and characterization of zinc oxide–neem oil–chitosan bionanocomposite for food packaging application. Int J Biol Macromol 74:76–84. https://doi.org/10.1016/J.IJBIOMAC.2014.11.036
19. Fu B, Liu Q, Liu M et al (2022) Carbon dots enhanced gelatin/chitosan bio-nanocomposite packaging film for perishable foods. Chin Chem Lett 33:4577–4582. https://doi.org/10.1016/J.CCLET.2022.03.048
20. Surendhiran D, Roy VC, Park JS, Chun BS (2022) Fabrication of chitosan-based food packaging film impregnated with turmeric essential oil (TEO)-loaded magnetic-silica nanocomposites for surimi preservation. Int J Biol Macromol 203:650–660. https://doi.org/10.1016/J.IJBIOMAC.2022.01.178
21. Ni Y, Shi S, Li M et al (2021) Visible light responsive, self-activated bionanocomposite films with sustained antimicrobial activity for food packaging. Food Chem 362:130201. https://doi.org/10.1016/J.FOODCHEM.2021.130201
22. Priyadarshi R, Kim S-M, Rhim J-W (2021) Carboxymethyl cellulose-based multifunctional film combined with zinc oxide nanoparticles and grape seed extract for the preservation of high-fat meat products. Sustain Mater Technol 29:e00325. https://doi.org/10.1016/J.SUSMAT.2021.E00325
23. Sadegh-Hassani F, Mohammadi Nafchi A (2014) Preparation and characterization of bionanocomposite films based on potato starch/halloysite nanoclay. Int J Biol Macromol 67:458–462. https://doi.org/10.1016/J.IJBIOMAC.2014.04.009

24. Yuan T, Zeng J, Wang B, Cheng Z, Gao W, Xu J, Chen K. Silver nanoparticles immobilized on cellulose nanofibrils for starch-based nanocomposites with high antibacterial, biocompatible, and mechanical properties. https://doi.org/10.1007/s10570-020-03567-y
25. Yu Z, Sun L, Wang W et al (2018) Soy protein-based films incorporated with cellulose nanocrystals and pine needle extract for active packaging. Ind Crops Prod 112:412–419. https://doi.org/10.1016/J.INDCROP.2017.12.031
26. Yang Z, Zhai X, Zhang C et al (2022) Agar/TiO_2/radish anthocyanin/neem essential oil bionanocomposite bilayer films with improved bioactive capability and electrochemical writing property for banana preservation. Food Hydrocoll 123:107187. https://doi.org/10.1016/J.FOODHYD.2021.107187
27. Yang W, Owczarek JS, Fortunati E et al (2016) Antioxidant and antibacterial lignin nanoparticles in polyvinyl alcohol/chitosan films for active packaging. Ind Crops Prod 94:800–811. https://doi.org/10.1016/J.INDCROP.2016.09.061
28. Sessini V, Arrieta MP, Kenny JM, Peponi L (2016) Processing of edible films based on nanoreinforced gelatinized starch. Polym Degrad Stab 132:157–168. https://doi.org/10.1016/J.POLYMDEGRADSTAB.2016.02.026
29. La DD, Nguyen-Tri P, Le KH et al (2021) Effects of antibacterial ZnO nanoparticles on the performance of a chitosan/gum arabic edible coating for post-harvest banana preservation. Prog Org Coat 151:106057. https://doi.org/10.1016/J.PORGCOAT.2020.106057
30. Punia Bangar S, Whiteside WS, Dunno KD et al (2022) Starch-based bio-nanocomposites films reinforced with cellulosic nanocrystals extracted from Kudzu (Pueraria montana) vine. Int J Biol Macromol 203:350–360. https://doi.org/10.1016/J.IJBIOMAC.2022.01.133
31. Kumar H, Ahuja A, Kadam AA et al (2022) Antioxidant film based on chitosan and tulsi essential oil for food packaging. Food Bioprocess Technol 1:1–14. https://doi.org/10.1007/S11947-022-02938-6/FIGURES/7
32. Kardam SK, Kadam AA, Dutt D (2021) Retention of cinnamaldehyde in poly(vinyl alcohol) films intended for preservation of faba beans through vapor-phase antimicrobial effect. Food Packag Shelf Life 29:100704. https://doi.org/10.1016/J.FPSL.2021.100704
33. Kumar A, Deshmukh RK, Gaikwad KK (2022) Quality preservation in banana fruits packed in pine needle and halloysite nanotube-based ethylene gas scavenging paper during storage. Biomass Convers Biorefinery 1:1–10. https://doi.org/10.1007/S13399-022-02708-6/FIGURES/4
34. Sharma H, Ahuja A, Sharma B et al (2023) Vapor phase antimicrobial active packaging application of chitosan capsules containing clove essential oil for the preservation of dry cakes. Food Bioprocess Technol 1–11. https://doi.org/10.1007/s11947-023-03151-9
35. Maftoonazad N, Ramaswamy H (2018) Novel techniques in food processing: bionanocomposites. Curr Opin Food Sci 23:49–56. https://doi.org/10.1016/J.COFS.2018.05.010
36. Fang Z, Bhandari B (2010) Encapsulation of polyphenols—a review. Trends Food Sci Technol 21:510–523. https://doi.org/10.1016/J.TIFS.2010.08.003
37. Souza MP, Vaz AFM, Correia MTS et al (2014) Quercetin-loaded lecithin/chitosan nanoparticles for functional food applications. Food Bioprocess Technol 7:1149–1159. https://doi.org/10.1007/S11947-013-1160-2/FIGURES/6
38. Priyadarshi R, Roy S, Ghosh T et al (2021) Antimicrobial nanofillers reinforced biopolymer composite films for active food packaging applications—a review. Sustain Mater Technol 32:e00353. https://doi.org/10.1016/J.SUSMAT.2021.E00353
39. Brandelli A, Brum LFW, dos Santos JHZ (2017) Nanostructured bioactive compounds for ecological food packaging. Environ Chem Lett 15:193–204. https://doi.org/10.1007/S10311-017-0621-7/FIGURES/2
40. De Azeredo HMC (2013) Antimicrobial nanostructures in food packaging. Trends Food Sci Technol 30:56–69. https://doi.org/10.1016/J.TIFS.2012.11.006
41. Nath D, Pal K, Sarkar P (2022) Nanoclay-based active food packaging systems: a review. Food Packag Shelf Life 31:100803. https://doi.org/10.1016/J.FPSL.2021.100803
42. Lazzara G, Cavallaro G, Panchal A et al (2018) An assembly of organic-inorganic composites using halloysite clay nanotubes. Curr Opin Colloid Interface Sci 35:42–50. https://doi.org/10.1016/J.COCIS.2018.01.002

43. Kurczewska J, Ratajczak M, Gajecka M (2021) Alginate and pectin films covering halloysite with encapsulated salicylic acid as food packaging components. Appl Clay Sci 214. https://doi.org/10.1016/J.CLAY.2021.106270
44. Roy S, Rhim J-W (2021) Effect of chitosan modified halloysite on the physical and functional properties of pullulan/chitosan biofilm integrated with rutin. Appl Clay Sci 211:106205. https://doi.org/10.1016/J.CLAY.2021.106205
45. Yang X, Zhang Y, Zheng D et al (2020) Nano-biocomposite films fabricated from cellulose fibers and halloysite nanotubes. Appl Clay Sci 190:105565. https://doi.org/10.1016/J.CLAY.2020.105565
46. Qiang X, Zhou S, Zhang Z et al (2018) Synergistic effect of halloysite nanotubes and glycerol on the physical properties of fish gelatin films. https://doi.org/10.3390/polym10111258
47. Mansour G, Zoumaki M, Marinopoulou A et al (2020) Characterization and properties of non-granular thermoplastic starch—clay biocomposite films. Carbohydr Polym 245. https://doi.org/10.1016/J.CARBPOL.2020.116629
48. Staroszczyk H, Gottfried K, Malinowska-Pańczyk E, Kołodziejska I (2017) Clay-filled starch films. Part I: Effect of clay kind and glycerol concentration on functional properties of composites. Starch Stärke 69:1500325. https://doi.org/10.1002/STAR.201500325
49. Huang R, Zhang X, Li H et al (2020) Bio-composites consisting of cellulose nanofibers and Na+ montmorillonite clay: morphology and performance property. Polymers (Basel) 12:1–13. https://doi.org/10.3390/polym12071448
50. Zheng M, Tajvidi M, Tayeb AH, Stark NM (2019) Effects of bentonite on physical, mechanical and barrier properties of cellulose nanofibril hybrid films for packaging applications. Cellulose 26:5363–5379. https://doi.org/10.1007/S10570-019-02473-2/TABLES/3
51. Ramos ÓL, Pereira RN, Cerqueira MA et al (2018) Bio-based nanocomposites for food packaging and their effect in food quality and safety. Food Packag Preserv 271–306. https://doi.org/10.1016/b978-0-12-811516-9.00008-7
52. Kalpana S, Priyadarshini SR, Maria Leena M et al (2019) Intelligent packaging: trends and applications in food systems. Trends Food Sci Technol 93:145–157. https://doi.org/10.1016/J.TIFS.2019.09.008
53. Teymouri Z, Shekarchizadeh H (2022) A colorimetric indicator based on copper nanoparticles for volatile sulfur compounds to monitor fish spoilage in intelligent packaging. Food Packag Shelf Life 33:100884. https://doi.org/10.1016/j.fpsl.2022.100884
54. Pirsa S, Sani IK, Mirtalebi SS (2022) Nano-biocomposite based color sensors: investigation of structure, function, and applications in intelligent food packaging. Food Packag Shelf Life 31:100789. https://doi.org/10.1016/j.fpsl.2021.100789
55. Jayakumar A, Radoor S, Kim JT et al (2022) Titanium dioxide nanoparticles and elderberry extract incorporated starch based polyvinyl alcohol films as active and intelligent food packaging wraps. Food Packag Shelf Life 34:100967. https://doi.org/10.1016/J.FPSL.2022.100967
56. Arora B, Bhatia R, Attri P (2018) Bionanocomposites: green materials for a sustainable future. New Polym Nanocompos Environ Remediat 699–712. https://doi.org/10.1016/B978-0-12-811033-1.00027-5
57. Alizadeh-Sani M, Mohammadian E, Rhim J-W, Jafari SM (2020) PH-sensitive (halochromic) smart packaging films based on natural food colorants for the monitoring of food quality and safety. Trends Food Sci Technol 105:93–144. https://doi.org/10.1016/j.tifs.2020.08.014
58. Pereira VA, de Arruda INQ, Stefani R (2015) Active chitosan/PVA films with anthocyanins from Brassica oleraceae (red cabbage) as time-temperature indicators for application in intelligent food packaging. Food Hydrocoll 43:180–188. https://doi.org/10.1016/J.FOODHYD.2014.05.014
59. Sun J, Jiang H, Wu H et al (2020) Multifunctional bionanocomposite films based on konjac glucomannan/chitosan with nano-ZnO and mulberry anthocyanin extract for active food packaging. Food Hydrocoll 107:105942. https://doi.org/10.1016/j.foodhyd.2020.105942
60. Shavisi N, Shahbazi Y (2022) Chitosan-gum Arabic nanofiber mats encapsulated with pH-sensitive Rosa damascena anthocyanins for freshness monitoring of chicken fillets. Food Packag Shelf Life 32:100827. https://doi.org/10.1016/j.fpsl.2022.100827

61. Koshy RR, Koshy JT, Mary SK et al (2021) Preparation of pH sensitive film based on starch/carbon nano dots incorporating anthocyanin for monitoring spoilage of pork. Food Control 126:108039. https://doi.org/10.1016/j.foodcont.2021.108039
62. Won K, Jang NY, Jeon J (2016) A natural component-based oxygen indicator with in-pack activation for intelligent food packaging. J Agric Food Chem 64:9675–9679. https://doi.org/10.1021/acs.jafc.6b04172
63. Llorens A, Lloret E, Picouet PA et al (2012) Metallic-based micro and nanocomposites in food contact materials and active food packaging. Trends Food Sci Technol 24:19–29. https://doi.org/10.1016/j.tifs.2011.10.001
64. Biji KB, Ravishankar CN, Mohan CO, Srinivasa Gopal TK (2015) Smart packaging systems for food applications: a review. J Food Sci Technol 52:6125–6135. https://doi.org/10.1007/s13197-015-1766-7
65. Raghuvanshi S, Khan H, Saroha V et al (2023) Potential of biofunctional magnetic nanoparticles in food safety: detection of toxins and foodborne pathogens. Nanotechnol Horizons Food Process Eng 353–399. https://doi.org/10.1201/9781003305408-14
66. Raghuvanshi S, Khan H, Saroha V et al (2023) Recent advances in biomacromolecule-based nanocomposite films for intelligent food packaging—a review. Int J Biol Macromol 253:127420. https://doi.org/10.1016/J.IJBIOMAC.2023.127420
67. Wu C, Zhu Y, Wu T et al (2019) Enhanced functional properties of biopolymer film incorporated with curcurmin-loaded mesoporous silica nanoparticles for food packaging. Food Chem 288:139–145. https://doi.org/10.1016/j.foodchem.2019.03.010
68. Li Y, Ying Y, Zhou Y et al (2019) A pH-indicating intelligent packaging composed of chitosan-purple potato extractions strength by surface-deacetylated chitin nanofibers. Int J Biol Macromol 127:376–384. https://doi.org/10.1016/j.ijbiomac.2019.01.060
69. Zhang X, Liu Y, Yong H et al (2019) Development of multifunctional food packaging films based on chitosan, TiO_2 nanoparticles and anthocyanin-rich black plum peel extract. Food Hydrocoll 94:80–92. https://doi.org/10.1016/j.foodhyd.2019.03.009
70. Lan W, Wang S, Zhang Z et al (2021) Development of red apple pomace extract/chitosan-based films reinforced by TiO_2 nanoparticles as a multifunctional packaging material. Int J Biol Macromol 168:105–115. https://doi.org/10.1016/J.IJBIOMAC.2020.12.051
71. Rahman S, Chowdhury D (2022) Guar gum-sodium alginate nanocomposite film as a smart fluorescence-based humidity sensor: a smart packaging material. Int J Biol Macromol 216:571–582. https://doi.org/10.1016/j.ijbiomac.2022.07.008
72. Jamróz E, Kulawik P, Kopel P et al (2019) Intelligent and active composite films based on furcellaran: structural characterization, antioxidant and antimicrobial activities. Food Packag Shelf Life 22:100405. https://doi.org/10.1016/J.FPSL.2019.100405
73. Alizadeh Sani M, Tavassoli M, Salim SA et al (2022) Development of green halochromic smart and active packaging materials: TiO_2 nanoparticle- and anthocyanin-loaded gelatin/κ-carrageenan films. Food Hydrocoll 124:107324. https://doi.org/10.1016/j.foodhyd.2021.107324
74. Ge Y, Li Y, Bai Y et al (2020) Intelligent gelatin/oxidized chitin nanocrystals nanocomposite films containing black rice bran anthocyanins for fish freshness monitorings. Int J Biol Macromol 155:1296–1306. https://doi.org/10.1016/j.ijbiomac.2019.11.101
75. Kong J, Ge X, Sun Y et al (2023) Multi-functional pH-sensitive active and intelligent packaging based on highly cross-linked zein for the monitoring of pork freshness. Food Chem 404:134754. https://doi.org/10.1016/j.foodchem.2022.134754
76. Koshy RR, Mary SK, Reghunadhan A et al (2022) Smart pH-sensitive films based on biomaterials: preparation from soy protein with hybrid fillers of CuO nanorods, cellulose nanofibrils, and Clitoria ternatea anthocyanin. JSFA Rep 2:535–547. https://doi.org/10.1002/jsf2.86
77. Amjadi S, Nazari M, Alizadeh SA, Hamishehkar H (2020) Multifunctional betanin nanoliposomes-incorporated gelatin/chitosan nanofiber/ZnO nanoparticles nanocomposite film for fresh beef preservation. Meat Sci 167:108161. https://doi.org/10.1016/j.meatsci.2020.108161

78. Moradi M, Tajik H, Almasi H et al (2019) A novel pH-sensing indicator based on bacterial cellulose nanofibers and black carrot anthocyanins for monitoring fish freshness. Carbohydr Polym 222:115030. https://doi.org/10.1016/j.carbpol.2019.115030
79. Mohammadalinejhad S, Almasi H, Moradi M (2020) Immobilization of Echium amoenum anthocyanins into bacterial cellulose film: a novel colorimetric pH indicator for freshness/spoilage monitoring of shrimp. Food Control 113:107169. https://doi.org/10.1016/j.foodcont.2020.107169
80. Alizadeh-Sani M, Tavassoli M, Mohammadian E et al (2021) pH-responsive color indicator films based on methylcellulose/chitosan nanofiber and barberry anthocyanins for real-time monitoring of meat freshness. Int J Biol Macromol 166:741–750. https://doi.org/10.1016/j.ijbiomac.2020.10.231
81. Duan M, Yu S, Sun J et al (2021) Development and characterization of electrospun nanofibers based on pullulan/chitin nanofibers containing curcumin and anthocyanins for active-intelligent food packaging. Int J Biol Macromol 187:332–340. https://doi.org/10.1016/j.ijbiomac.2021.07.140
82. Zhai X, Li Z, Shi J et al (2019) A colorimetric hydrogen sulfide sensor based on gellan gum-silver nanoparticles bionanocomposite for monitoring of meat spoilage in intelligent packaging. Food Chem 290:135–143. https://doi.org/10.1016/j.foodchem.2019.03.138
83. Tirtashi FE, Moradi M, Tajik H et al (2019) Cellulose/chitosan pH-responsive indicator incorporated with carrot anthocyanins for intelligent food packaging. Int J Biol Macromol 136:920–926. https://doi.org/10.1016/j.ijbiomac.2019.06.148
84. Asadi S, Pirsa S (2020) Production of biodegradable film based on polylactic acid, modified with lycopene pigment and TiO_2 and studying its physicochemical properties. J Polym Environ 28:433–444. https://doi.org/10.1007/s10924-019-01618-5
85. Ma W, He S, Ma H et al (2020) Silver nanoparticle exposure causes pulmonary structural damage and mitochondrial dynamic imbalance in the rat: protective effects of sodium selenite. Int J Nanomed 15:633–645. https://doi.org/10.2147/IJN.S232986
86. Hans M, Bansal R, Nayik GA et al (2022) Potential risks, health safety features, and public acceptance of nanoparticles in packaging. Nanotechnol Interv Food Packag Shelf Life 283–294. https://doi.org/10.1201/9781003207641-16
87. Ali SA, Gooda SM, Aboul Naser AF et al (2022) Chromosomal aberrations, DNA damage, and biochemical disturbances induced by silver nanoparticles in mice: role of particle size and natural compounds treatment. 27:349–360. https://doi.org/10.1080/1354750X.2022.2046856
88. Lorite GS, Rocha JM, Miilumäki N et al (2017) Evaluation of physicochemical/microbial properties and life cycle assessment (LCA) of PLA-based nanocomposite active packaging. LWT Food Sci Technol 75:305–315. https://doi.org/10.1016/j.lwt.2016.09.004
89. Huang JY, Li X, Zhou W (2015) Safety assessment of nanocomposite for food packaging application. Trends Food Sci Technol 45:187–199. https://doi.org/10.1016/j.tifs.2015.07.002
90. Sharma R, Jafari SM, Sharma S (2020) Antimicrobial bio-nanocomposites and their potential applications in food packaging. Food Control 112:107086. https://doi.org/10.1016/j.foodcont.2020.107086
91. Azeredo HMC, Otoni CG, Corrêa DS et al (2019) Nanostructured antimicrobials in food packaging-recent advances. https://doi.org/10.1002/biot.201900068. www.biotechnology-journal.com
92. Störmer A, Bott J, Kemmer D, Franz R (2017) Critical review of the migration potential of nanoparticles in food contact plastics. Trends Food Sci Technol 63:39–50. https://doi.org/10.1016/J.TIFS.2017.01.011
93. Perera KY, Jaiswal S, Jaiswal AK (2022) A review on nanomaterials and nanohybrids based bio-nanocomposites for food packaging. Food Chem 376:131912. https://doi.org/10.1016/J.FOODCHEM.2021.131912

Migration and Safety Issues in Food Packaging Bionanocomposites

Tanja S. Petrović, Ana S. Salević-Jelić, Senka Z. Popović, Nevena M. Hromiš, and Danijela Z. Šuput

Abstract Biopolymers have great potential for use in the production of food packaging. They are renewable, biodegradable, widely available, non-toxic, and inexpensive, but possess poor mechanical, thermal, and barrier properties which limit their application on a large scale. Therefore, bionanocomposites aim to improve functional characteristics by incorporating small amounts of nanofillers into the bio-based matrix. In addition, these composites can be tailored to enable active and intelligent packaging which can help prolong the shelf life of food and ensure traceability throughout the supply chain. Depending on their size and morphology, commonly used nanofillers are divided into three groups: one-dimensional 1D nanomaterials (nanofibers, nanotubes, and nanocrystals), two-dimensional 2D nanomaterials (layered nanomaterials) and three-dimensional 3D nanomaterials (nanoparticles). Although these substances are generally recognized as safe (GRAS), there is a potential risk that they could migrate from the packaging material into the food and have a negative impact on consumer health. Therefore, this chapter discusses the nature and properties of the main nanofillers used to manufacture bionanocomposites, migration issues, and factors affecting their potential release from bionanocomposites. In addition, an overview of the regulations and viewpoints of food safety authorities and scientific perspectives on the migration of nanoparticles into food is provided.

Keywords Bionanocomposites · Nanofillers · Food packaging materials · Migration · Food safety

---

T. S. Petrović (✉) · A. S. Salević-Jelić
Faculty of Agriculture, Institute of Food Technology and Biochemistry, University of Belgrade, Nemanjina 6, 11080 Zemun-Belgrade, Serbia
e-mail: tpetrovic@agrif.bg.ac.rs

S. Z. Popović · N. M. Hromiš · D. Z. Šuput
Faculty of Technology Novi Sad, University of Novi Sad, Bulevar Cara Lazara 1, 21000 Novi Sad, Serbia

M. Alle et al. (eds.), *Functional Nanomaterials and Nanocomposites for Biodegradable Food Packaging*, Smart Nanomaterials Technology,
https://doi.org/10.1007/978-981-97-9084-5_12

1 Introduction

Nanotechnology offers great potential for use in different areas of the food sector, particularly in food packaging. In the middle of the twentieth century, researchers were intensively engaged in the study of various natural structures in living organisms in which organic and inorganic components are linked together and arranged hierarchically, from the nano to the macro-size. These structures include pearls and shells, ivory, bone, enamel, and dentin in teeth, whereupon nanocomposite materials have been developed [1].

Based on the realization that bionanocomposites can replace synthetic polymers with natural, renewable, abundant, and inexpensive biopolymers, materials science, engineering, and nanotechnology combine in a multidisciplinary manner with the aim of expanding the application possibilities of biopolymer materials. The term bionanocomposite refers to structures composed of a bio-based polymer matrix and organic/inorganic fillers that have at least one nanoscale (1–100 nm) dimension [1]. Due to their nano size and their large surface area to volume ratio, nanofillers exhibit specific properties such as adsorption capacity and biological effectiveness [2, 3].

Nowadays, there is a growing demand for minimally processed foods with extended shelf life, which has led to increased interest in developing new packaging materials with desirable barrier properties against gases, water vapor, and flavors [3]. When using biopolymers, the reduction of water vapor permeability is crucial due to their polar nature. However, bionanocomposites can overcome this disadvantage by containing nanofillers that can significantly improve the thermal, barrier, and mechanical properties of bio-based materials. Nanofillers can either be used as direct antioxidant and antimicrobial agents or as carriers for the protection and controlled release of other active ingredients. In addition, these materials can be used as nanosensors (temperature, gas, moisture, freshness, pathogen detection, etc.) that inform consumers about the condition and safety of food products [4]. In this way, the range of applications for biopolymer materials is expanding at the expense of conventional plastics.

The major concern with the use of nanofillers is their nano-size, as the physical and chemical characteristics of these small materials can be very distinct from those of their macroscopic counterparts, with potential risk to human health and the environment [5]. Many factors can affect migration, such as the characteristics of the nanofillers and packaging materials, chemical composition of food, environmental conditions, surface area of packaging in contact with food, and contact time [6]. Furthermore, these risks arise from the absence of sufficient understanding of the interactions of nanofillers with the human body at the cellular and molecular levels. Thus, the safety assessment of nanofillers in food packaging can be of great importance [7, 8]. In addition, more attention should be paid to the development and harmonization of regulations, guidelines, and standards for the use of bionanocomposites based on comprehensive research and studies.

Hence, the purpose of this chapter is to provide information on the migration of nanofillers from bionanocomposite materials used in food packaging and on factors influencing the migration and toxicological behavior of these materials. Additionally, safety, regulatory, and legal issues, as well as scientific perspectives on the migration of nanosubstances are discussed.

2 Bionanocomposites for Food Packaging

The commercialization of nanotechnologies, has led to various applications of bionanocomposites in food packaging, opening up a promising field in the food industry. Bionanocomposites are moving into the spotlight due to the ongoing search for innovative, environmentally friendly, green, and high-performance packaging materials. Biopolymers can be obtained from plant and animal sources and isolated from microorganisms such as polyhydroxyalkonoates (PHA) and bacterial cellulose (BC). Bio-based matrix used for the preparation of bionanocomposites is diverse and include polysaccharides (starch, cellulose, and its derivatives, agar, alginate, chitosan, carrageenan, etc.), proteins (myofibrillar proteins, wheat gluten, lectins, gelatin, collagen, whey protein, corn zein, soybean, and sunflower protein), synthetic biopolymers (polylactic acid—PLA, polycaprolactone—PCL, polyhydroxyalkanoates—PHA), and biopolymer composites and bi-layer films: chitosan/sodium alginate film, gelatin/chitosan film, polyurethane/chitosan film, and others [1, 9].

The commonly used nanofillers in bionanocomposites are different, and their development is continuous. According to their dimensions and morphology, they are divided into three groups: one-dimensional 1D nanomaterials (nanofibers, nanotubes, nanocrystals), two-dimensional 2D nanomaterials (layered nanomaterials), and three-dimensional 3D nanomaterials (nanoparticles). One-dimensional (1D) nanomaterials are generally defined as linear structures less than 100 nm in diameter that form wires, tubes, rods, fibers, and bands. The preparation of 1D nanostructure mainly includes hydrothermal, vapor deposition, sol–gel, template, and electrospinning methods [10]. Examples of this group of nanomaterials for food packaging include nanofibers (cellulose and chitin nanofibers), nanotubes (halloysite nanotubes (HNT), carbon nanotubes (CNT): single wall and multiple walls, titanium dioxide nanotubes (TNT)) and nanocrystals (cellulose and chitin nanocrystals, or these nanocrystals oxidized by TEMPO). It is known that carbon nanotubes can significantly improve various properties of composite materials: electrical, mechanical, thermal, physical, and photocatalytic properties. The extraordinary advantages of carbon nanotubes can only be exploited if the nanotubes are appropriately distributed in the composite materials (Fig. 1).

However, the appropriate distribution of carbon nanotubes is difficult to achieve due to the formation of clusters and aggregates in most solvents [11]. Titanium dioxide nanotubes (TNT) are characterized by stability, ethylene scavenging activity, photocatalytic activity, as well as biocompatibility. TNT's photocatalytic activity leads to antimicrobial activity. Under UV light, TNTs generate reactive oxygen species which can cause bacterial DNA fragmentation, protein denaturation, and cell wall interruption. TNTs modified with CuO extend the photocatalytic activity to

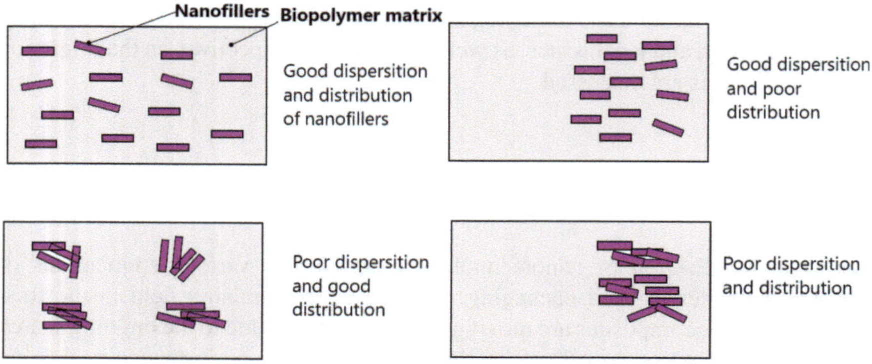

Fig. 1 Schema of possible dispersion and distribution of nanofillers in bio-based matrix

visible light. It was reported that the addition of CuO nanoparticles into carrageenan can lead to improved antibacterial and mechanical strength and water vapor barrier properties of the obtained film [12]. Carbon-coated titanium dioxide nanotubes combine a high surface-to-volume ratio of TiO_2 nanotubes with an enhanced interaction with organic compounds of the carbon coating in synergy [12, 13].

Two-dimensional nanomaterials (2D nanomaterials) are built of one to a few atom layers. They are produced using one of two approaches: the top-down method, in which synthesis starts from bulk materials and the layers are peeled off to obtain ultra-thin 2D layers, or the bottom-up method, in which atoms are bonded together to create an ultra-thin 2D layer of nanomaterials. These materials include graphene and its derivatives, metal–organic frameworks (MOFs), boron nitride (BN), and layered double hydroxides (LDH). They are characterized by anisotropy in the three-dimensional view, which enables them to align in different directions. This anisotropy can either reduce the gas molecules mobility in the bionanocomposite material or increase the diffusion path, resulting in improved gas barrier. In addition to high anisotropy, 2D nanomaterials are also characterized by functional properties such as large surface area, large diameter-length ratio, high mechanical strength and flexibility, ability to intercalate different species, the possibility of surface functionalization, excellent light transmittance [9, 14].

Different nanomaterials in combination result in hybrid nanomaterials that have been developed to take advantage of the good properties of the individual materials. Representatives of these materials include Ag/graphene nanohybrids and clay/TiO_2 nanohybrids [15].

3D nanomaterials are materials with all three dimensions in the nanoscale range. Inorganic 3D nanomaterials, metallic and nonmetallic nanoparticles include sulfur, silver, titanium dioxide (TiO_2), zinc oxide (ZnO), calcium oxide (CaO), silicon dioxide (SiO_2), magnesium-oxide (MgO), and copper sulfide (CuS). Their stability combined with antibacterial properties makes them ideal for use in active packaging films that extend the shelf life of products [16]. In this group, silver nanoparticles showed the most pronounced antimicrobial effect against a wide spectrum of microorganisms, including bacteria, as well as viruses, yeasts, and molds [17]. By enriching Fe_3O_4 nanoparticles with Ag^+ ions, a more environmentally friendly active nanomaterial with strong antimicrobial activity was developed for the production of carrageenan-based active packaging films [18]. Nano-TiO_2 and nano-ZnO particles have a photocatalytic activity which allows them to degrade ethylene and maintain fruit freshness of the fruit by activating the antimicrobial activity of the reaction products. Silica nanoparticles (SiO_2) with a large surface area to volume ratio have been shown to increase the stability of active molecules in the formation of active packaging. The addition of Ca^{2+} ions can cause a strong cross-linking reaction [9, 19].

Thermal properties of packaging materials are crucial when exposed to extreme temperatures, during food preservation and storage. Reported benefits of using nanoparticles as nanofillers in bionanocomposite films include increased thermal stability and improved mechanical properties. Reports on tensile strength vary in the literature, with some reports observing a decrease in tensile strength, likely due to nanoparticle agglomeration phenomena that can lead to the loosening of the bionanocomposite structure [19–22]. Barrier properties, which are important for protecting foods from photodegradation and oxidation during storage and transportation, are also improved by the incorporation of nanoparticles, as reported in several studies [23–26]. This was likely due to the hydrophobic nature of the nanoparticles and their interactions with the biopolymers, as well as the development of zigzag trajectories through low concentrations of the nanoparticles. However, the barrier properties of bionanocomposites may also deteriorate due to the aggregation of nanoparticles at high concentrations. As the concentration of nanoparticles increases, formed agglomerates destroy the polymer matrix, resulting in a loose structure and allowing gas molecules to penetrate the bionanocomposite film [20, 27, 28]. Nanosensors are being developed for intelligent food packaging that not only detects food contamination or spoilage (by detecting microbial contaminants, pesticides, and toxins in products) but can also dose antimicrobial agents of a nanoscale to extend the shelf life of food [29].

Bionanocomposites can be effectively improved by the addition of non-toxic organic nanoparticles. Examples of organic nanoparticles that have been successfully applied in bionanocomposites include: chitosan nanoparticles, nanocellulose, curcumin, essential oil nanoemulsions, and melanin nanoparticles [9].

3 Migration of Nanofillers

Migration occurs when small molecules from packaging materials unexpectedly migrate into food [30]. The formulation of bionanocomposites mainly uses biodegradable, edible, or biocompatible polymers, and thus their potential risk in this regard is low. Concerns exist regarding the conversion of bulk to nano-size and an insufficient understanding of the outcoming effects [31]. In this context, a risk assessment of the potential migration of these substances into packaged foods and beverages and the influence of released compounds on human health, animals, and

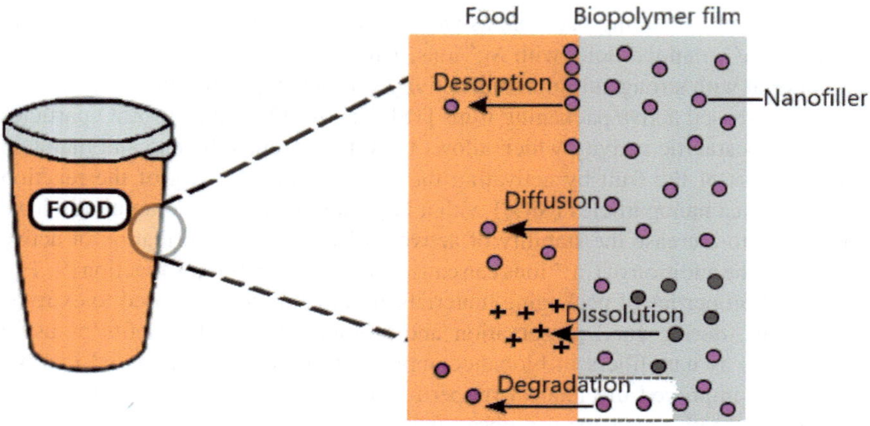

Fig. 2 Four possible mechanisms of migration (desorption, diffusion, dissolution, and degradation) of nanofillers from packaging materials into the food were suggested by Noonan et al. [33]

the surroundings is of immense importance [32]. Figure 2 illustrates the release of nanofillers from the polymer matrix by four potential mechanisms: desorption, diffusion, dissolution, and matrix degradation, as suggested by Noonan et al. [33]. Migration by desorption appears through the unforced migration of nanofillers that adhere to the interface between the polymer and the food. Diffusion is the migration of nanofillers positioned inside the nanocomposite material into the food or other areas of low concentration, controlled by their entropically driven mass. Nanofillers located inside the bionanocomposite or on its surface can migrate through dissolution processes by easily transforming into ions, which are then dispersed in the food. The migration of nanofillers by matrix degradation would occur in the case of transformation and decomposition of the biopolymer matrix so that the nanofillers are no longer firmly bound to the interior of the matrix.

In addition, foods in contact with nanocomposite packaging may experience undesirable sensory changes, such as 'off' odors and flavors caused by rancidity induced by the migration of TiO_2 into the lipid food [34]. Therefore, the risk assessment of bionanocomposite packaging materials presents a unique challenge for food safety [35]. The best way to perform migration testing is to use real food samples. However, due to the complex composition and small amounts of nanofillers used in the production of bionanocomposites (typically around 2% by volume), it is challenging to directly measure the amount of substances that migrate from bionanocomposites into food [7]. The optional method can be the use of food simulants to measure the migration of various packaging materials into food. The migration of nanofillers into food or food simulants can be evaluated using standard migration tests, which are standardized methods in Europe and the U.S. (FDA) for evaluating migration from food contact materials (FCMs) and are performed with various food simulant solutions. The European Commission (EC) has established regulations specifying conditions for migration testing [36]. Distilled water, acetic acid (3% v/v), ethanol

(10–50% v/v), and vegetable oil are usually indicated as aqueous, acidic, and fatty food simulants, respectively, depending on the type of food product in contact with the packaging. Although these tests were developed for plastic materials, they can also be applied to other materials [36]. However, several issues can complicate the determination and interpretation of the results, such as the uncertainty of analytical techniques for the detection of nanoparticles in food or food simulants. Standard migration tests are based on contact between packaging materials and food simulants under specific time and temperature conditions, selected according to the type of food packaged. These methods are used to measure the 'overall' or 'specific migration' which is compared with the respective thresholds. The overall migration limit (OML) is the maximum allowable amount (mg/kg) of all non-volatile substances that can be released from a material or article into food simulants. These substances do not pose a major risk to consumer health but are still undesirable, while the specific migration limit (SML) defines the migration (mg/kg) of a specific, single substance that can pose a risk to consumer health even in low concentrations and therefore requires strict control. During the migration test, only those parts of the sample should come into contact with the simulant that is intended to come into contact with food during actual use. In the migration testing, a worst-case scenario for migration and a highly probable scenario, different factors such as storage time, storage temperature, nanofiller content, and nanofiller size, should be investigated for their effect on the migration of nanofiller. It should be indicated that natural light can induce the degradation of polymer matrix by photooxidation reactions which accelerate migration [37]. In addition, other components of packaging material can also either accelerate or delay the migration of nanosubstances [36]. In general, most migration studies have used food simulants but there are insufficient studies on the migration of nanoparticles into real food.

The discussion on the potential risks of nanoparticles for consumers requires a thorough insight into their properties, such as particle dispersion, changes in the main matrix, and the nature of the particle-polymer interface [8, 38]. Due to the structural properties of nanofillers, their detection in a complex matrix such as food and food simulants can be highly ambiguous [39, 40]. Nanofillers exhibit poor dispersion stability when dispersed in an unsuitable medium. Detecting the migration of nanomaterials in food matrices requires more sensitive analytical techniques. This is due to the complexity of nanomaterials and the fact that they represent a very small fraction of the bulk food or simulant, with the amount in samples after migration testing being in the trace or ultra-trace range. Therefore, the detection of these substances requires a combined methodology, as no single technique can provide all the detailed information. The nature of the surrounding matrix largely determines whether the nanofillers remain stable in the dispersion or are "lost" through particle dissolution or sedimentation.

Spectroscopic, microscopic, and quantitative analytical techniques are useful for these investigations. Spectroscopic techniques such as UV–Vis spectroscopy and X-ray diffraction spectroscopy can be used for the qualitative determination of nanofillers. Transmission electron microscopy (TEM), scanning electron microscopy

(SEM), and atomic force microscopy (AFM) can be used to determine the properties of nanoparticles such as size, shape, structure, dispersion, and coagulation state. Spectroscopic techniques such as X-ray diffraction (XRD) and UV–Vis spectroscopy can also be used to characterize and analyze nanomaterials (qualitative determination). Dynamic Light Scattering (DLS), single particle Inductive Couple Plasma Mass Spectroscopy (sp-ICP-MS), Field Flow Fractionation (FFF), Field Flow Fractionation-Inductive Couple Plasma Mass Spectroscopy (FFF-ICP-MS) can also be used for size determination. For quantitative determination, there are only a few techniques that enable sensitive and selective detection and characterization of nanofillers. In addition, the migration of NPs can be analyzed using inductively coupled plasma mass spectrometry (ICP-MS), inductively coupled plasma atomic emission spectrometry (ICP-AES), and inductively coupled plasma optical emission spectrometry (ICP-OES), which are three important techniques for quantification and elemental analysis [5, 41]. The advantages of these methods include their high selectivity, sensitivity, and accuracy. These methods can detect nanofillers in quantities as low as 0.1–10 ppm [8].

The European Commission has published a Union list of approved substances for use in the manufacture of polymeric food contact materials [36, 42]. The nanofillers on the list include, for example, nano-clay, titanium nitride, nano-silver, silanated silicon dioxide, titanium oxide, zinc oxide, and iron oxide.

3.1 Factors Affecting Migration

Many factors can influence the migration of nanomaterials into food, as shown in Fig. 3. These factors can be related to the properties of the nanofillers [43] and packaging materials [44], food characteristics [45], and environmental conditions [46]. However, they may also be related to the position of the nanofillers in the packaging material, the size of the contact area, the interaction between the nanofillers and the FCMs, the kinetics and thermodynamics of the migration process, and the contact time [6, 33, 47].

For example, higher temperatures stimulate greater mobility of molecules in packaging, and lower food pH increases the solubility of nanofillers in aqueous solution, leading to increased migration [48]. Low molecular weight polymers, which have a larger free volume, also accelerate the migration rate and diffusivity of nanofillers [49]. Furthermore, there is an inverse relationship between the migration rate of the system and the size of the nanofillers [50]. Some food ingredients, especially fat, can be absorbed into packaging materials when they combine with the polymer matrix, increasing the space between the polymer chains, which accelerates the migration rate [8]. Due to their high surface activity in relation to their volume, nanomaterials can trigger undesirable chemical reactions. This would be unfavorable if their presence in the packaging promotes the formation of unintended reactions and by-products during processing [51]. Nanofillers can also influence the migration of other packaging components (monomer residues, additives, impurities) by either accelerating

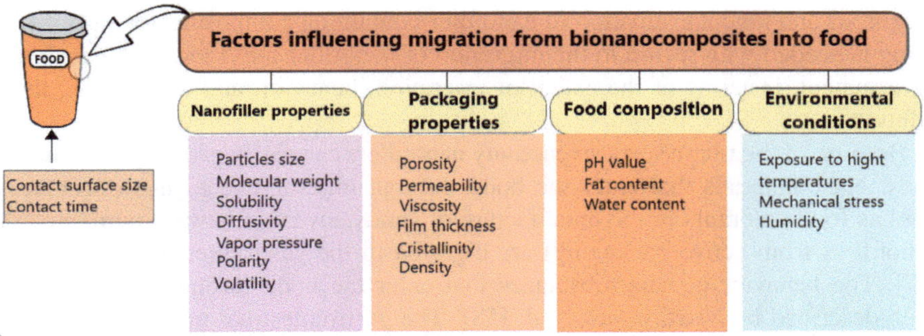

Fig. 3 The main factors influencing the migration of nanofillers from bionanocomposites into food products

or slowing down their migration. Food preservation techniques can also provoke the migration of nanofillers, and it has been published that the use of microwave heating caused a structural change in FCMs, intensifying the migration of silver ions [52].

3.2 *Toxicological Aspects*

In most cases, the effects of nanofillers on human health and the environment are not yet fully understood, leading to concerns about their safety. The main concern is the physiological exposure of humans to engineered nanocomposites and the nanofillers that may migrate from the packaging. In the case of migration from packaging or FCMs, it is possible that nanofillers, such as nanoparticles, can easily penetrate cellular barriers due to their three-dimensional nano-size and exceptional physicochemical reactivity, which could lead to damage or even toxicity [53, 54]. The possible formation of NIAS (not-intentionally added substances) during nanocomposite processing should also be investigated [31].

The behavior of nanofillers released from bionanocomposites into the environment and their possible bioaccumulation/bioconcentration in the food chain have also not been sufficiently studied [55]. In addition, the high surface-to-volume ratio of nanofillers may increase their reactivity and trigger interactions with other materials during disposal and recycling [56] or affect microbial populations when disposed of as waste [2]. Due to different physical, chemical, and biological properties and the small size of nanofillers they may differ greatly compared to their macroscale counterparts, so that findings from toxicity studies and migration limits for the conventional form of larger particles cannot be directly transferred to analogous nanoparticles. Therefore, the use of substances with "generally recognized as safe" (GRAS) status in the development of bionanocomposites may be questioned as their risk assessment was conducted without a nanoscale safety evaluation [8]. Consequently, a rigorous and urgent evaluation of nanofiller-containing composites for food packaging is needed with regard to the migration of nanofiller into food products or the environment, in terms of the extent of migration, behavior, and effects, i.e., their safety.

Because of their high surface reactivity nanofillers can easily penetrate the human body. Nanomaterials that enter the body through ingestion (e.g., nanosubstances used as food colorants in the manufacture of candy, sweets, chewing gum, etc., or nanofillers from active packaging) are digested in the gastrointestinal tract—GIT [57]. The behavior of nanosubstances that enter the body through ingestion has been described by Martirosyan et al. [58]. The gastrointestinal tract (GIT) and its mucous membranes should play the role of a selective barrier to systemic exposure to substances, including particulates. In this case, nanosubstances may remain in the intestinal lumen, possibly with the potential to interact with the GIT surfaces or the inhabitants of the lumen (e.g. the gut microbiota), but are essentially completely excreted from the body via the feces [5]. Furthermore, it is possible for nano-sized molecules to enter the body through various pathways, spread throughout all body organs, and damage human cells by altering mitochondrial function, increasing membrane permeability, and leading to toxic effects and chronic diseases such as various inflammations, allergies, bronchitis, asthma, cardiovascular problems and cancer [59]. Depending on their composition and structure, nanomaterials can generate toxicity in cells via a variety of different mechanisms. This toxicity is usually influenced by their physical characteristics such as size, shape, surface area, density, and chemical composition. The most important factor contributing to toxicity is the ability of nanomaterials to generate ROS, which can interact with lipids, proteins, or nucleic acids and damage cell membranes, organelles, and the cell nucleus [60–62].

Therefore, their toxicological properties need to be determined, which requires physicochemical characterization and in vitro and in vivo testing [63, 64]. It is also necessary to determine whether and how much of the nanofillers migrate into the packaging when they come into contact with food and the effects of their uptake on GIT and other organs. However, available information is limited, and further research in this area is urgently needed.

4 Safety and Regulations Issues

Given the positive impact and promising potential of nanocomposite packaging materials alongside the (eco)toxicological concerns, it is crucial to establish regulations, legislation, and guidelines on their application and safety. This would be a driving force for promoting cutting-edge research, further development, public acceptance, market introduction, and commercialization of these materials. The Food and Agricultural Organization (FAO) of the United Nations, the World Health Organization (WHO), the European Union (EU), the Food and Drug Administration (FDA) of the United States, as well as various governmental and nongovernmental organizations globally, have made efforts to establish regulations and legislation on nanotechnology for food applications [41]. However, no regulations and legislation exclusively address bionanocomposites for use in food packaging. Still, some existing regulations may apply to these materials. The Food and Agriculture Organization (FAO) of the United Nations and the World Health Organization (WHO) paid attention to the considerable opportunities concerning nanotechnology applications in developing various innovative products in the food and agriculture sectors, including food packaging and contact materials. This is reflected in the FAO/WHO technical paper that presents national and international initiatives and activities on risk analysis, including risk assessment, management approaches, and strategies for dealing with potential hazards of nanotechnology-related products, including nanomaterials [65].

The International Organization for Standardization (ISO) has created Technical Committees (ISO/TC 229) for standardization dedicated to nanotechnologies [66]. The standardization focuses on understanding and controlling the properties of the nanoscale substances, materials, and processes, aiming to enable the design and development of enhanced materials devices, and systems. Accordingly, it aims to develop standards for nomenclature, instrumentation, methodology, reference materials, and simulations, and to establish health, safety, and environmental practices based on science.

The European Food Safety Authority (EFSA) released "Guidance on risk assessment of the application of nanoscience and nanotechnologies in the food and feed chain: Part 1, human and animal health" [67]. Among the various areas of nanomaterial applications, this guidance also covers nanocomposites for the application as FCM, which implies that the exposure to a nanomaterial occurs via migration or transfer from food contact material into food. It addressed detailed physicochemical testing and characterization of nanomaterials, assessment of a contact to a nanomaterial according to the experimental information on the migration into food, and hazard identification and characterization of nanomaterials. The guidance also emphasizes that migrating species should be qualitatively and quantitatively determined, alongside the determination of their form (nanoparticulate or solubilized/degraded).

The European Commission set up Regulation (EU) No 10/2011 on plastic materials and articles intended to come into contact with food [36]. Aiming to ensure the safe use of food contact materials for human health, this regulation provides specific migration limits of certain substances, i.e., their maximum permitted amounts that migrate from a material into a food product or media specified as food simulants. If the risk assessment does not imply lower limits, an overall limit of maximum migration, (10 mg of migrated substances/1 dm^2 of food contact surface area) shall apply for good manufacturing practice. Concerning a unit of cubic packaging with 1 kg of food, the set limit of overall migration corresponds to the migration of 60 mg/1 kg of food. This migration limit is set according to the assumption that a person of 60 kg body weight consumes daily 1 kg of food packaged in a cubic package with a surface area of 6 dm^2 from which substances migrate. However, this regulation accentuates that nanoparticles may exhibit different physical and chemical properties than those on a larger scale, and thus the risk assessment of these substances has to be performed by the Authority (the European Food Safety Authority) case-by-case. The requirement to assess nanoparticles' risk case-by-case is supported by Commission Regulation (EC) No. 450/2009 on active and intelligent materials and articles intended to come into contact with food [68]. On the other hand, the concept of a functional barrier for active and intelligent materials is not specified for nanoparticles. Accordingly, Regulation (EU) No. 10/2011 [36] specified that the authority should assess the risk of a migrated substance considering the most adverse use and toxicity. The risk assessment should include the substance properties and purity, possible reactions, and degradation products formed during the use. As an outcome, the authorization would establish the specification concerning the use of the substance, restrictions, and migration limits, i.e., safe use of the resulting material.

In recent years, the amendments and annexes to the Regulation, based on scientific evidence, have introduced the use of substances manufactured in nanoform. Thus, the use and incorporation of some nanoparticles have been approved for food contact materials [69–71]. For example, the nanoforms of copolymers methacrylic acid, ethyl acrylate, n-butyl acrylate, methyl methacrylate, and butadiene are considered safe for consumers as an additive at levels up to 15% (w/w) in non-plasticized polylactic acid (PLA) for long-term contact with all categories of food at room temperature or lower [69]. In addition, nanoparticles of synthetic amorphous silicon dioxide, silanated, are authorized for materials in the form of aggregates with sizes >100 nm [70]. These nanoparticles can be used as fillers, resulting in improved thermal, mechanical, gas, and light barrier properties of the materials [2]. The Authority has also approved zinc oxide nanoparticles, uncoated and coated with [3-(methacryloxy)propyl]trimethoxysilane, in plasticizer-free polymers. A specific migration limit (0.05 mg/kg) was set up in the case of the coated zinc oxide nanoparticles [70]. These nanoparticles could be employed as UV light-absorbing agents for plasticizer-free polymers [2].

Montmorillonite clay modified with hexadecyltrimethylammonium bromide is an approved consumer-safe constituent up to 4% (w/w) in PLA food contact materials for water stored at ambient temperature or lower. These nanoparticles (<100 nm) dispersed in PLA are fully embedded and are not expected to migrate [71]. Another additive approved as safe for consumers in nanoform for food contact materials is titanium dioxide surface-treated with fluoride-modified alumina. The Authority allowed the use of these nanoparticles at levels up to 25% (w/w), supported by compliance verification [71].

The regulation of the United States specifies that the status of materials intended for food contact applications depends directly on each contained substance, i.e., its regulatory status [72]. In this regard, any substance expected to migrate from a food contact material into food has to be covered by a regulation listed in Title 21 Code of Federal Regulations; the conditions to have GRAS status; an authorization letter; request for regulation exception; or a statement on the effectiveness of food contact substance. In particular, the substance manufacturer has to ensure the compliance of the composite formulations and food contact materials with authorization concerning the substance's identification, specifications on its properties, as well as limitations for use. The United States Food and Drug Administration (FDA) has also established guidance for the regulation of products associated with the use of nanotechnology that does not address or presuppose the ultimate regulatory outcomes for particular nanotechnology products [73]. This suggests that explicit matters related to nanotechnology articles covering safety, efficacy, impact on public health, or regulatory aspects should be considered case-by-case at the initial development stage based on the specific scientific and regulatory matter and the FDA's existing review processes.

There is also a standard established by the People's Republic of China on materials intended for food contact applications [74]. This standard also emphasizes the importance of controlling and standardizing the migration of substances from materials in contact with food. It specifies allowed additives for food contact use, along with the corresponding requirements of their use, the maximum usage quantity, and both specific and total migration limits of food contact materials into food or food simulants. Similar to the EU regulation, the quantity of a migrating substance could be expressed in mg/kg of food or food simulant medium, either mg/dm^2 of the area being in contact between food contact material and food or food simulant. However, this standard did not include specific regulations for a nanoform.

The Korean Ministry of Food and Drug Safety has set standards and provisions dedicated to various food contact items and articles, including utensils, containers, and packaging [75], intended to promote the safe distribution of these items, and the safety of consumer health. To this end, the methods of migration testing are given describing the procedures and migrant specifications. Of note, the migration of individual substances or overall migration is specified. Thus, apart from giving specific migration limits for certain substances, a maximum migration limit into the food of 30 mg/L is also specified. Interestingly, the Korean standards also include specifications for migration from active and intelligent materials. However, as with the regulations and standards discussed previously, migration from nanofillers is not covered.

5 Migration Studies

Migration of nanofillers into packaged food products is one of the key difficulties confronted by nano-incorporated active packaging [76]. The nanofillers embedded in the packaging material are generally safe for humans and do not pollute the environment. However, if they migrate into food, it could adversely affect both human health and the ecosystem [41]. Environmental exposure to nanofillers via packaging is the subject of limited research [77]. In the review paper, Part et al. [78] described the potential pathway of nanofillers in waste, with the potential conclusion that waste degradation mechanisms are not affected, although their presence may have an impact on microbial communities. However, there is a huge knowledge gap in this field.

During the use, storage, or disposal of materials, a migration process takes place. Desorption, diffusion, dissolution, and decomposition of the packaging material can lead to migration of nanofillers into food [33, 79]. Therefore, before using the nanocomposite in food packaging, it is essential to determine the degree of migration of the nanofiller from the packaging into the food and provide a means to limit this migration [56]. Most studies related to migration have targeted nanofillers and synthetic polymer matrices, which is clear from the reviews by Garcia et al. [2] and Paidari et al. [59], who summarized the applications of nanofillers to food packaging and the migration studies of nanofillers incorporated into polymer packaging materials.

The limited number of studies on nanofiller migration from bionanocomposite matrices is due to the newness of using nanotechnology to modify food packaging materials [8]. Furthermore, the difficulty of identifying nanofillers in composites and the limited availability of qualitative and quantitative analytical methods make conducting such investigations more difficult [80]. The techniques used to evaluate the migration level of nanofillers must be precise, accurate, and sensitive to very low nanofiller concentrations.

Table 1 presents the most relevant investigations in the field of nanofillers application in biopolymer matrices, with data on the test procedure (simulant, test conditions, applied analytical method) highlighting the main results on migration levels.

NPs in bio-nanocomposites spread, desorb, evaporate, or have been transmitted into food from bionanocomposite materials by matrix disintegration [90]. Studies on nanofiller migration have found that when added to polymers, nanofillers have a tendency to agglomerate and stay firmly attached to the matrix, which prevents them from migrating. This appears to be common for non-ionizing elements such as titanium (Ti). However, since ion migration is required to achieve the antimicrobial effect of elements such as zinc (Zn), their controlled removal from the active packing must guarantee that the migrating levels remain within acceptable limits [2].

Some studies suggest that nanofillers or their residues may migrate into packaged foods during storage, although it is unclear whether these releases can be regarded as real migration or residual nanofillers release, that are attached to surfaces of food-contact materials. Further research is needed to fill these gaps in the field of migration and to understand the mechanisms underlying the migration of nanofillers [91].

On the other hand, many experimental studies fail to give a convincing answer to the potential migration of nanofillers from food packaging materials into food, suggesting that the reason for this omission is a deficiency of appropriate analytical methods to identify nanofillers with smaller dimensions and lower contents. It is advised that investigations that come to the conclusion that there is no evidence of migration disclose the assessment's detection limit, which ought to take into account both particle mass and particle size [92]. Furthermore, nanofiller migration is hard to anticipate with predictive models that only consider diffusion-based migration. There is currently insufficient information on the possible migration mechanisms of the detected nanofillers [93].

Stormer et al. [94] reported that there are conflicting and even contradictory conclusions regarding the likelihood of migration of nanofillers from polymers, which could be the result of flawed experimental design. Governments and public bodies, therefore, need to address these issues. Since the migration of nanofillers

Table 1 Migration studies on bionanocomposites

Nanomaterial	Matrix	Test procedure	Analytical method	Migration amount	References
TiO_2 TiO_2/Ag (TiO_2:1%; 5%)	Poly-lactic acid (PLA)	Simulants: 95% E 3% AA Test conditions: 25 °C/ 40 days	ICP-AES	• For 3% AA: 2.19 (for PLA/ TiO_2 1%), 2.36 (for PLA/ TiO_2 1%/Ag), 3.12 (for PLA/ TiO_2 5%) and 3.5 ug/kg (for PLA/ TiO_2 5%/Ag) • For 95% E: 0.593 (for PLA/ TiO_2 1%), 0.72 (for PLA/ TiO_2 1%/Ag), 0.80 (PLA/ TiO_2 5%) and 0.99 ug/kg (for PLA/ TiO_2 5%/Ag)	[81]
TiO_2 (0.1% w/v)	Polyvinyl alcohol/ chitosan (PVA/CHI)	Simulants: distilled water, 3% AA, 10% E, olive oil Test conditions: room temperature/ 1–11 h	ICP-MS	• Film samples decomposed in the AA and E • Migration ratio for olive oil 5.40 × 10^{-3}%	[82]

(continued)

Table 1 (continued)

Nanomaterial	Matrix	Test procedure	Analytical method	Migration amount	References
TiO_2 and ZnO (1–5%)	Polybutylene adipate-co-terephthalate/ thermoplastic starch (PBAT/TPS)	Simulants: distilled water, 10% E and 3% AA Test conditions: 5 ± 2 °C/ 10 days	ICP-OES	• OM values for TiO_2: in all simulants in the range 0.2–1.3 mg/ dm^2 • OM values for ZnO in distilled water were lower compared to the control, while 10% E and 3% AA had higher values • OM values for 5% ZnO in 3% AA exceeded 10 mg/dm^2 limit • SM values for TiO_2: max was 0.1 mg/ dm^2 for distilled water • SM values for ZnO: max was 3.5 mg/ dm^2 for 3% AA	[83]
TiO_2 TiO_2 + Ag	Poly-lactic acid (PLA)	Simulant: hexane Test conditions: 40 °C/ 25 days	ICP-AES	• The max Ti migration from PLA/TiO_2: 112.92 µg/kg PLA/TiO_2 + Ag: 127.38 µg/ kg • The max Ag migration: 135.58 µg/kg	[84]

(continued)

Table 1 (continued)

Nanomaterial	Matrix	Test procedure	Analytical method	Migration amount	References
Al (atomic layer deposition of Al_2O_3)	Poly-lactic acid (PLA)	Simulants: distilled water, 10% E, 3% AA Test conditions: room temperature/ 10 days	ICP-MS	• Al migration was • Small in water and E • Migration in 3% AA was 0.5 mg/dm^2 which fits to Al_2O_3 • Migration of 0.9 mg/dm^2	[85]
TiO_2 (0–20 wt%)	Poly-lactic acid (PLA)	Simulant: 50% E Test conditions: 40 °C/ 45 days	ICP-AES	For PLA/T20 film on day 45th migration was 0.54 ± 0.04 mg/ kg	[86]
	Chitosan	Simulant: 3% AA Test conditions: 10–60 min	UV–VIS spectra	Traces (absorbance maintained at a low level (<0.2), demonstrating CDs composite film insignificant levels of migration	[87]
TiO_2	Chitosan	Simulants: 95% E, olive oil Test conditions: 40 °C/ 10 days for 95% E 5 °C/ 10 days for olive oil	ICP-MS	• 1.97×10^{-4}% to 4.46×10^{-4}% of Ti migrated into the ethanol • 1.47×10^{-4}% to 5.44×10^{-4}% of Ti migrated into the olive oil	[76]

(continued)

Table 1 (continued)

Nanomaterial	Matrix	Test procedure	Analytical method	Migration amount	References
Cellulose nanocrystals and Ag NP	Poly-lactic acid (PLA)	Simulants: 10% E and isooctane Test conditions: 40 °C/ 10 days for 10% E 20 °C/ 2 days for isooctane	ICP-MS	• OM values: not a single sample went over the OM limit (60 mg/kg) • Max SM values for 10% E: 87 ± 7 μg/kg for PLA/5sCNC/1Ag • Max SM values for isooctane: 23 ± 10 μg/kg for PLA/5sCNC/1Ag	[88]
Clay (5% Cloisite30B)	Poly-lactic acid (PLA)	Simulant: 95% E Test conditions: 40 °C/ 10 days	ICP-MS and AF4	• Blank simulant: 0.4 mg/dm^2 • Pure PLA: 1.7 ± 0.6 mg/dm^2 • PLA/5% Cloisite®30B: 6.7 ± 0.5 mg/dm^2	[89]

E = ethanol; AA = acetic acid; ICP-AES = Inductively coupled plasma atomic emission spectroscopy; ICP-MS = Inductively coupled plasma mass spectrometry; ICP-OES = Inductively Coupled Plasma Optical Emission spectroscopy; UV–VIS spectra = UV–Visible spectroscopy; AF4 = asymmetric flow field flow fractionation; OM = overall migration; SM = specific migration; PLA/$TiO_2$1% = PLA film with 1 wt% nano-TiO_2; PLA/$TiO_2$1%/Ag = PLA film with 1 wt% nano-TiO_2 and 0.5 wt% nano-Ag; PLA/$TiO_2$5% = PLA film with 5 wt% nano-TiO_2; PLA/$TiO_2$5%/Ag = PLA film with 5 wt% nano-TiO_2 and 0.5 wt% nano-Ag; PLA/TiO_2 = PLA film with 3 wt% nano-TiO_2; PLA/TiO_2 + Ag = PLA film with 2 wt% nano-TiO_2 plus 1 wt% nano-Ag; PLA/T20 = PLA with 20 wt% content nano-TiO_2; PLA/5sCNC/1Ag = PLA with 5 wt% s-CNC and 1 wt% Ag; PLA/5% Cloisite®30B = PLA with 5% Cloisite30B (a derivatized montmorillonite clay)

from packaged foods (nanocomposite systems) poses the possibility of health risks to consumers, the final authorization of nanocomposites used in food packaging must be based on the results of their migration studies [50].

6 Conclusion

Nanotechnology has opened up great possibilities for use in the food industry, especially in the realm of food packaging. Bionanocomposites are being developed to produce environmentally friendly and sustainable packaging materials with enhanced functional properties. Biocompatible or biodegradable films frequently used to produce bionanocomposites include polysaccharides, proteins, bioplastics (polylactic acid—PLA, polycaprolactam—PCL, polyhydroxyalkanoates—PHA), biopolymer composites and bi-layer films. The addition of nanofillers, such as nanofibers, nanotubes, nanocrystals, layered nanomaterials, or nanoparticles, to a bio-based matrix, can significantly enhance the thermal stability, barrier properties, mechanical strength, antimicrobial characteristics, antioxidant activity, and biosensing properties of resulting bionanocomposites.

However, the introduction of these nano-sized molecules may pose a risk to consumer health, as nanofillers from bionanocomposite materials may migrate into food as well as enter the environment, leading to environmental pollution. The potential migration of nanosubstances is influenced by various factors such as the properties of nanofillers and packaging materials, food composition, environmental conditions, contact time, and contact surface size. Since food is a heterogeneous and very complex matrix, it is very difficult to determine the migration of nanofillers into real food. Therefore, migration tests usually use simulant solutions that mimic different types of food. Detection of nano-sized molecules in these tests is challenging and requires the use of various spectroscopic, microscopic, and quantitative analytical techniques to obtain thorough information.

The growing concern about the use of bionanocomposites relates to the potential toxicity of nanofillers, as they can easily penetrate the human body. Therefore, further migration studies are needed to determine the quality of bionanocomposites and to guarantee the safety of both the packaging material and the food.

This potential release in food and the toxicity of nanofillers has raised concerns among regulatory authorities (FAO, WHO, EU, FDA) and global governmental and nongovernmental organizations, which have sought to establish regulations and guidelines for the application of nanotechnology in the food industry and for bionanocomposites in food packaging. To date, there are no regulations and legislation that deal exclusively with nanofillers. In the EU, it is recommended to conduct risk assessments for these compounds on an individual case basis. Similarly, in the United States, the regulatory status of nano-based products should also be determined on an individual case basis in accordance with specified scientific and regulatory issues and existing FDA review procedures. While the People's Republic of China and the Republic of Korea have standards and specifications for food contact materials, there are currently no specific regulations for nanofillers.

In general, there are few studies on the potential migration of nanofillers into food, due to the lack of appropriate analytical methods to identify substances with trace or ultra-trace levels in samples after migration testing. However, the available information is scarce, and further research in this area is of vital importance.

References

1. Darder M, Aranda P, Ruiz-Hitzky E (2007) Bionanocomposites: a new concept of ecological, bioinspired, and functional hybrid materials. Adv Mater 19(10):1309–1319. https://doi.org/10.1002/adma.200602328
2. Garcia CV, Hwa G, Tae J (2018) Metal oxide-based nanocomposites in food packaging: applications, migration, and regulations. Trends Food Sci Technol 82:21–31. https://doi.org/10.1016/j.tifs.2018.09.021
3. Chausali JN, Saxena J, Prasad R (2022) Recent trends in nanotechnology applications of bio-based packaging. J Agric Food Res 100257. https://doi.org/10.1016/j.jafr.2021.100257
4. Burris KB, Stewart JAC (2012) Fluorescent nanoparticles: sensing pathogens and toxins in foods and crops. Trends in Food Sci Technol. 28(2):143–152
5. Holgate ST (2010) Exposure, uptake, distribution and toxicity of nanomaterials in humans. J Biomed Nanotechnol 6(1):1–19
6. Kato LS, Conte JC (2021) Safety of plastic food packaging: the challenges about non-intentionally added aubstances (NIAS) discovery, identification and risk assessment. Polymer 13(13):2077. https://doi.org/10.3390/polym13132077
7. Honarvar Z, Hadian Z, Mashayekh M (2016) Nanocomposites in food packaging applications and their risk assessment for health. Electron Phys 8(6):2531–2538. https://doi.org/10.19082/2531
8. Huang J-Y, Li X, Zhou W (2015) Safety assessment of nanocomposite for food packaging application. Trends Food Sci Technol 45:187–199. https://doi.org/10.1016/j.tifs.2015.07.002
9. Xue W, Zhu J, Sun P, Yang F, Wu H, Li W (2023) Permeability of biodegradable film comprising biopolymers derived from marine origin for food packaging application: a review. Trends Food Sci Technol 136:295–307. https://doi.org/10.1016/j.tifs.2023.05.001
10. Yuan S, Zhang Q (2021) Application of one-dimensional nanomaterials in catalysis at the single-molecule and single particle scale. Front Chem 9:812287
11. Zaib Q, Ahmad F (2019) Optimization of carbon nanotube dispersions in water using response surface methodology. ACS Omega 4(1):849–859
12. Ezati P, Riahi Z, Rhim J-W (2021) Carrageenan-based functional films integrated with CuO-doped titanium nanotubes for active food-packaging applications. ACS Sustainable Chem Eng 9(28):9300–9307. https://doi.org/10.1021/acssuschemeng.1c01957
13. Papanicolaou GC, Portan DV (2015) Carbon and titanium dioxide nanotube polymer composite manufacturing-characterization and interphase modeling. In: Beaumont PWR, Soutis C, Hodzic A (eds) Structural integrity and durability of advanced composites. Woodhead Publishing Elsevier Ltd., Cambridge, pp 735–761. https://doi.org/10.1016/B978-0-08-100137-0.00027-4
14. Baig N (2023) Two-dimensional nanomaterials: a critical review of recent progress, properties, applications, and future directions. Compos Part A Appl 165:107362. https://doi.org/10.1016/j.compositesa.2022.107362
15. Bayan R, Karak N (2020) Polymer nanocomposites based on two-dimensional nanomaterials. In: Raju K, Barua S (eds) Two-dimensional nanostructures for biomedical technology. Elsevier, pp 249–279. https://doi.org/10.1016/B978-0-12-817650-4.00008-5
16. Priyadarshi R, Roy S, Ghosh T Biswas D, Rhim JW (2022) Antimicrobial nanofillers reinforced biopolymer composite films for active food packaging applications—a review. SM&T 32:e00353. https://doi.org/10.1016/j.susmat.2021.e00353
17. Carbone M, Donia DT, Sabbatella G, Antiochia R (2016) Silver nanoparticles in polymeric matrices for fresh food packaging. J King Saud Univ Sci 28(4):273–279. https://doi.org/10.1016/j.jksus.2016.05.004
18. Saedi S, Shokri M, Priyadarshi R, Rhim JW. Silver ion loaded 3-aminopropyl trimethoxysilane-modified Fe_3O_4 nanoparticles for the fabrication of carrageenan-based active packaging films. Colloids Surfaces B: Biointerfaces 208:112085. https://doi.org/10.1016/j.colsurfb.2021.112085

19. Dong W, Su J, Chen Y, Xu D, Cheng L, Mao L et al (2022) Characterization and antioxidant properties of chitosan film incorporated with modified silica nanoparticles as an active food packaging. Food Chem 373(A):131414. https://doi.org/10.1016/j.foodchem.2021.131414
20. Wu W, Liu L, Goksen G, Demir D, Shao P (2022) Multidimensional (0D-3D) nanofillers: fascinating materials in the field of bio-based food active packaging. Food Res Int 157:111446. https://doi.org/10.1016/j.foodres.2022.111446
21. Mukherjee A, Panda S, Jeelani PGh, Mossa A-T, Chidambaram R (2023) Biodegradable polymers/silica nanocomposites: applications in food packaging. In: Sharma A, Vijayakumar PS, Prabhakar Er PK, Kumar R (eds) Nanotechnology applications for food safety and quality monitoring. Academic Press, London, pp 395–414. https://doi.org/10.1016/B978-0-323-85791-8.00001-X
22. Radusin T, Ristić I, Pilić B, Duraccio D, Novaković A (2018) Silicium-based nanocomposite materials for food packaging applications. In: Cirillo G, Kozlowski MA, Spizzirri UG (eds) Composites materials for food packaging. Scrivener Publishing LLC, pp 175–208. https://doi.org/10.1002/9781119160243.ch5
23. Indumathi M, Rajarajeswari G (2019) Mahua oil-based polyurethane/chitosan/nano ZnO composite films for biodegradable food packaging applications. Int J Biol Macromol 124(1):163–174. https://doi.org/10.1016/j.ijbiomac.2018.11.195
24. Arfat YA, Benjakul S, Prodpran T, Sumpavapol P, Songtipya P (2014) Properties and antimicrobial activity of fish protein isolate/fish skin gelatin film containing basil leaf essential oil and zinc oxide nanoparticles. Food Hydrocoll 41:265–273. https://doi.org/10.1016/j.foodhyd.2014.04.023
25. Roy S, Shankar S, Rhim J-W (2019) Melanin-mediated synthesis of silver nanoparticle and its use for the preparation of carrageenan-based antibacterial films. Food Hydrocoll 88:237–246. https://doi.org/10.1016/j.foodhyd.2018.10.013
26. Roy S, Rhim J-W, Jaiswal L (2019) Bioactive agar-based functional composite film incorporated with copper sulfide nanoparticles. Food Hydrocoll 93:156–166. https://doi.org/10.1016/j.foodhyd.2019.02.034
27. Ashfaq A, Khursheed N, Fatima S, Anjum Z, Younis K (2022) Application of nanotechnology in food packaging: pros and cons. J Agric Food Res 7:100270. https://doi.org/10.1016/j.jafr.2022.100270
28. Siripatrawan U, Kaewklin P (2018) Fabrication and characterization of chitosan-titanium dioxide nanocomposite film as ethylene scavenging and antimicrobial active food packaging. Food Hydrocoll 84:125–134. https://doi.org/10.1016/j.foodhyd.2018.04.049
29. Onyeaka H, Passaretti P, Miri T, Al-Sharify ZT (2022) The safety of nanomaterials in food production and packaging. Curr Res Food Sci 5:763–774. https://doi.org/10.1016/j.crfs.2022.04.005
30. Torres A, Guarda A, Moraga N, Romero J, Galotto MJ (2012) Experimental and theoretical study of thermodynamics and transport properties of multilayer polymeric food packaging. Eur Food Res Technol 234(4):713–722. https://doi.org/10.1007/s00217-012-1683-1
31. Jiménez A, Vargas M, Chiralt A (2016) Antimicrobial nanocomposites for food packaging applications: novel approaches. Novel approaches of nanotechnology in food. Elsevier Inc., pp 347–386. https://doi.org/10.1016/B978-0-12-804308-0/00011-X
32. Dimitrijević M, Karabasil N, Bošković M, Teodorović V, Vasilev D, Đorđević V et al (2015) Safety aspects of nanotechnology applications in food packaging. Procedia Food Sci 5:57–60. https://doi.org/10.1016/j.profoo.2015.09.015
33. Noonan GO, Whelton AJ, Carlander D, Ducan TV (2014) Measurement methods to evaluate engineered nanomaterials release from food contact materials. Compr Rev Food Sci F 13:679–692. https://doi.org/10.1111/1541-4337.12079
34. de Azeredo HM (2013) Antimicrobial nanostructures in food packaging. Trends Food Sci Technol 30(1):56–69. https://doi.org/10.1016/j.tifs.2012.11.006
35. Huang Y, Chen S, Bing X, Gao C, Wang T, Yuan B (2011) Nanosilver migrated into food-simulating solutions from commercially available food fresh containers. Packag Technol Sci 24(5):291–297. https://doi.org/10.1016/j.tifs.2015.07.002

36. European Commission (2011) Commission Regulation (EU) No 10/2011 of 14 January 2011 on plastic materials and articles intended to come into contact with food. Off J Eur Union [Internet] L 12/1:1–89. https://eur-lex.europa.eu/LexUriServ/LexUriServ.do?uri=OJ:L:2011:012:0001:0089:EN:PDF
37. Kumar AP, Depan D, Singh Tomer N, Singh RP (2009) Nanoscale particles for polymer degradation and stabilization—trends and future perspectives. Prog Polym Sci 34:479–515. https://doi.org/10.1016/j.progpolymsci.2009.01.002
38. Arora A, Padua GW (2010) Review: nanocomposites in food packaging. J Food Sci 75(1):43–49. https://doi.org/10.1111/j.1750-3841.2009.01456.x. PMID: 20492194
39. Singh S, Stephan C, Westerhoff P et al (2014) Measurement methods to detect, characterize, and quantify engineered nanomaterials in foods. Compr Rev Food Sci Food Safe 13:693–704. https://doi.org/10.1111/1541-4337.12078
40. Peters RJB, Hoekstra EJ (2019) Migration of nanomaterials from food contact materials (Chapter 9). In: Suman M (ed) Food contact materials analysis: mass spectrometry techniques. Royal Society of Chemistry. Food Chemistry Function and Analysis No 10, p 226. ISSN: 2398-0664
41. Perera KY, Jaiswal S, Jaiswal AK (2022) A review on nanomaterials and nanohybrids based bio-nanocomposites for food packaging. Food Chem 376:131912. https://doi.org/10.1016/j.foodchem.2021.131912
42. Chaudhry Q, Castle L, Bradley E, Blackburn J, Aitken R, Boxall A (2008) Assessment of current and projected applications of nanotechnology for food contact materials in relation to consumer safety and regulatory implications. Final report Project Ă3. Food Standards Agency. http://www.food.gov.uk/sites/default/files/ă3.pdf
43. Simon P, Chaudhry Q, Bakos D (2008) Migration of engineered nanoparticles from polymer packaging to food-a physicochemical view. J Food Nutrit Res 47:105–113
44. Bhunia K, Sablani, Tang J, Rasco B (2013) Migration of chemical compounds from packaging polymers during microwave, conventional heat treatment, and storage. Compr Rev Food Sci F 12:523–545. https://doi.org/10.1111/1541-4337.1202
45. Metak AM, Nabhani F, Connolly SN (2015) Migration of engineered nanoparticles from packaging into food products. LWT 64:781–787. https://doi.org/10.1016/j.lwt.2015.06.001
46. Bott J, Franz R (2018) Investigation into the potential migration of nanoparticles from laponite-polymer nanocomposites. Nanomaterial 8(9):723. https://doi.org/10.3390/nano8090723
47. Heydari-Majd M, Ghanbarzadeh B, Shahidi-Noghabi M et al (2019) Kinetic release study of zinc from polylactic acid based nanocomposite into food simulants. Polym Test 76:245–260. https://doi.org/10.1016/j.polymertesting.2019.03.040
48. Song H, Li B, Lin QB, Wu HJ, Chen Y (2011) Migration of silver from nanosilverpolyethylene composite packaging into food simulants. Food Addit Contam Part A 28:1758–1762. https://doi.org/10.1080/19440049.2011.603705
49. Schmidt B, Katiyar V, Plackett D, Larsen EH, Gerds N, Koch CB, Petersen JH (2011) Migration of nanosized layered double hydroxide platelets from polylactide nanocomposite films. Food Addit Contam Part A 28(7):956–966. https://doi.org/10.1080/19440049.2011.572927
50. Cushen M, Kerry J, Morris M, Cruz-Romero M, Cummins E (2014) Evaluation and simulation of silver and copper nanoparticle migration from polyethylene nanocomposites to food and an associated exposure assessment. J Agric Food Chem 62:1403–1411. https://doi.org/10.1021/jf404038y
51. Bradley EL, Castle C, Chaudhry Q (2011) Applications of nanomaterials in food packaging with a consideration of opportunities for developing countries. Trends Food Sci Tech 22:604–610. https://doi.org/10.1016/j.tifs.2011.01.002
52. Echegoyen Y, Nerín C (2013) Nanoparticle release from nano-silver antimicrobial food containers. Food Chem Toxicol 62:16–22. https://doi.org/10.1016/j.fct.2013.08.014
53. Sharma R, Jafari SM, Sharma S (2020) Antimicrobial bio-nanocomposites and their potential applications in food packaging. Food Control 112:107086. https://doi.org/10.1016/j.foodcont.2020.107086

54. Ahmad A, Qurashi A, Sheehan D (2023) Nano packaging—progress and future perspectives for food safety, and sustainability. Food Packag Shelf Life 35:100997. https://doi.org/10.1016/j.fpsl.2022.100997
55. Gomes V, Souza L, Fernando AL (2016) Nanoparticles in food packaging: biodegradability and potential migration to food—a review. Food Packag Shelf Life 8:63–70. https://doi.org/10.1016/j.fpsl.2016.04.001
56. Rhim J, Park H, Ha C (2013) Bio-nanocomposites for food packaging applications. Prog Polym Sci 38(10–11):1629–1652. https://doi.org/10.1016/j.progpolymsci.2013.05.008
57. Enescu D, Cerqueira MA, Fucinos P, Pastrana LM (2019) Recent advances and challenges on applications of nanotechnology in food packaging. A literature review. Food Chem Toxicol 134:110814. https://doi.org/10.1016/j.fct.2019.110814
58. Martirosyan A, Polet M, Bazes A, Sergent T, Schneider YJ (2012) Food nanoparticles and intestinal inflammation: a real risk? In: Szabo I (ed) Inflammatory bowel disease. In Tech: Rijeka, Croatia (Chapter 8), pp 259–282. https://doi.org/10.5772/52887
59. Paidari S, Tahergorabi R, Anari ES, Nafchi AM, Zamindar N, Goli M (2021) Migration of various nanoparticles into food samples: a review. Foods 10:2114. https://doi.org/10.3390/foods10092114
60. Wu HH, Yin JJ, Wamer WG, Zeng MY, Lo YM (2014) Reactive oxygen species related activities of nano-iron metal and nano-iron oxides. J Food Drug Anal 22:86–94. https://doi.org/10.1016/j.jfda.2014.01.00710.3390/ijms17040509
61. Gaillet S, Rouanet JM (2015) Silver nanoparticles: their potential toxic effects after oral exposure and underlying mechanisms-a review. Food Chem Toxicol 77:58–63. https://doi.org/10.1016/j.fct.2014.12.019
62. Sarma VK, Siskova KM, Zboril R, Gardea-Torresdey JL (2014) Organic-coated silver nanoparticles in biological and environmental conditions: fate, stability and toxicity. Adv Coll Inter Sci 204:15–34. https://doi.org/10.1016/j.cis.2013.12.002
63. Wyser Y, Adams M, Avella M, Carlander D, Garcia L, Pieper G et al (2016) Outlook and challenges of nanotechnologies for food packaging. Packag Technol Sci 29:615–648. https://doi.org/10.1002/pts.2221
64. Oberdörster G, Maynard A, Donaldson K, Castranova V, Fitzpatrick J, Ausman K et al (2005) Principles for characterizing the potential human health effects from exposure to nanomaterials: elements of a screening strategy. Part Fibre Toxicol 2:8. https://doi.org/10.1186/1743-8977-2-8
65. FAO/WHO, Food and Agriculture Organization of the UNH (2013) FAO/WHO paper: state of the art on the initiatives and activities relevant to risk assessment and risk management of nanotechnologies in the food and agriculture sectors. Geneva
66. International Organization for Standardization (ISO) (2005) ISO/TC 229 Nanotechnologies
67. EFSA Scientific Committee, Hardy A, Benford D, Thorhallur Halldorsson MJJ, Knutsen HK, More S, Naegeli H, Co HN, Ricci A, Rychen G, Schlatter JR, Vittorio Silano RS, Turck D, Younes M, Chaudhry Q, Francesco Cubadda DG, Oomen A, Weigel S, Melpo Karamitrou RS (2018) Guidance on risk assessment of the application of nanoscience and nanotechnologies in the food and feed chain: Part 1, Human and animal health. EFSA J 16(May)
68. European Commission (2009) Commission regulation (EC) No 450/2009 of 29 May 2009 on active and intelligent materials and articles intended to come into contact with food. Off J Eur Union (135):3–11. https://eur-lex.europa.eu/LexUriServ/LexUriServ.do?uri=OJ:L:2009:135:0003:0011:EN:PDF
69. European Commission (2017) Commission regulation (EU) 2017/752 of 28 April 2017 amending and correcting regulation (EU) No 10/2011 on plastic materials and articles intended to come into contact with food. Off J Eur Union L 113/18
70. European Commission (2016) Commission regulation (EU) 2016/1416 of 24 August 2016 amending and correcting regulation (EU) No 10/2011 on plastic materials and articles intended to come into contact with food. Off J Eur Union L 230/22
71. Commission E (2020) Commission Regulation (EU) 2020/1245 of 2 September 2020 amending and correcting regulation (EU) No 10/2011 on plastic materials and articles intended to come into contact with food. Off J Eur Union 16(10):1–17

72. Food and Drug Administration (FDA) (2018) Determining the regulatory status of components of a food contact material [Internet]. https://www.fda.gov/food/packaging-food-contact-substances-fcs/determining-regulatory-status-components-food-contact-material
73. Food and Drug Administration (FDA) (2014) Guidance for industry considering whether an FDA-regulated product involves the application of nanotechnology. (June):1–14
74. National Health and Family Planning Commission TPR of National Food Safety standard for uses of additives in food contact materials and their products (2016)
75. Ministry of Food and Drug Safety (MFDS) of Republic of Korea. Standards and specificationsfor utensils, containers and packages (2021)
76. Enescu D, Dehelean A, Gonçalves C, Cerqueira MA, Magdas DA, Fucinos P, Pastrana LM (2020) Evaluation of the specific migration according to EU standards of titanium from chitosan/metal complexes films containing TiO_2 particles into different food simulants. A comparative study of the nano-sized vs. micro-sized particles. Food Packag Shelf Life 26:100579. https://doi.org/10.1016/j.fpsl.2020.100579
77. Karimi M, Sadeghi R, Kokini J (2018) Human exposure to nanoparticles through trophic transfer and the biosafety concerns that nanoparticle-contaminated foods pose to consumers. Trends Food Sci Technol 75:129–145. https://doi.org/10.1016/j.tifs.2018.03.012
78. Part F, Berge N, Baran P, Stringfellow A, Sun W, Bartelt-Hunt S, Mitrano D, Li L, Hennebert P, Quicker P, Bolyard SC, Huber-Humer M (2018) A review of the fate of engineered nanomaterials in municipal solid waste streams. Waste Manag 75:427–449. https://doi.org/10.1016/j.wasman.2018.02.012
79. Nile SH, Baskar V, Selvaraj D, Nile A, Xiao JB, Kai GY (2020) Nanotechnologies in food science: applications, recent trends, and future perspectives. Nano-Micro Lett 12(1):Article 45. https://doi.org/10.1007/s40820-020-0383-9
80. Han W, Yu Y, Li N, Wang L (2011) Application and safety assessment for nano-composite materials in food packaging. Sci Bull 56:1216–1225. https://doi.org/10.1007/s11434-010-4326-6
81. Li W, Zhang C, Chi H, Li L, Lan T, Han P, Chen H, Qin Y (2017) Development of antimicrobial packaging film made from poly (lactic acid) incorporating titanium dioxide and silver nanoparticles. Molecules 22(7):1170. https://doi.org/10.3390/molecules22071170
82. Lian Z, Zhang Y, Zhao Y (2016) Nano-TiO_2 particles and high hydrostatic pressure treatment for improving functionality of polyvinyl alcohol and chitosan composite films and nano-TiO_2 migration from film matrix in food simulants. Innov Food Sci Emerg Technol 33:145–153. https://doi.org/10.1016/j.ifset.2015.10.008
83. Phothisarattana D, Harnkarnsujarit N (2022) Migration, aggregations and thermal degradation behaviors of TiO_2 and ZnO incorporated PBAT/TPS nanocomposite blown films. Food Packag Shelf Life 33:100901. https://doi.org/10.1016/j.fpsl.2022.100901
84. Li W, Li L, Zhang H, Yuan M, Qin Y (2018) Evaluation of PLA nanocomposite films on physicochemical and microbiological properties of refrigerated cottage cheese. J Food Process Preserv 42:e13362. https://doi.org/10.1111/jfpp.13362
85. Vähä-Nissi M, Pitkänen M, Salo E, Sievänen-Rahijärvi J, Putkonen M, Harlin A (2015) Atomic layer deposited thin barrier films for packaging. Cellul Chem Technol 49:575–585. http://www.cellulosechemtechnol.ro/pdf/CCT7-8(2015)/p.575-585.pdf
86. Yang C, Zhu B, Wang J, Qin Y (2019) Structural changes and nano-TiO_2 migration of poly(lactic acid)-based food packaging film contacting with ethanol as food simulant. Int J Biol Macromol 139:85–93. https://doi.org/10.1016/j.ijbiomac.2019.07.151
87. Wen F, Li P, Yan H, Su W (2023) Turmeric carbon quantum dots enhanced chitosan nanocomposite films based on photodynamic inactivation technology for antibacterial food packaging. Carbohydr Polym 311:120784. https://doi.org/10.1016/j.carbpol.2023.120784
88. Fortunati E, Peltzer M, Armentano I, Jiménez A, Kenny JM (2013) Combined effects of cellulose nanocrystals and silver nanoparticles on the barrier and migration properties of PLA nano-biocomposites. J Food Eng 118:117–124. https://doi.org/10.1016/j.jfoodeng.2013.03.025

89. Schmidt B, Petersen JH, Koch CB, Plackett D, Johansen N, Katiyar V, Larsen E (2009) Combining asymmetrical flow field-flow fractionation with light-scattering and inductively coupled plasma mass spectrometric detection for characterization of nanoclay used in biopolymer nanocomposites. Food Addit Contam Part A Chem Anal Control Expo Risk Assess 26:1619–1627. https://doi.org/10.1080/02652030903225740
90. Azizi-Lalabadi M, Garavand F, Jafari SM (2021) Incorporation of silver nanoparticles into active antimicrobial nanocomposites: release behavior, analyzing techniques, applications and safety issues. Adv Colloid Interface Sci 293:102440. https://doi.org/10.1016/j.cis.2021.102440
91. Khajavi MZ, Mohammadi R, Ahmadi S, Farhoodi M, Yousefi M (2019) Strategies for controlling release of plastic compounds into foodstuffs based on application of nanoparticles and its potential health issues. Trends Food Sci Technol 90:1–12. https://doi.org/10.1016/j.tifs.2019.05.009
92. Jokar M, Pedersen GA, Loeschner K (2017) Six open questions about the migration of engineered nano-objects from polymer-based food-contact materials: a review. Food Addit Contam Part A Chem 34 (3):434–450. https://doi.org/10.1080/19440049.2016.1271462
93. Biswas R, Alam M, Sarkar M, Haque MdI, Hasan MdM, Hoque M (2022) Application of nanotechnology in food: processing, preservation, packaging and safety assessment. Heliyon 8:e11795. https://doi.org/10.1016/j.heliyon.2022.e11795
94. Stormer A, Bott J, Kemmer D, Franz R (2017) Critical review of the migration potential of nanoparticles in food contact plastics. Trends Food Sci Technol 63:39–50. https://doi.org/10.1016/j.tifs.2017.01.011

Converting Challenges for the Novel Materials in Food Packaging Production

Igor Karlovits and Gregor Lavrič

Abstract The conversion of different natural or synthetic polymers into productive intermediary materials like plastic film rolls or cardboard sheets or rolls, as well as laminated multiply packaging solutions, require several technical requirements. The newly developed packaging materials besides conforming to the food safety standards, must also fulfil several other requirements. They must be mechanically resistant, resilient in different logistic conditions and supply chain management handlings, easily recyclable, preferably obtained from renewable resources, and easily convertible in the packaging processes. These requirements are often determined by the food production processes (for example, high-temperature pasteurization or some form of thermal or other energy processing), existing packaging machines, and other machines involved in the packaging process like the printing presses and converting units. As one material must withstand all the production requirements, these are very important to be taken into consideration during the development of novel packaging materials. The chemical modification of the surface materials to achieve better barrier properties must not deprive the surface groups that are reacting during the conversion stage of the packaging and hamper pacakging production or content filling operations. In this chapter, we shall discuss the main operational processes for food packaging materials and the main material properties like tensile strength, porosity, and coefficient of friction to processing properties like printability, sealability, and glueability which can alter the basic protective and other functions of the packaging. The topics are grouped by material types (biopolymer or cellulose-based materials for flexible and semi-rigid materials) with typical food packaging products. Also, standardized methods for testing these materials are presented.

Keywords Packaging materials · Converting · Printing · Sealing

I. Karlovits (✉) · G. Lavrič
Pulp and Paper Institute, Ljubljana, Slovenia
e-mail: igor.karlovits@danfoss.com

I. Karlovits
Danfoss Trata doo, Danfoss A/S, Ljubljana, Slovenia

M. Alle et al. (eds.), *Functional Nanomaterials and Nanocomposites for Biodegradable Food Packaging*, Smart Nanomaterials Technology,
https://doi.org/10.1007/978-981-97-9084-5_13

1 Introduction

The development of the product's packaging must secure all functionalities of the food distribution systems mostly containment, protection, and preservation of the food itself. Packaging as an integral part of this system needs to connect the food science contributing factors (food preparation operations) as well as the processes of product manufacturing, distribution, and economics of the whole supply chain (Fig. 1).

Nowadays when materials scientists develop new types of (nano)materials for different food protection functionalities (UV protection, mechanical strengths, antimicrobial and antioxidant properties), additional steps that make total product development need to be considered. Just one improved material property is often the reason for a drastic change in the production process and converting steps (additional heating or cooling of the materials), but scientists must ensure that the newly developed solutions are adjusted to existing packaging materials specifications and machinery.

Marketing and advertising functions of the packaging were important in the last few decades (with improved food shelf protection), but the recent shift towards more sustainable packaging options is influencing the whole food packaging value chain. Based on a new holistic definition, sustainable food packaging is defined as an optimized, quantified, and validated solution, which takes into consideration the balance of social, economic, ecological, and safe implementations of the circular value chain, based on the entire history and life cycle of the food product-package unit [1]. In this chapter, we shall describe the most common technological processes and the physical and chemical requirements of the nano or other food-based packaging solutions.

One of these examples is described by Parisi et al. [2] where it is common to use plasticizers (oleates, polyglycols, and so on) in plastic materials if products have to be produced with acceptable ductility or flexibility and improved mechanical properties. The mineral fillers (carbon black, zinc oxide, fibreglass, and so on) may be added to

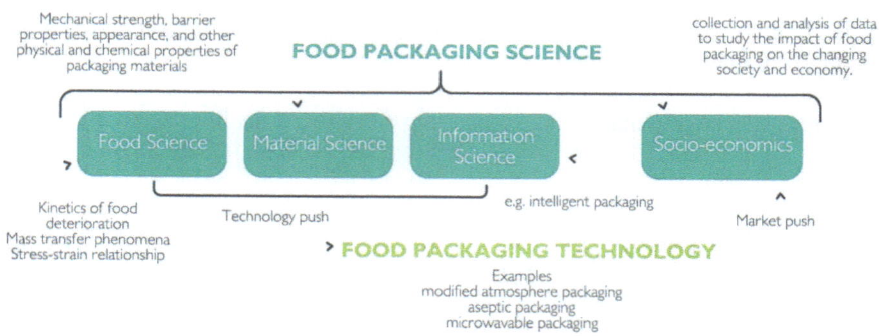

Fig. 1 Integration of food packaging with food production and distribution

plastic mixtures to solve these challenges as this lowers the mechanical damage and insulation problems. The oxidation and UV exposure can be limited with the use of antioxidant agents (phenols and hydroquinones, organic phosphites, acrylonitriles). Another problem can be the accumulation of electrostatically charged water particles. For this reason, antistatic substances (amines, alkyl phosphates, amides) may reduce this phenomenon.

The question of friction and superficial damage during manufacturing and successive steps (delivery, storage, food packaging) is also very important. For this reason, lubricating substances (fats, polyglycols, silicones, waxes) are necessary [2]. This is also important for cellulosic materials where the fatty acids and waxes, as well as the used mineral fillers directly influence the production process in the paper mills (roll vibration and handling) as well as processing in the packaging lines.

2 Packaging Types and Processes

As there are many types of packaging with specific processes, we shall group them based on the mechanical stiffness where we shall discern between:

1. Flexible packaging solutions: films, foils pouches, and bags made eather from synthetic polymer or biobased raw materials and their combination.
2. Semi-rigid packaging: corrugated and paperboard used to form cases, cartons, and trays.
3. Rigid packaging: glass, plastic, metal, and composite bottles, cans, and tubes.

We shall mainly discuss the first two groups with composite bottles from group 3, but the technological processes for glass and metal packaging are out of the scope of this chapter. Sometimes for a specific product like a mayonnaise tube with a rigid cap and flexible body made of polymer or composite solutions, it is hard to make an exact distinction between semi-rigid, rigid, and flexible, but the majority of food packaging products go through similar bulk processes.

2.1 Flexible Materials Processing

Single films, coextruded films, and coated and laminated films in reel form are often used to make plastic bags, sachets, pouches, and overwraps. Plastic bags are made by folding, cutting, and sealing with welded seams which are also cut in the same operation. Pouches are usually made from laminates. They may be formed on the packing machine either from one reel by folding, or from two reels and sealing, inside face to inside face on three sides before filling and closing [3]. These operations also apply to cellulose-based packaging where the sealing functions are mostly achieved with lamination and coating of the paper reels. The needs of the flexible packaging converter are summed in Table 1 [4].

Table 1 End user needs by the converting stakeholders for polymer flexible packaging

End user need	Material property
Stiffness	Material selection and layer or multilayer thickness
Heat seal performance	Properties of heat seal material (heat transfer ratio, melting point, etc.)
Barrier requirements	Material selection and layer thickness
Durability	Toughness and interlayer adhesion (z-tensile strength)
Printability	Surface energy and ink type
Packing line speeds	Heat seal material, seal initiation temperature, hot tack, possibility for welding
Slip properties	Coefficient of friction, bulk packages
Post filling steps	Material temperature ranges

One decision is whether lamination or coextrusion should be used to make the multiple-layer structure. Lamination involves combining two films with an adhesive or polymer melt (extrusion lamination). In coextrusion, layers are combined while in the melt during the film-converting process. Coextrusion has several attributes including bringing the desired properties of different materials into one structure, and reducing emissions from solvent lamination processes. However, it is to be noted that water-based or solventless adhesives are predominantly used today in adhesive lamination. Nevertheless, the coextrusion helps in processing multiple materials in one pass to economize on process steps, cost, and time. Coextrusion is inherently a more flexible process than monolayer extrusion but is more complex: more extruders, more complicated flow geometries, greater potential for flow issues, higher cost of capital investment, and higher scrap costs if the waste generated during the process cannot be recycled. These factors must be weighed by the manufacturer when determining what process to use [4].

Barrier requirements for specific food types are met by the selection of materials. For example, for a high oxygen barrier, the converter may choose between several materials, such as aluminium foil, metallized films, AlO_x- or SiO_x-coated film, polyvinylidene chloride (PVDC), ethylene vinyl alcohol (EVOH), and polyamine (PA). For lower values, some forms of biopolymers, starch films, or nanocellulose solutions have already been tested, but they may require different processes other than polymer lamination or extrusion.

Here, the cost, the ability to meet the other requirements of the package, and the compatibility with the converter's process will dictate which material to use with a focus on food shelf life and food contact safety. For example, if transparency is required, then aluminium foil or metallized films would not be considered. The barrier is also influenced by the structure design. The permeation is typically inversely proportional to the thickness. However, increasing thickness improves the barrier, but increases cost. Multiple barrier layers may be beneficial. Protecting barrier layers, such as aluminium foil, from stress cracking during use may also be an effective strategy. Currently, many companies aim for mono-material solutions due to the high waste separation and recycling costs and other challenges of these kinds of multilayer materials.

The bending stiffness of the film is affected by the modulus of each of the layers, their thickness, and their position within the structure. Stiffer materials near the outside layers of the film have a greater effect on increasing stiffness, than near the centre. As structures are downsized to meet cost or sustainability targets, these design features become increasingly important to maintain the strength and durability of the package, as well as consumer acceptance regarding their perception of quality. The chemistry of the sealant polymer affects sealability, particularly when dealing with powdery or greasy contamination in the seal area.

The durability of the package film when exposed to puncture, impact, scratch, and other threats is influenced by the durability of the layers within the package. The adhesion between the layers may also play a role. Controlled delamination during an impact may help dissipate energy and prevent failure of the package.

Printability of flexible packaging films requires adequate surface energy, as most polymer surfaces are naturally hydrophobic, and resist being wetted. Characteristically, these plastics are chemically inert, nonporous surfaces with low surface energy, which makes printing adhesion nearly impossible. Low molecular weight materials, such as silicones, mould release, anti-slip agents, and process additives, inhibit the ink's ability to flow and achieve intimate contact essential for adhesion. These steps are often solved using some sort of surface treatment like flame, corona, or plasma where the surface energy of the surfaces is altered, or changing the micro-roughness, which can enable better ink adhesion of polymer surfaces.

Packaging line speed is a function of how fast the package will seal. This is determined by how fast the heat arrives at the interface (film thickness and seal bar temperature) and the characteristics of the sealant. Sealants with low-heat seal initiation temperature and high hot tack strength allow for faster line speeds. The slip properties of the film are determined by the coefficient of friction of the film, which in turn, is often controlled using additives in the polymer.

The ability of the film to withstand post-filling steps such as sterilization, retort, or blast freezing will depend on the temperature, which must be within the use ranges of the materials. Retort, for example, typically involves exposing the filled package to 121 °C for 30 min or more under pressure. Materials that melt below this temperature may not be suitable [4].

2.1.1 Heat Sealing

Heat sealing is used for joining thermoplastic films, typically less than 0.5 mm thick. There are two main types of heat sealing: hot-bar welding and impulse welding. Hot-bar welding and impulse welding have found applications in several industrial sectors but are most widely used in the packaging industry for sealing bags, films, and containers made from thermoplastics. This includes plastic pot and tray sealing for food packaging applications [5]. It is an important packaging processing step as the seal area represents the most problematic part of food packaging for controlling the moisture and gas permeation which can hamper the food product quality. Hydrogen bonds, polar and ionic interactions are molecular forces that can come into play, depending on the thermoplastic materials that are used in the seal layer. Bonds between identical polymers, referred to as autohesion, are formed in pouch applications (e.g., horizontal and vertical form-fill-seal packages). Besides the polymers being used, there are three main processing factors that influence the heat-sealing integrity: temperature, time, and pressure. For temperature processing, we must determine the melt temperature of the packaging polymer. Low temperature will produce a poor seal which can peel, where too high temperature will degrade the sealing by changing the properties of the used polymer material. In the case of combining two non-identical materials, the processing melting point needs to be adjusted to the melting point of higher melting material, while not thermally degrading the lower melting polymer. The time factor, equally important if prolonged, can cause thinning of the material. Materials should be cooled to the point where they can be handled without stretching or tearing. The pressure must be minimal both with equal distribution, as uneven pressure also causes poor transfer of heat from the platens to the material. Using the minimum pressure to force the melted polymer together results in quality welds.

One specific case is, for example, blister packaging where we have sealing of thermoformed material which is usually rigid with a flexible polymer that can be multiplied. To achieve the correct sealing properties the press plate must reach a temperature that will result in the proper amount of heat transferring through the layers to ensure that both adhesive layers reach their activation energies. Another simple way of looking at the situation is to use a circuit equivalent system, where each layer provides resistance to heat transfer through the package [6]. This resistance would have the following setup:

$$Q = \frac{\Delta T}{\Sigma R} = \frac{\Delta T}{R_{\text{Paper}} + R_{\text{Adhesive}} + R_{\text{Plastic}}} \tag{1}$$

where

Q: The heat is transferred through the system,

ΔT: time difference,

R: resistance to heat pressure.

The resistance equivalent sheds light on one area of difficulty when considering the conduction of heat through multiple surfaces in direct contact with one another. This difficulty is how heat is transferred across the contact points between the materials. However, due to differing surface metrologies, it is highly unlikely that the two materials are in full contact. Instead, pockets of vacuum or fluid can form in between two materials, thus increasing the thermal resistivity of the overall system. These cavities result in the need to consider new resistance terms besides those of the materials themselves, where there may not only be conduction occurring, but convection and/or radiation as well [6].

Table 2 ASTM-based standardized test methods for heat sealing properties testing

Standard name	Code	Output of testing
Standard test method for seal strength of flexible barrier materials	ASTM F88/ F88M-21	Seal strength (N mm^{-1}) and seal failure modes (adhesive peel, seal break)
Standard test methods for hot seal (hot tack) of thermoplastic polymers and blends compromising the sealing surfaces of flexible webs	ASTM F 1921/ F19M-21	Hot tack strength (N mm^{-1} and seal failure modes)
Standard test methods for detecting gross leaks in packaging by internal pressurization (bubble test)	ASTM 2096-11	Leak tightness: yes/no (250 μm defect size + leak location)
Standard test methods for pressure decay test for flexible packages with and without restraining plates in packaging by internal pressurization	ASTM 2095-07	Leak tightness: yes/no + pressure decay (N mm^{-2})
Test method for non-destructive detection of leaks in packaging by vacuum decay method	ASTM 2338-09	Leak tightness: yes/no + 125 μm defect size Pressure decay (N mm^{-2})
Standard test method for detecting leaks in nonporous packaging or flexible barriers by dye penetration	ASTM F 3039-15	Leak tightness: yes/no + (50 μm defect size)

For classic polymer materials, low-density polyethylene (LDPE) and linear low-density polyethylene (LLDPE) are both commonly used seal materials. These semi-crystalline materials are flexible and can be sealed at relatively low temperatures, as the respective T_m of LDPE and LLDPE range between 98–115 °C and 118–125 °C. On the other hand, there were some studies conducted regarding the heat sealability of biodegradable and/or compostable polymeric materials like poly-D-lactic acid (PDLA) and polylactic acid (PLA), which are semi-crystalline materials with a T_m between 170 and 180 °C and a T_g around 55–600 °C. As PLA has some downsides like brittleness, it needs to be blended with such other polymers as poly(butylene adipate-co-terephthalate) (PBAT), poly(caprolactone) (PCL), or starch and/or using additives.

For testing the heat-sealing properties there are several ASTM standards that are widely used in the industry are presented in Table 2.

2.1.2 Ultrasonic Welding of Packaging Materials

Ultrasonic welding of packaging materials is one option for joining and closing the plastic and paper materials without the application of additional adhesive material. Weld tools (anvil or sonotrodes) focus the ultrasonic energy on the point at which the welding should take place, and friction heat is generated. The sealing time is between 100 and 200 ms. Unlike heat sealing, the heat required in ultrasonic technology is only generated within the materials being welded. The tools remain cold and support the dissipation of heat. The hot-tack resistance (strength of the seam directly after welding without cooling) is also significantly higher than with other welding methods.

The potential of ultrasonic welding for bonding 100% lignocellulosic papers has been recently investigated [7]. For papers containing high lignins and hemicellulose contents (50% in total), the adhesion strength generated by welding is almost equivalent to the one obtained by glueing. The increase in temperature during welding triggers the flow of lignins and hemicelluloses, which form a matrix around the fibres at the interface. However, as suggested by the mechanics of this phenomenon, it is not possible to weld papers obtained from chemical pulp containing low hemicelluloses and lignin contents. As quality paperboards are mainly delignified and contain high cellulose content to improve mechanical properties, some kind of thermoplastic coatings can help with this kind of welding. One of the more recent works is presented by Monot et al. [8], which indicates that welding of coated folding box boards is possible, provided that coating weight and binder content are high enough.

The ultrasonic welding of polymer materials gives a wide range of packaging products, such as flexible films, capsules and tea bag spots, valves, and zipper bag solutions, as well as cups and blister packs sealing. The ultrasonic sealing process can help to join fine films or laminates. The ultrasonic waves cause the molecules of the layers positioned on top of one another to vibrate. The friction leads to local heat development at the contact points of both layers. There, the materials are joined, and a weld joint is formed. Once the joint is cooled down, the connection is almost as solid as the starting materials [9]. Some of the basic underlying principles were investigated by Bach et al. [10] where they found some industry ranges for the typical frequencies for ultrasonic welding, i.e., 20, 30, 35, and 40 kHz. Amplitudes used for plastic welding are in the range of 25–60 mm. For the sealing of thin packaging films, amplitudes of 10–30 mm are usually used, depending on the film properties and equipment setup. To ensure the transmission of the ultrasonic vibrations and energy dissipation, the horn presses the films against a counter tool, which is called the anvil. In the case of ultrasonic sealing of films, the anvil additionally has the function of an energy conductor. For the intermittent ultrasonic sealing process with force control, there are four main process steps. In the first step, the horn attaches the films and presses them against the anvil with the predefined sealing force. In the second step, the vibration progresses with increasing amplitude up to its set-up value. In step three, the temperature rises to the crystalline melting point, and the sealing material begins to flow out. Depending on the properties of the sealing material and the process parameters, the horn displacement reaches a stationary plateau, which means a balanced condition between heating in the inner fusion zone and cooling at the border of the fusion zone. In this step, the melt flow decreases or even stops. By switching off the vibration, one ensures that the horn sinks into the molten material and reaches the final displacement set. After the effective sealing time, the seam cools down in the last step (step four). On reaching the recrystallization point of the material, the seam becomes almost completely resilient. This typically occurs within several milliseconds. Hence, if the application, that is the machine speed, allows an adequate cooling time, problems with open seams after sealing, because of the product load, may no longer arise when using the ultrasonic sealing process.

In this context, the Hot-Tack property of the packaging films does not influence the seam quality as much as in conduction sealing. In research, Coralie-Koo-Sin-Lin [11] introduced a new sealing process based on High-Intensity Focused Ultrasound (HIFU), which improved previous challenges with the polymer films, and studied the acoustic and thermal effects, as well as their interrelationships, which affected polymer films exposed to HIFU to implement successfully this new process on a form fill and seal machine. The weldability of PLA films was studied by Stoehr et al. [12] who determined that the addition of a plasticizer resulted in a wider process window, that is, it was possible to weld plasticized PLA films at lower amplitude and the higher welding force compared with PLA films without a plasticizer.

2.1.3 Coefficient of Friction

The coefficient of friction, commonly referred to as CoF, is a ratio of the force required to move one surface over another to the total force applied normally to those surfaces. Friction is an important material property not just during the conversion but also at the end-use of almost every packaging material. In printing the surface, for example, the cellulose-based material influences the good runnability of the paper/boards in the printing process. This applies also to all roll-to-paper transportation systems like automatic teller machines, office printers, or any paper web based printing systems. In packaging during transport, high friction is important for sack papers and line boards in corrugated containers, so any damage to goods is avoided. CoF is broken down into two readings, static and kinetic. Static CoF is the ratio of force required to move one surface over another from the instant the motion starts. Whereas, kinetic CoF is the same ratio once the surface is already in motion. CoF values range from 0.00 to 1.00, with lower CoF values indicating lesser resistance to sliding, or higher "slip."

- CoF < 0.25 is considered low CoF—high-slip
- CoF > 0.45 indicates high CoF—low/non-slip

CoF is reported in two values: static (μs)—the force needed to begin movement, and kinetic (μk)—the force required to maintain movement. Generally, Kinetic CoF is used for roll films. Static CoF is of greater concern for stacked or palletized items [13].

For example, in wrapping with foils the wrap speed determines the required friction values (dynamic and static). The faster the machine, the lower the CoF specifications should be. Table 3 is s general guideline for choosing both kinetic and static CoF specifications.

For vertical form fill seal machines, in particular, packaging film interacts with a machine component called pull belts. The pull belts press against the outside of the film, sandwiching it between the belts and the forming tube, helping to draw it through the packaging system. To optimize performance, there must be enough friction on the outside of the film to overcome the friction on the inside. In other

Table 3 General properties of flexible films kinetic and static CoF for wrapping

Type of packaging	CoF specification range	
	Kinetic	Static
Above 300 IPM	15–20	25–30
100–300 IPM	20–25	30–35
Below 100 IPM	23–30	35–40
Manual wrapping	Depending on specific requirements	

IPM = Inches per minute

words, the inside of the film must be 'slippery' enough to be able to slide over the forming tube assembly (FTA) when the pull belts draw the film through the machine.

The packaging material processed through a vertical form fill seal (VFFS) machine must be within an ideal range for CoF to be compatible with the machinery. Below are industry-accepted ranges for the ideal coefficient of friction values:

Inside CoF: 0.20–0.30.

Outside CoF: 0.300–0.400.

'Perfect film' will have a CoF difference of 0.100 from inside to outside, with the outside being the higher number. However, film suppliers indicate that 'perfect film' is nearly impossible to achieve in practice. The most important thing is to make sure the film is always more slippery on the inside than the outside [13].

Any deviations from the tolerances can cause serious problems in the production.

Customized CoFs are achieved by adding a "slip agent" to a film resin during production. The traditional approach to reducing CoF involves adding a compound that is incompatible with the film resin and will migrate to the surface of the film over time. Non-migratory slip agents offer benefits in the area of thermal stability and consistency but can affect film clarity.

CoF can be affected by several factors including anti-block additives, corona treatment, antistats, inks, varnishes, and adhesives. Because laboratory testing cannot simulate every element of any packaging process, CoF should not be considered an indication of system-specific performance.

The most commonly used industry standards for measuring the coefficient of friction are:

- ASTM D1894—Standard test method for static and kinetic coefficients of friction of plastic film and sheeting.
- ASTM D202—Standard test methods for sampling and testing untreated paper used for electrical insulation.
- ASTM D4918—Standard test method for the coefficient of static friction of uncoated writing and printing paper by use of the inclined plane method (Withdrawn 2010).
- ISO 8295—Plastics—Film and sheeting—Determination of the coefficients of friction.

- TAPPI T815—Coefficient of static friction (slide angle) of packaging and packaging materials (including shipping sack papers, corrugated and solid fiberboard).
- TAPPI T816—Coefficient of static friction of corrugated and solid fiberboard.

For cellulosic materials, the influence of surface species present on the fibres has been studied by a few authors [14]. When wood extractives, a heterogeneous mixture of low molecular weight aliphatic substances such as fatty acids and sterols, are present on the paper surface they are found to reduce the surface energy and also the coefficient of friction [15]. Higher molecular weight resin acids, stearyl ethers, and glycerides, however, appear to increase the friction despite producing low-energy surfaces. These conflicting results may be associated with the boundary-lubricating properties of the extractives when present on fibre surfaces [16, 17]. For example, surface sizing agents like alkyl ketene dimers (AKD) have a detrimental effect on paper friction, resulting in a lowering of the coefficient of friction, as well, for example, recycling of the paper. Hoyland and Neill [18] determined that the side chain length of the AKD molecule has a significant effect on friction, the longer the side chain the greater the decrease in the friction coefficients. The addition of AKD decreases the surface energy of the paper, as well as the friction coefficients. The decrease in the friction coefficients had a 'step' relationship to the decrease in the surface energy of the paper. It was found that there was a critical surface energy value, below which there was a dramatic reduction in the friction coefficients.

2.2 *Semi-rigid Packaging Processing*

The semi-rigid packaging where cellulose cardboard and corrugated board dominate, can come either from the pulp in the form of fibre moulded packaging (where moulding and drying steps are included), or in sheets of cardboard and corrugated boards which are printed, die-cutted and glued or deep drawn with mechanical tools in 3D shapes of packaging solutions. The typical steps for semi-rigid packaging are summarized in Table 4 [19].

Only a few of the different forming processes are used in cellulose biocomposites (paper and board-like materials) converting technology—mainly compressive forming (e.g., embossing), tensile/compressive forming (e.g., drawing), and particularly forming by bending (e.g., corrugating).

For example, folding is one of the main forming processes of cellulose-based materials where its elastic and plastic properties play a significant role. As elastic behaviour is a result of simple deformation of the fibre network, plastic deformation occurs if single fibre–fibre bonds rupture under the action of tensile stress. Tensile stress is directly connected to the fibre processing, fibre type, and auxiliary materials which are used in paper and board production. To prepare folding lines, it is usual

Table 4 Processes of semi-rigid (paperboard type composites) converting

Processing step	The function of the applied converting step
Forming	Changing the shape and form of the composite
Separating	Local destruction of the cohesion in material
Joining	Applying cohesion to certain surfaces on the material
Combining	Changing the usage of the materials
Transporting	Processes in packaging machinery that change location, amount, and position in the processed material
Printing	Information transfer to the surface of the processed material

to crease the material, which in turn locally decreases bending stiffness. For cellulosic materials main influencing factors for stiffness property are: grammage, thickness, bulk or density, single/multi-layer structure. Besides these, the tear strength, tensile strength, and moisture content play an important role. Especially in cellulose-based materials with high hydrophilic tendencies, moisture content can hamper the processing steps as well as the end-user experience and functionality. While adding nanocellulose to the fibre mixture can improve board stiffness initially, the extreme moisture content in the supply chain can additionally lower the stiffness due to these additives.

Embossing is an example of compressive forming and is often applied. The low-yield point of paper and board under compression is utilized. To change the shape of the material, overcoming a certain forming resistance of the material is important. The necessary force (F) depends on the load (A) and the specific forming resistance (σ_0) of the material. In embossing, the relief pattern can be 0.15–2.5 mm high or deep. If the relief has only small differences in levels, the possible machine speed will be higher, and stacking the finished sheets will also be easier. The cellulose nanocomposite board or paper material has to withstand a high degree of elongation without cracking (loss of local cohesion), but at the same time retain a permanent deformation after the impression process—basically to be easily formed. If the surface is coated, the coating also needs to be highly elastic to avoid any leak or reduction in barrier properties (for grease, oil or oxygen). Whereas, the basic materials are controlled through compressibility and thickness variation.

2.2.1 Cutting or Die Cutting

In the conversion of paperboard, the die-cutting and creasing operations are usually performed simultaneously in the flat-bed die-cutting station, which can be offline or in-line with the printing press. Pressure cutting involves the separation of the material to be cut using a mostly wedge-shaped straight or circular knife, working against a plane anvil perpendicular to it, with or without a horizontal component of movement. Paper or carton board is compressed by the knife with the following effects:

(1) elongation of the top layers of the material;
(2) displacement of material by the flanks of the wedge-shaped knife;
(3) strong compression under the edge of the knife.

Therefore, the result of the deformation is a strong elongation of the upper layers of the material where fibre structure and inter-fiber bonding are important. Also, to press the knife into the material, friction forces as well as the forces to displace the material must be overcome. It can be estimated that approximately half of the load is used for friction and displacement and not for cutting [19]. It can be concluded that a low knife angle, along with low friction between the knife and the material will result in a more accurate cut. Additional important characteristics are the strength properties of the paperboard, i.e., tearing strength and tensile strength. The nick strength is proportional to tensile strength and grammage. The moisture content level is essential for both runnability and quality during the die-cutting process. Too high moisture level will make the paperboard stronger, and tougher to cut. On the other hand, if the moisture level is too low, it will make the paperboard more brittle and difficult to safely transfer. Difficulties in cutting through can be caused by the paperboard thickness, moisture variation, or tool wear and adjustments [20]. More contemporary systems also employ laser cutting techniques where the cellulosic material (paperboard or other biocomposite) used in laser cutting should be as lightly coated as possible for two reasons. The first is that the lighter the coating is, the faster the laser cutting works. The second is that the operation leaves a slightly brownish discolouration on the reverse side of the paperboard, and this is more noticeable on a heavily coated paperboard. However, this discolouration can be covered by printing [20].

Creasing is an process which facilitates the folding operation. During creasing the paperboard is weakened along well-defined folding lines, which then act as hinges for folding packaging and graphical products. It is very demanding to fold a paperboard with a good result without creasing. The paperboard is creased by being pressed into a channel or groove in the make-ready. The generated forces deform the paperboard in a predetermined way and the deformation is permanent. The result is a reduction in the bending resistance of the crease. The paperboard is therefore weaker along the crease than elsewhere. During the creasing operation, the paperboard sheet is bent in four narrow zones and each of these the paperboard must endure high tensile forces or compression forces. So again z-tensile strength and ply or internal bonding of materials is one of the crucial factors in achieving good creasing.

2.2.2 Joining Processes

The aim of joining processes is to cause structural cohesion in the material to be processed in well-defined areas. Joining processes with the paper converting industry is the use of adhesive bonding which can be either gluing or sealing. Surface energy is a key parameter in adhesive technology concerning adhesion as well as surface strength properties, and porosity for specific glueing processes.

For paper-polymer laminates, heat sealing is predominant for processes in the erecting unit. The closure of the package is very often done with hot melt glue because the counter-pressure in the contact areas for the glueing is often insufficient for heat sealing. Also, other purposes like fitting a window on a paperboard blank, involves glueing cellulose acetate or oriented polypropylene film (or any other bio-based film) to the reverse side of the paperboard. For this purpose, tacky dispersion glues and hot melt glues are preferable.

Frictional connections are in use in paper converting as

- stapling and sewing,
- screwing and riveting,
- embossing.

Stapling is used in bookbinding and as an element of joining in the corrugated board industry. The strength of stapled connections depends on

- the strength of the material to be processed, particularly tear strength;
- the strength of staples, particularly bending resistance;
- the details of load impact and arrangement of staples concerning the direction of load impact.

Sewing is sometimes used to seal paper sacs filled with, for example, dry animal food, crops, or similar dry products. Single-needle as well as double-needle seams can be used for this purpose.

Screwing and riveting is seldom used in paper converting. Sometimes application to converted products is fixed by rivets. In paperboard containers where the top cover is hinged to the bottom part, rivets are used as hinges. Embossing describes a special form of connection of two paper webs without adhesive, solely by applying pressure. Examples are multilayer tissue paper products such as handkerchiefs, kitchen rolls, and toilet paper. Special design of embossing tools is used to produce paper filters for coffee and tea. The tool consists of two interlocked cogwheels. If the two layers of filter paper are to be connected past the nip of the tool it is corrugated and connected by high pressure particularly acting in the flanks of the cogs [19].

Finally, two examples of connections by positive locking should be mentioned:

- closure of envelopes by inserting flaps into slots;
- closure of folding boxes by inserting flaps.

2.2.3 Combining Materials

An operation where semi-rigid materials like cellulose boards and paperboards are combined is somewhat similar to extrusion or lamination processes for flexible packaging, the difference is the nature of materials. While in flexible polymer materials, both plies have similar film properties and are in the form of rolls or as a heated substance for extrusion; combining semi-rigid materials often involves a combination of sheets, cellulosic paper, or cardboard rolls and different types of materials which are combined. The most common combination is the polymer-cellulose combination where the cellulose is in the semi-rigid form, and where we can use synthetic coating, impregnation, and lamination. In lamination, the combining is done through the use of intermediate adhesive or the melting of the foil through dry or wet lamination. In dry lamination, after the applied adhesive is dried and thermally reactivated by the laminator, the combining is finished and no permeability of materials is needed due to the internal cohesion of the applied gluing. In lamination, the permeability of at least one of the materials is a precondition so if both have a very low porosity and low surface energy problems with delamination can arise.

2.3 *Rigid Material Processing*

The composite rigid packaging is mostly injection moulded or thermoformed with additional operations like printing and closure system applications by welding. The basic materials are bioplastics polythelene teretphalate (PET), polyethylene (PE), polyvinyl chloride (PVC), and most recently fibre polymer combinations like fibre-based bottles. The main processes are presented in Table 5 [21].

In recent years many producers have tried to switch to fibre-based solutions where fiber mixture (most commonly wood pulp) is used instead of some polymer. Due to the poor barrier properties surface enhancement in the form of barrier properties is also needed. In a work by Su et al. [22] differences between fiber and expanded polystyrene (EPS) packaging solution are presented in Table 6.

On the other hand, to endow some special properties and value, like smooth surface, waterproofing, etc., some additives, such as colloidal rosin, wax emulsions, and alum are commonly added to make hydrophilic fibre into hydrophobic (water-repellent) products. A more detailed summary of additives for water-repellent moulded pulp products can be seen in Table 7 [23].

Table 5 Processing steps of rigid packaging

Process	Container types	Main polymers used
Thermoforming	Tapered cups, tubs, trays and lids	Polystyrene (PS), polypropylene (PP), amorphous PET (APET), and polyvinyl chloride (PVC), also bio-based versions
Injection moulding	Cups, tubs, trays, pails, tubes, caps, and closures	Polystyrene (PS), polypropylene (PP), PET, and ISBM preforms
Extrusion blow moulding	Bottles, bottles with integral handles, jars, and drums	Polyethylene (PE), PP, and PVC
Injection blow moulding	Cosmetic jars, bottles, and vials	Polystyrene, polyethylene, and polypropylene
Injection stretch blow moulding (ISBM)	Bottles and jars, carbonated drink bottles	Dominant PET, some PP

Table 6 Moulded fiber versus expanded polystyrene (EPS)

	Moulded fiber	EPS
Material	Plant fiber	Petroleum-based polystyrene
Environment	Sustainable that is fully recyclable and biodegradable	Non-biodegradable materials facing bans in many countries
Economic	Stable cost bases	Not stable, prices are correlated to rising oil and resin prices
Protection	Outstanding vibration and cushioning	Good vibration and cushioning
Shipping and storage	Product nests	Product rarely nests
Temperature tolerance	Unaffected by extreme temperature	Temperature affects flexibility

Table 7 Various base materials and additives for water-resistance features for moulded pulp products

Base material	Additive
Recycled paper pulp	Diatomite and bentonite, pore structures with great adsorption performance
Softwood pulp	Low molecular weight and highly cationic coagulant followed by a high molecular weight anionic flocculent
Unbleached kraft pulp	Palm oil from empty fruit bunch (EFB) and alkyl ketene dimer (AKD)

One of the more interesting processes is the production of the so-called paper bottles, where the inflatable core assessed fibre moulding process is used. In a paper [24] it is described that the forming process begins with the injection of the pulp inside a split mould cavity (Fig. 2a). The set of moulds forms a negative replica of the bottle shape to be produced. The mould is set up such as the pulp enters from the neck section and the bottle is formed upside down. The pulp is kept inside a tank and continuously stirred to avoid precipitation of fibres. It is then fed into the forming mould with the help of a pump. The tool is provided with larger holes for efficient removal of the 'free water'. A wired mesh in the shape of a bottle is placed at the inner lining of the mould. The dimension of mesh is such that it allows water to pass through but it doesn't allow the fibers to go out, thereby forming the shape. With the pulp being pumped in, a pressure differential is created across the walls of the tool using vacuum suction. This process continues for a short period, to ease the formation of the bottle shape. Once the shape has been formed, the vacuum suction is turned off and the wet bottle is removed from the mould. This wet bottle is then transferred to a drying mould for the drying process (Fig. 2b). The drying mould is also a split mould made of porous material. The porous mould is preheated to the temperature of 140 °C. As the wet paper bottle comes in contact with the hot tool, the drying process starts. During the drying process, an inflatable core/insert is inserted and inflated inside the mould. The core is made up of a rubber material. The core

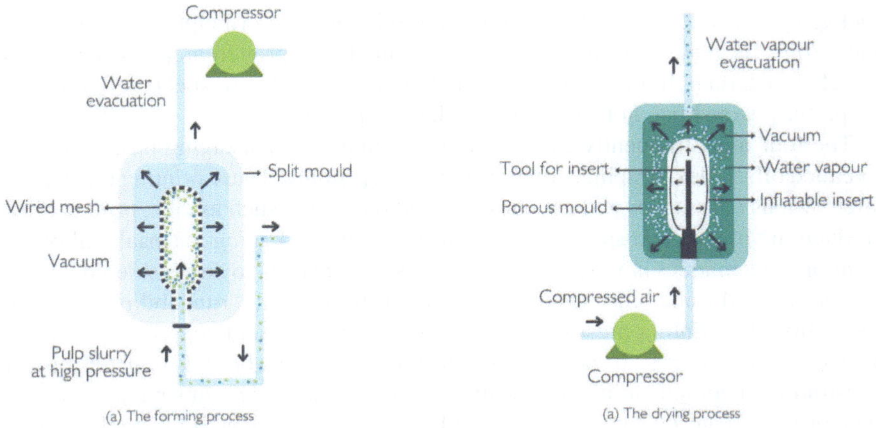

Fig. 2 Paper bottle production process

serves two purposes: (a) compaction of the wet fibres and (b) avoiding the shrinkage of paper during the drying process. A pressure differential is created across the walls of the tool using vacuum suction. Once the drying process is completed the split mould opens up and the dried paper bottle is obtained.

While in these products, the bottle shell is produced with a mixture of cellulose fibres, the main nanocomposite challenge and the main characteristic is the resistance to water penetration in the hydrophobicity of classic synthetic polymers. The materials must be preferably recyclable and biodegradable, and if this requirement is not fulfilled by the inner lining, larger amounts of fibres are wasted. In this case, a full mono-material liquid packaging solution is more sustainable. A detailed study of the pulp materials and coatings and the potential of biopolymers like polylactic acid, polyhydroxyalkanoates, and their nanocomposites are further explored [25]. Moreover, the emerging biomaterial such as cellulose fatty ester, cellulose nanopaper, waterborne acrylic emulsion, and natural wax-based dispersion for coating on paper packaging are detailed.

3 Printability and Runnability of Flexible and Semi-rigid Packaging Materials

Printing is an important process that enriches the majority of packaging materials nowadays. By printing, we add important and sometimes even necessary information, in the form of pictures and text [26].

When designing the packaging it must be assured that there is no significant impact on the recycling or potential biodegradability of the packaging material or product. Therefore, it is even more important to carefully plan the design of the packaging (as well as the printing process) following the principles of eco-design and design for recycling. In practice, this means fewer printed areas, wider use of recycled materials, limiting the use of additional/spot colours, special effects such as (partial) varnishing, hot foil stamping, lamination, etc. [27].

The four most frequently used printing techniques in packaging production are so-called offset, flexographic, gravure, and inkjet printing. These printing techniques offer various advantages and are chosen based on factors such as cost, print quality, production volume, and specific requirements of the packaging material. All of the mentioned techniques have specific characteristics that need to be considered if good printability and runnability need to be ensured. In printing, "runnability" refers to the ability of a printing press or machine to run smoothly and consistently without interruptions or issues. It encompasses factors such as stable ink transfer, proper registration of colours, minimal downtime for maintenance or adjustments, and efficient printing material handling. "Printability" on the other hand refers to the quality and suitability of a substrate or material for the selected printing process. In summary, runnability focuses on the smooth operation of the printing press, while printability pertains to the characteristics of the substrate or material that affect the quality of the printed output [26, 28].

3.1 Flexographic Printing

Flexographic printing, also known as flexo printing, is a widely used technique in the packaging industry. It involves using flexible printing forms made of (photo)polymer materials. These forms have raised printed areas that transfer ink onto various flexible materials such as paper, cardboard, plastic films, and labels. Liquid (water or solvent-based) inks are stored in ink chambers and transferred to the printing form through anilox rollers. The inked printing plate then transfers the image onto the substrate as it passes through the press. Drying or curing systems are employed to ensure proper ink adhesion and prevent smudging [29] (Fig. 3).

In flexographic printing, achieving good printability relies on several important factors. One crucial consideration is the surface energy of the substrate. Surface energy refers to the inherent capability of a material's surface to allow ink adhesion. The key aspects that need to be considered for printability in flexographic printing are as follows:

- Ink Selection: Choosing the appropriate ink type is crucial to ensure compatibility with the substrate's surface energy. Ink formulations should have properties that enable proper adhesion and minimize issues such as ink smearing or poor drying/curing.
- Substrate Surface: The surface of the substrate plays a significant role in printability. It should have an appropriate surface energy to facilitate ink adhesion. Surface treatments, such as corona treatment or primers, can modify the surface energy of substrates to enhance printability.

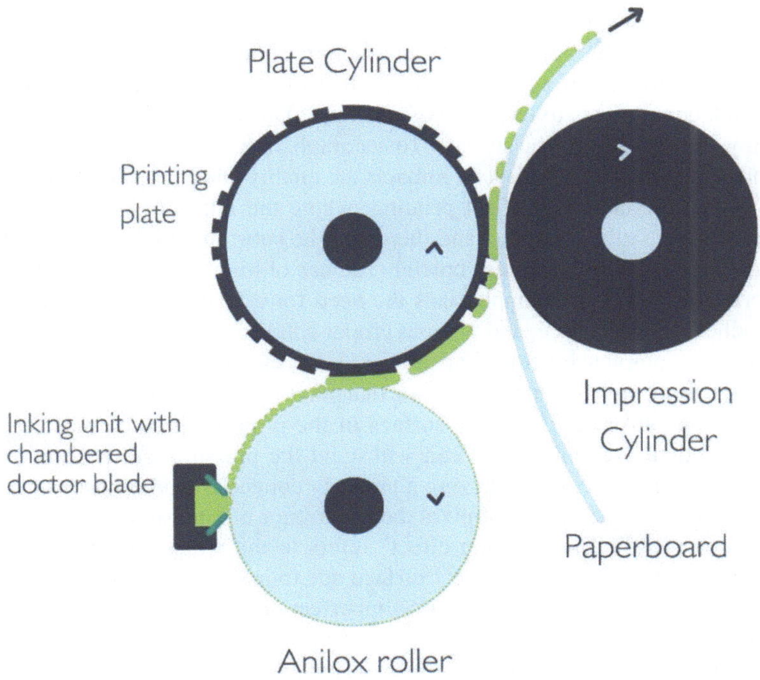

Fig. 3 Principle of flexography printing [28]

- Ink adhesion: Proper ink adhesion is vital for achieving high print quality. When the substrate's surface energy matches the ink's surface tension, it promotes better wetting and adhesion. Adequate ink adhesion ensures the printed image remains intact and durable.
- Anilox roller selection: Anilox roller plays a crucial role in controlling ink transfer. Selecting anilox rollers with the right cell geometry and volume is essential for achieving proper ink coverage and maintaining printability. The cell structure affects ink transfer and distribution on the printing plates [26, 28, 29].

3.2 *Offset Printing*

So-called wet offset printing is a widely used printing technique in the commercial printing industry. It offers high-quality, sharp, and vibrant prints and is mostly suitable for large print runs. The offset printing process involves several key steps. Oil-based ink is transferred from ink rollers to a rubber blanket, which is the speciality of offset printing. For this, dampening rollers must wet the non-printing areas of the printing form with a dampening solution, repelling ink from these areas. The inked image on the rubber blanket is then transferred to the printing surface, typically paper or other substrate, as it passes between the blanket cylinder and an impression cylinder. This transfer creates a high-quality, offset reproduction of the original image.

In offset printing, it is particularly important to achieve proper adhesion of the printing ink components to the printing substrate. This ensures that the ink adheres uniformly and securely to the surface. To accomplish this, a lot of attention is given to the binding process, as it directly impacts the quality and longevity of the printed output. Given the nature of offset printing, where the ink is transferred from the printing form to a rubber blanket and then onto the substrate, the interplay between the ink and the substrate becomes crucial. The use of highly viscous and sticky oil-based printing inks further emphasizes the need for effective bonding. These inks tend to adhere to the surface, and without proper adhesion, issues such as smudging, smearing, or incomplete ink transfer can arise, compromising the overall print quality. Moreover, it is not only the ink formulation that plays a vital role but also the condition of the printing substrate itself. The surface of the printing material must be well-connected and strong, ensuring it can withstand the pressure and friction exerted during the printing process. By having a properly connected and robust surface, the substrate resists damage and minimizes the occurrence of undesired phenomena like the "picking effect". The "picking effect" refers to the undesired lifting of small particles or fibres from the substrate's surface due to inadequate adhesion and weak surface integrity. This can result in visible imperfections, rough textures, or even tiny voids in the printed image, diminishing the overall visual appeal and clarity of the final product. Therefore, through effective curing and strengthening of the printing substrate, the risk of such issues is mitigated, ensuring a smooth and flawless printing process [26, 28].

3.3 Gravure Printing

Gravure printing, also known as intaglio printing, is another widely used printing technique in the commercial printing industry. It is particularly suitable for high-volume print production and is known for its ability to produce sharp, detailed, and consistent prints. The gravure printing process involves several distinct steps that contribute to its exceptional quality. Initially, the image to be printed is etched or engraved onto a cylindrical plate made of metal, typically copper or chrome. This plate contains recessed cells (alveoli) that hold the ink. The depth and size of these cells determine the amount of ink transferred onto the substrate. During printing, the plate cylinder is partially submerged in an ink fountain, where it picks up a generous amount of ink, which in most cases is solvent-based. Simultaneously, a doctor blade ensures that the surface of the plate is clean and free of excess ink. Next, the substrate, such as paper or film, passes between the plate cylinder and a pressure cylinder. As the pressure cylinder applies force to the substrate, it pushes it against the plate cylinder, and the inked cells come into contact with the substrate's surface. The ink is transferred from the plate to the substrate due to the pressure exerted, resulting in the reproduction of the engraved image with remarkable accuracy and detail. Due to the nature of gravure printing, it is best suited for long print runs as it involves significant setup time and cost for plate preparation. However, once the setup is complete, the process can efficiently produce large quantities of high-quality prints consistently.

In gravure printing, the printing substrate/material is practically the only compressible component in the system. Therefore, the substrate must be appropriately compressible while maintaining good mechanical properties to prevent permanent damage from the high pressures applied during the printing nip. Additionally, the material needs to be as smooth as possible. Surface roughness can cause irregularities in the form of the so-called "missing dots" effect, which manifests as missing elementary or raster dots in the print. It is also essential to ensure that no foreign substances pass from the printing substrate into the printing system, such as paper particles or coatings in the case of paper printing. Surface particles, typically resulting from cutting fibre-based materials (paper, cardboard) with dull blades, can clog the small cavities within the system, leading to poor print quality and cleaning challenges associated with such systems. Furthermore, attention must be given to the selection of printing substrates to minimize any undesired interactions with the ink and ensure optimal print performance. Factors such as ink absorption, drying time, and substrate compatibility play significant roles in achieving consistent and high-quality prints. Gravure printing relies on the precise control of various parameters, including ink viscosity, doctor blade pressure, and cylinder alignment, to achieve optimal results. Regular maintenance and cleaning of the printing equipment are crucial to prevent any buildup or contamination that could affect print quality [28].

3.4 Digital Printing

Digital printing encompasses various techniques, including inkjet printing and electro-photography (also known as laser printing). Each method offers distinct advantages and processes.

3.4.1 Inkjet Printing

Inkjet printing is a popular digital printing technique widely used in the commercial printing industry. It offers versatility and is suitable for various applications, including short print runs, personalized printing, and on-demand production. The inkjet printing process involves the following steps:

- Ink delivery: Ink cartridges or tanks containing liquid ink are connected to the inkjet printer. The ink, which is typically water-based or solvent-based, is selectively dispensed through microscopic nozzles. The inkjet printer precisely controls the placement and volume of ink droplets according to the image's pixel pattern.

- Substrate interaction: The substrate, such as paper, fabric, or other materials, is fed into the printer. As the substrate passes through the printing area, the printheads deposit the ink droplets onto its surface. The ink droplets adhere to the substrate, forming the desired image.
- Drying or curing: After printing, the ink on the substrate needs to dry or cure, depending on the ink type. Water-based inks typically require time to air dry, while solvent-based inks may require additional drying methods or exposure to radiations, e.g., UV light, for curing.

Inkjet printing offers several advantages, such as high resolution, vibrant colour reproduction, and the ability to print on a wide range of substrates. It is well-suited for short to medium print runs, personalized printing, variable data printing, and prototypes.

3.4.2 Electro-photographic Printing (Laser Printing)

Electro-photographic printing, commonly known as laser printing, is another prevalent digital printing technique. It utilizes a combination of electrostatic principles and toner particles to produce high-quality prints. The electro-photographic printing process involves the following steps:

- Charging: The surface of a cylindrical drum or belt, typically made of a photoconductive material, is uniformly charged with static electricity using a corona wire or roller. The charged drum becomes sensitive to light.
- Image formation: A laser beam scans across the drum's surface, selectively discharging areas to create the desired image pattern. The laser beam is controlled by the digital image data.
- Toner application: Toner particles, which are fine, dry powder composed of pigments and resins, are attracted to the discharged areas on the drum's surface. The toner is transferred from a storage unit, known as a toner cartridge, to the drum using electrostatic forces.
- Image transfer: The toner image on the drum is transferred onto the substrate, typically paper, by applying an electrostatic charge to the paper. The toner particles adhere to the paper's surface, forming the final printed image.
- Fusing: The printed substrate passes through a pair of heated rollers called the fuser unit. The heat and pressure melt the toner particles, bonding them to the paper, and producing a durable, permanent image.

Electro-photographic printing offers advantages such as high-speed printing, consistent image quality, and sharp text reproduction. Electro-photography does face challenges when it comes to charging wider areas, as it can result in striped patterns that can negatively affect print quality. As a result, inkjet printing is more commonly used for industrial applications where consistent and precise image reproduction is crucial. However, it's important to note that there are many other printing techniques available, although they may be less prevalent in certain contexts.

3.5 Critical Parameters Affecting Printing Processes in Flexible and Semi-rigid Packaging Materials

The printing of flexible and semi-rigid packaging materials is influenced by several critical parameters that have a pervasive impact across various materials and printing processes, including climate conditions such as temperature and humidity, surface energy, porosity, and surface roughness or smoothness.

The utilization of natural fibre-based materials in the domain of flexible and semi-rigid packaging presents distinct challenges due to their inherent hydrophilic nature, which enables the absorption and release of moisture. This characteristic has a profound impact on the printing process and overall print quality, posing significant challenges in terms of printing performance and alignment. Achieving optimal conditions is crucial to ensure dimensional stability, appropriate print adhesion, and desired output.

The ability of paper to absorb and release moisture has a direct influence on its printing behaviour. Optimal dimensional stability of paper is achieved within a relative air humidity range of 35–50%. Maintaining suitable printing conditions is paramount, with the recommended temperature range for paper printing being between 21 and 29 °C, accompanied by a relative humidity of 40–45%. It is essential to strike a balance between moisture management and temperature-humidity interplay to ensure successful printing operations.

Another crucial factor that significantly influences the runnability of all printing processes is the friction of the material. Adequate friction is indispensable for facilitating the smooth transfer of the material through the printing machine. Inadequate friction can impede the effective handling and transportation of the material by various rollers. Maintaining the appropriate level of friction is vital to ensure optimal printing performance and prevent issues during the printing process. Friction (static friction) is strongly influenced by the surface smoothness. Rough surfaces can cause excessive friction, while excessively smooth surfaces exhibit the 'glass plate effect' due to their large contact area. Achieving an optimal balance in surface smoothness is essential to control static friction during the printing process.

Measuring material friction in the field of printing involves the utilization of various methods to assess the surface characteristics and frictional properties of the materials. Additionally, surface smoothness and roughness can be measured using different techniques. In this scientific text, we will explore some of the commonly employed methods for measuring material friction and the assessment of surface smoothness and roughness.

Coefficient of friction (CoF) measurement: The CoF is a widely used parameter to quantify the frictional behaviour of materials. It is typically measured using a tribometer, which applies a controlled force to the material and measures the resulting frictional force. The CoF is calculated as the ratio of the frictional force to the applied force. This method provides valuable information about the material's resistance to sliding or slipping during the printing process.

Inclined plane method: The inclined plane method is employed to evaluate the angle at which a material starts sliding due to the applied force. The angle of inclination is gradually increased until the material begins to move. This method allows the determination of the angle of repose, which indicates the material's resistance to motion and its frictional properties.

Surface smoothness measurement: Surface smoothness is an important parameter that affects material friction. It can be measured using different techniques, including:

1. Contact profilometry: It involves using a profilometer to scan the material's surface and measure its topography. The profilometer utilizes a stylus or a laser to trace the surface and record height variations. This method provides data on parameters such as surface roughness average (Ra), root mean square roughness (Rq), and peak-to-valley height (Rz), which quantitatively describe surface smoothness.
2. Optical interferometry: It involves techniques, such as white light interferometry or coherence scanning interferometry, are employed to measure surface smoothness and roughness at a microscopic level. These methods utilize interference patterns generated by reflected light to precisely measure the surface height variations and calculate parameters like roughness average (Sa) and spatial frequency characteristics.
3. Atomic Force Microscopy (AFM): AFM is a high-resolution imaging technique that can be utilized to assess surface roughness and smoothness at a nanoscale level. It uses a sharp probe to scan the material's surface, measuring the forces between the probe and the surface. AFM provides detailed three-dimensional surface topography information and allows the determination of parameters like roughness average (Ra) and peak height.

The interaction between the printing ink and the printing material is a significant parameter that influences both the runnability and printability of various printing processes. Different printing techniques employ distinct types of printing inks, each with its characteristics and requirements.

Flexographic printing utilizes liquid-based inks that are primarily water/solvent-based or UV-curable. The ink interacts with the printing material, which can be porous or non-porous, affecting the ink absorption and drying process. Porous materials, such as paper or cardboard, allow the ink to penetrate and be absorbed, while non-porous materials, like plastic films, require specialized surface treatments or primers to enhance ink adhesion. These treatments are designed to modify the surface energy of the substrate, allowing the ink to adhere more effectively. Those are:

- Corona treatment: Corona treatment involves exposing the surface of the substrate to a high-frequency electrical discharge. This discharge creates micro-level etching and oxidation, increasing the surface energy of the material. The increased surface energy promotes better wetting and adhesion of the ink.
- Flame treatment: In flame treatment, a controlled flame is applied to the surface of the material. The heat and oxidation generated by the flame modify the

surface properties of the substrate, increasing its surface energy and improving ink adhesion.

- Plasma treatment: Plasma treatment involves exposing the substrate to a low-pressure plasma, which consists of ionized gas. The plasma interacts with the material surface, resulting in chemical reactions and physical changes that increase surface energy. Plasma treatment can be particularly effective for enhancing ink adhesion on various materials.
- Chemical primers: Chemical primers are coatings applied to the substrate surface to enhance ink adhesion. These primers contain specific chemical compounds that create a receptive surface for the ink. The primer molecules bond with both the substrate and the ink, promoting better adhesion.

It is important to note that the desired surface energy level varies depending on the specific printing components involved. The rule of increasing surface energy establishes the minimum surface energy levels required for effective ink adhesion in various components of flexographic printing. According to this rule, the printing ink should have a surface energy of at least 30 mN/m, the anilox roller of at least 34 mN/m, the printing form of at least 36 mN/m, and the printing material should have a surface energy equal to or greater than 40 mN/m.

Gravure printing employs liquid inks that are typically solvent based. The ink is transferred from recessed cells engraved on a cylindrical printing plate onto the printing material. The interaction between the ink and the material is influenced by the surface characteristics and porosity of the material. Drying in gravure printing is usually achieved through a combination of evaporation and heat.

Digital printing utilizes various types of inks, including aqueous, solvent-based, UV-curable, and latex inks. The interaction between the ink and the printing material varies depending on the technology used, such as inkjet or electrophotographic printing. Inks designed for digital printing are formulated to optimize compatibility with specific substrates, ensuring good adhesion and colour vibrancy. Drying methods in digital printing vary based on the ink formulation, with UV curing and heat-assisted drying being common techniques.

Different ink types, including oil-based, and UV-curable inks, are used in offset printing. Oil-based inks are composed of pigments or dyes dispersed in a petroleum-based solvent or vegetable oil. These inks require a drying process that involves the evaporation of solvents, leaving behind a thin ink film on the printing material. The drying process in oil-based inks is typically aided by the absorption of the solvents into the paper or by heat-assisted drying methods. UV-curable inks cure rapidly using UV light [30].

References

1. Dörnyei KR, Uysal-Unalan I, Krauter V, Weinrich R, Incarnato L, Karlovits I et al (2023) Sustainable food packaging: an updated definition following a holistic approach

2. Parisi S (2013) Food industry and packaging materials: performance-oriented guidelines for users. Smithers Rapra, Shawbury
3. Coles R, Mcdowell D, Kirwan MJ (2003) Food packaging technology. CRC Press, Cop, Boca Ratón, Florida
4. Troughton MJ (2008) Handbook of plastics joining: a practical guide. William Andrew, Norwich
5. Morris BA (2017) The science and technology of flexible packaging: multilayer films from resin and process to end use. Elsevier Ltd., Oxford; William Andrew
6. Cantwell M, Cantwell M, Cardwell J, Davison B, Davison GC (2015) Heat sealing fundamentals, testing, and numerical modeling [Internet]. Worcester Polytechnic Institute. https://core.ac.uk/download/pdf/212990336.pdf
7. Regazzi A, Viguié J, Harthong B, Dumont P, Imbault D, Peyroux R et al (2019) Ultrasonic welding of 100% lignocellulosic papers. J Mater Sci 54(19):12938–12950
8. Monot C, Viguié J, Charlier Q, Bras J, Guérin D, Harthong B et al (2021) Ultrasonic welding of folding boxboards. BioResources 16(3):5766–5779
9. Ultrasonic sealing—welding of packaging | Herrmann Ultraschall [Internet]. www.herrmannultraschall.com. https://www.herrmannultraschall.com/en/welding-using-ultrasonics/welding-of-packaging/ultrasonic-sealing. Accessed 2 Oct 2023
10. Bach S, Thürling K, Majschak JP (2011) Ultrasonic sealing of flexible packaging films—principle and characteristics of an alternative sealing method. Packag Technol Sci 25(4):233–248
11. Koo Sin Lin C, Oehm L, Majschak J (2020) Innovative focused ultrasound-based sealing method of flexible packaging films—principles and characteristics. Packag Technol Sci
12. Stoehr N, Baudrit B, Haberstroh E, Nase M, Heidemeyer P, Bastian M (2014) Properties and weldability of plasticized polylactic acid films. J Appl Polym Sci 131(12):n/a–n/a
13. Ohl D (2020) What is coefficient of friction (COF), and why is it important in packaging? [Internet]. Viking Masek. https://vikingmasek.com/packaging-machine-resources/packaging-machine-blog/what-coefficient-friction-cof-and-why-it-important-packaging#How%20Do%20You%20Measure%20Coefficient%20of%20friction?. Accessed 22 May 2023
14. de Silveira G, Hutchings IM (1997) Determination of the friction of paper and board. In: Baker CF (ed) The fundamentals of papermaking materials. Transactions of the XIth fundamental research symposium, Cambridge. FRC, Manchester, pp 1329–1353. https://doi.org/10.15376/frc.1997.2.1329
15. Back EL (1991) Paper-to-paper and paper-to-metal friction. In: International paper physics conference. TAPPI Press, pp 49–65
16. Tabor D (1974) Triction, adhesion and boundary lubrication of polymers. In: Lee L-H (eds) polymer science and technology, vol 5A. Advances in polymer friction and wear. Plenum Press, pp 5–28
17. Hutchings IM (1992) Tribology: friction and wear of engineering materials. Edward Arnold, London, pp 51–57
18. Hoyland RW, Neill MP (1997) The effect of AKD on paper friction—prepared contribution. In: Baker CF (ed) The fundamentals of papermaking materials. Transactions of the XIth fundamental research symposium, Cambridge. FRC, Manchester, pp 1617–1631. https://doi.org/10.15376/frc.1997.3.1617
19. Holik H (2013) Handbook of paper and board. Wiley
20. Iggesund H (no date) Embossing and debossing. https://www.iggesund.com/services/knowledge/graphics-handbook/finishing/embossing-and-debossing/. Accessed 29 June 2023
21. Hannay F, Rapra Technology Limited (2002) Rigid plastics packaging: materials, processes and applications. Rapra Technology Ltd., Shawbury, Shrewsbury
22. Su Y, Yang B, Liu J, Sun B, Cao C, Zou X et al (2018) Prospects for replacement of some plastics in packaging with lignocellulose materials: a brief review. BioResources 13(2)
23. Zhang Y, Duan C, Bokka SK, He Z, Ni Y (2022) Molded fiber and pulp products as green and sustainable alternatives to plastics: a mini review. J Bioresour Bioproducts [Internet] 7(1):14–25. https://www.sciencedirect.com/science/article/pii/S2369969821000803

24. Saxena P, Bissacco G, Meinert KÆ, Bedka FJ (2020) Mold design and fabrication for production of thermoformed paper-based packaging products. J Manuf Process 58:311–321
25. Ahuja A, Samyn P, Rastogi VK (2022) Paper bottles: potential to replace conventional packaging for liquid products. Biomass Convers Bioref
26. Kipphan H (2001) Springerlink (Online Service). Handbook of print media: technologies and production methods. Springer, Berlin, Heidelberg
27. Gürlich U, Kladnik V, Pavlovic K (2022) Circular packaging design guideline
28. Wilson LA (1997) What the printer should know about paper. Gatf Press
29. Iggesund H (no date) Flexography, Holmen Iggesund. https://www.iggesund.com/services/knowledge/reference-manual/printing-and-converting-performance/flexography/. Accessed 29 June 2023
30. Simon J, Langenscheidt A (2020) Curing behavior of a UV-curable inkjet ink: distinction between surface-cure and deep-cure performance. J Appl Polym Sci 137(40):49218

MIX
Papier aus verantwortungsvollen Quellen
Paper from responsible sources
FSC® C105338

If you have any concerns about our products,
you can contact us on
ProductSafety@springernature.com

In case Publisher is established outside the EU,
the EU authorized representative is:
Springer Nature Customer Service Center GmbH
Europaplatz 3, 69115 Heidelberg, Germany

Printed by Libri Plureos GmbH
in Hamburg, Germany